Lecture Notes in Computer Science 8080

Commenced Publication in 1973
Founding and Former Series Editors:
Gerhard Goos, Juris Hartmanis, and Jan van Leeuwen

Editorial Board

Traian Muntean Dimitrios Poulakis
Robert Rolland (Eds.)

Algebraic Informatics

5th International Conference, CAI 2013
Porquerolles, France, September 3-6, 2013
Proceedings

Springer

Volume Editors

Traian Muntean
Aix-Marseille University
ERISCS Research Group
13288 Marseille CEDEX 9, France
E-mail: traian.muntean@univ-amu.fr

Dimitrios Poulakis
Aristotle University of Thessaloniki
Department of Mathematics
54124 Thessaloniki, Greece
E-mail: poulakis@math.auth.gr

Robert Rolland
Aix-Marseille University
ERISCS Research Group
and IML Research Laboratory
13288 Marseille CEDEX 9, France
E-mail: robert.rolland@acrypta.fr

ISSN 0302-9743
e-ISSN 1611-3349
ISBN 978-3-642-40662-1
e-ISBN 978-3-642-40663-8
DOI 10.1007/978-3-642-40663-8
Springer Heidelberg New York Dordrecht London

Library of Congress Control Number: Applied for

CR Subject Classification (1998): F.4, I.1.3, F.1.1, F.4.1, F.4.3, F.4.2

LNCS Sublibrary: SL 1 – Theoretical Computer Science and General Issues

Typesetting: Camera-ready by author, data conversion by Scientific Publishing Services, Chennai, India

Printed on acid-free paper

Springer is part of Springer Science+Business Media (www.springer.com)

Preface

This volume of LNCS is devoted to CAI 2013, the 5th International Conference on Algebraic Informatics, organized under the auspices of Aix-Marseille University, ERISCS Research Group and IML Research Laboratory. The conference, held at IGESA Center, Porquerolles Island, during September 3–6, intended to cover topics in mathematical aspects of computing such as algebraic specifications and algorithms, algebraic coding theory, algebraic aspects of cryptography, computational number theory, formal power series, algebraic semantics, finite and infinite computations, algebraic characterization of logical theories, process algebra, program construction and refinements, acceptors and transducers for discrete structures, decision problems, term rewriting, abstract machines and systems, hybrid automata composition.

These topics involve considerable interaction between various theoretical mathematical disciplines, including the theory of computation, algebraic models of computing, and data coding, for future critical computer applications in industry.

This volume contains five invited lectures and 19 contributed papers, out of 24 submissions from eight countries, which were presented at the conference.

The papers cover a broad range of topics of recent research and ongoing work in the field featuring

- Data models and coding theory
- Fundamental aspects of cryptography and security
- Algebraic and stochastic models of computing
- Logic and program modelling

We are grateful to all members of the Program Committee for the evaluation of the submissions and the valuable suggestions from referees who assisted in this work.

We also thank all authors for having submitted high-quality papers.

Special thanks are due to Alfred Hofmann, Editorial Director at Springer, who helped us to publish the proceedings of CAI 2013 in the LNCS series, as well as to Anna Kramer for the excellent cooperation.

The sponsors of CAI 2013 are also gratefully acknowledged.

May 2013

Traian Muntean
Dimitrios Poulakis
Robert Rolland

Organization

CAI 2013 was jointly organized by the ERISCS research group and IML research Laboratory at Aix-Marseille University.

Steering Committee

Symeon Bozapalidis	Thessaloniki
Zoltán Ésik	Szeged
Werner Kuich	Vienna
Arto Salomaa	Turku

Chairs

Traian Muntean	Marseille
Dimitrios Poulakis	Thessaloniki
Robert Rolland	Marseille

Program Committee

Stéphane Ballet	Marseille
Alexis Bonnecaze	Marseille
Symeon Bozapalidis	Thessaloniki
Bruno Courcelle	Bordeaux
Zoltán Ésik	Szeged
Tero Harju	Turku
Kristin Lauter	Microsoft Research Seattle
Andreas Maletti	Stuttgart
Bruno Martin	Nice Sophia Antipolis
Traian Muntean	Marseille
Alexander Okhotin	Turku
Jean-Éric Pin	Paris
Dimitrios Poulakis	Thessaloniki
Anna Rimoldi	Trento
George Rahonis	Thessaloniki
François Rodier	Marseille
Robert Rolland	Marseille
Franz Winkler	Linz

Referees

R. Aragona
S. Ballet
P. Barreto
A. Bonnecaze
S. Bozapalidis
J.-L. Cayrel
B. Courcelle
C. Ene
Z. Ésik
S. Galbraith
T. Garefalakis
J. Hansen
T. Harju
H. Hisil
A. Kalampakas
D. Katz
D. Kuperberg
G. Landsmann
K. Lauter
A. Maletti
E. Mandrali
B. Martin
E. Martinez Moro
S. Mesnager
D. Moody
L. de Moura
L. Mugwaneza
T. Muntean
A. Okhotin
A. Papistas
J.-É. Pin
D. Poulakis
A. Rimoldi
G. Rahonis
F. Rodier
R. Rolland
I. Sakho
M. Sala
I. Szabolcs
L. Vaux
S. Vladuts
F. Winkler
A. Zeh

Sponsoring Institutions

Aix-Marseille University
ERISCS research group; Aix-Marseille University
IML research Laboratory; Aix-Marseille University and CNRS
Laboratoire d'Excellence Archimède; Aix-Marseille University and CNRS
Laboratoire d'Excellence AMIES; CNRS, UPMC and Grenoble 1 University

Table of Contents

Invited Speakers

Contributed Papers

Euclidean Model Checking: A Scalable Method for Verifying Quantitative Properties in Probabilistic Systems

Gul Agha*

Department of Computer Science
University of Illinois, Urbana, Illinois, USA
agha@illinois.edu
http://osl.cs.illinois.edu

We typically represent the global state of a concurrent system as the cross-product of individual states of its components. This leads to an explosion of potential global states: consider a concurrent system with a thousand actors, each of which may be in one of 5 states. This leads to a possible $5^{1}000$ global states. Obviously, it is not feasible to exhaustively search the state space in such systems. In fact, actors often have an even larger number of states (than say 5), although these states may be abstracted to fewer states.

Our work is motivated by the following observation. In large concurrent systems, we are often interested in probabilistic guarantees on the behavior of the system. This suggests the possibility of sampling the behavior. In the real world, engineers often use monte carlo simulations to analyze systems. This process can be made more rigorous by expressing the desired properties of a system in a formal logic such as *continuous stochastic logic* (CSL). We have earlier proposed using an approach we call *statistical method checking* to verify properties expressed in a sublogic of CSL [11]. This work was extended to verify properties involving *unbounded untils* in [12]. The methods are implemented in a tool called VESTA [12] which has been used in a number of applications. A parallel version of the tool, called pVeSTA, has also been implemented [2].

In this lecture, I will focus on an alternate method for addressing the problem of large state spaces. For many purposes, it may not be necessary to consider the global state as a cross-product of the states of individual actors. We take our inspiration from statistical physics where macro properties of a system may be related to the properties of individual molecules using probability distributions on the states of the latter. Consider a simple example. Suppose associated with each state is the amount of energy a node consumes when in that state (such

* The work reviewed in this presentation was done primarily in collaboration with YoungMin Kwon, Koushik Sen, Vijay Korthikanti, and Mahesh Viswanathan. The research has been supported in part by the Defense Advanced Research Projects Agency (DARPA) under Award No. F33615-01-C-1907, by ONR Grant N00014-02-1- 0715, by NSF under grant CNS 05-09321, by the AFRL and the AFOSR under agreement number FA8750-11-2-0084, and by the Army Research Office under Award No. W911NF- 09-1-0273.

T. Muntean, D. Poulakis, and R. Rolland (Eds.): CAI 2013, LNCS 8080, pp. 1–3, 2013.

an associated value mapping is called the *reward function* of the state). Now, if we have a frequency count of the nodes in each state, we can estimate the total energy consumed by the system. This suggests a model where the global state is a vector of probability mass functions (pmfs). In the above example, the size of the vector would be 5, one element for each possible state of a node. Each element of the vector represents the probability that any node is in the particular state corresponding to entry.

Given transitions between the global states, we can also compute how much energy has been consumed up to some point in time. Note that using such a global state assumes a certain symmetry (at least as a statistical approximation). However, we have also explored cases where there may be more than one type of node in a system (with its own associated Markovian behaviors). We have defined a temporal logic *iLTL* which can be used to write specifications where the global state of system is defined by such pmf vectors [6]. These vectors represent points in (convex subspace of) an Euclidean space. The evolution of a concurrent system can then be modeled as trajectories in this Euclidean space.

The behavior of many large concurrent system is Markovian and often modeled using *Discrete Time Markov Chains* (DTMC). The DTMC governs transforms of pmf vectors representing the state evolution. We have developed methods to check properties of systems governed by DTMCs expressed using iLTL formulas [9], [10]. Given the nature of the model, we term the method *Euclidean Model Checking* [10]. The method has been implemented in a tool [8] and used to verify properties such communication bandwidth, maximum expected queue lengths, energy consumption in sensor networks, pharmacokinetic models [5], and software reliability of many threaded concurrent software [7]. More general models may be considered, for example, *Markov Decision Processes* (MDPs) where there is no commitment to a single Markov matrix governing the transformations [4]. However this makes the verification problem in general intractable, except in special cases such as MDPs with a unique compact invariant set of distributions [3].

References

1. Second International Conference on the Quantitative Evaluaiton of Systems (QEST 2005), Torino, Italy, September 19-22. IEEE Computer Society (2005)
2. AlTurki, M., Meseguer, J.: PVESTA: A parallel statistical model checking and quantitative analysis tool. In: Corradini, A., Klin, B., Cîrstea, C. (eds.) CALCO 2011. LNCS, vol. 6859, pp. 386–392. Springer, Heidelberg (2011)
3. Chadha, R., Korthikanti, V.A., Viswanathan, M., Agha, G., Kwon, Y.: Model checking mdps with a unique compact invariant set of distributions. In: QEST, pp. 121–130. IEEE Computer Society (2011)
4. Korthikanti, V.A., Viswanathan, M., Agha, G., Kwon, Y.: Reasoning about mdps as transformers of probability distributions. In: QEST, pp. 199–208. IEEE Computer Society (2010)
5. Kwon, Y., Kim, E.: Specification and verification of pharmacokinetic models. In: Advances in Computational Biology. Advances in Experimental Medicine and Biology, vol. 680, pp. 463–472. Springer (2010)

6. Kwon, Y., Agha, G.: Linear inequality LTL (*iLTL*): A model checker for discrete time markov chains. In: Davies, J., Schulte, W., Barnett, M. (eds.) ICFEM 2004. LNCS, vol. 3308, pp. 194–208. Springer, Heidelberg (2004)
7. Kwon, Y., Agha, G.: A markov reward model for software reliability. In: IPDPS, pp. 1–6. IEEE (2007)
8. Kwon, Y., Agha, G.A.: iltlchecker: A probabilistic model checker for multiple dtmcs. In: [1], pp. 245–246
9. Kwon, Y., Agha, G.A.: Verifying the evolution of probability distributions governed by a dtmc. IEEE Trans. Software Eng. 37(1), 126–141 (2011)
10. Kwon, Y., Agha, G.A.: Performance evaluation of sensor networks by statistical modeling and euclidean model checking. ACM Transactions on Sensor Networks (2013)
11. Sen, K., Viswanathan, M., Agha, G.: Statistical model checking of black-box probabilistic systems. In: Alur, R., Peled, D.A. (eds.) CAV 2004. LNCS, vol. 3114, pp. 202–215. Springer, Heidelberg (2004)
12. Sen, K., Viswanathan, M., Agha, G.A.: Vesta: A statistical model-checker and analyzer for probabilistic systems. In: [1], pp. 251–252

Quantitative Analysis of Randomized Distributed Systems and Probabilistic Automata*

Christel Baier

Technische Universität Dresden, Faculty of Computer Science, Germany

The automata-based model checking approach for randomized distributed systems relies on an operational interleaving semantics of the system by means of a Markov decision process (MDP) and a formalization of the desired event E by an ω-regular linear-time property, e.g., an LTL formula. The task is then to compute the greatest lower bound for the probability for E that can be guaranteed even in worst-case scenarios. Such bounds can be computed by a combination of polynomially time-bounded graph algorithm with methods for solving linear programs. See e.g. [7,4,3].

In the classical approach, the "worst-case" is determined when ranging over all schedulers that decide which action to perform next. In particular, all possible interleavings and resolutions of other nondeterministic choices in the system model are taken into account.

As in the nonprobabilistic case, the commutativity of independent concurrent actions can be used to avoid redundancies in the system model and to increase the efficiency of the quantitative analysis. This motivates the use of partial-order reduction to construct and analyze a smaller sub-MDP that is equivalent to the original MDP for stutter-invariant ω-regular properties. Although the main concepts are the same as in the non-probabilistic case, there are certain phenomena that are specific for the probabilistic case and require additional conditions for the reduced model to ensure that the worst-case probabilities are preserved [2,6].

Related to this observation is also the fact that the worst-case analysis that ranges over all schedulers is often too pessimistic and leads to extreme probability values that can be achieved only by schedulers that are unrealistic for parallel systems. This motivates the switch to more realistic classes of schedulers that respect the fact that the individual processes only have partial information about the global system states. Such classes of partial-information schedulers yield more realistic worst-case probabilities, but computationally they are much harder since the semantic model of randomized systems with partial-information

* This work was in part funded through the CRC 912 Highly-Adaptive Energy-Efficient Computing (HAEC), the EU under FP7 grant 295261 (MEALS), the DFG/NWO-project ROCKS, the cluster of excellence cfAED (center for Advancing Electronics Dresden) and the DFG project QuaOS.

T. Muntean, D. Poulakis, and R. Rolland (Eds.): CAI 2013, LNCS 8080, pp. 4–5, 2013.

schedulers is closely related to probabilistic automata over words. Indeed, a wide range of verification problems that impose conditions on all partial-information schedulers turns out to be undecidable [5,1].

References

1. Baier, C., Größer, M., Bertrand, N.: Probabilistic -automata. Journal of the ACM 59(1) (2012)
2. Baier, C., Größer, M., Ciesinski, F.: Partial order reduction for probabilistic systems. In: 1st International Conference on Quantitative Evaluation of Systems (QEST), pp. 230–239. IEEE Computer Society (2004)
3. Baier, C., Größer, M., Ciesinski, F.: Model checking linear-time properties of probabilistic systems. In: Droste, M., Kuich, W., Vogler, H. (eds.) Handbook of Weighted Automata. Monographs in Theoretical Computer Science, pp. 519–570. Springer (2009)
4. Courcoubetis, C., Yannakakis, M.: The complexity of probabilistic verification. Journal of the ACM 42(4), 857–907 (1995)
5. Giro, S., D'Argenio, P.R., Fioriti, L.M.F.: Partial order reduction for probabilistic systems: A revision for distributed schedulers. In: Bravetti, M., Zavattaro, G. (eds.) CONCUR 2009. LNCS, vol. 5710, pp. 338–353. Springer, Heidelberg (2009)
6. Größer, M.: Reduction Methods for Probabilistic Model Checking. Dissertation, Technische Universität Dresden (2008)
7. Vardi, M.Y., Wolper, P.: An automata-theoretic approach to automatic program verification. In: 1st IEEE Symposium on Logic in Computer Science (LICS), pp. 332–345. IEEE Computer Society Press (1986)

On Elliptic Curve Paillier Schemes

Marc Joye

Technicolor, France
marc.joye@technicolor.com

Abstract. In 1999, Paillier proposed an elegant cryptosystem from the integers modulo N^2 where N is an RSA modulus. Paillier public-key encryption scheme enjoys a number of interesting properties, including a homomorphic property: the encryption of two messages allows anyone to derive the encryption of their sum. This reveals useful in cryptographic applications such as electronic voting. In this talk we review several generalizations of the original Paillier scheme to the elliptic curve setting. Using similar ideas, we then present a new elliptic curve scheme which is semantically secure in the standard model. Interestingly, the new encryption scheme does not require to encode messages as points on an elliptic curve and features a partial homomorphic property.

T. Muntean, D. Poulakis, and R. Rolland (Eds.): CAI 2013, LNCS 8080, p. 6, 2013.

Proofs of Storage: Theory, Constructions and Applications

Seny Kamara

Microsoft Research
senyk@microsoft.com

Abstract. Proofs of storage (PoS) are cryptographic protocols that allow a client to efficiently verify the integrity of remotely stored data. To use a PoS, the client sends an encoded version of its data to the server while keeping a small amount of state locally. At any point in time, the client can then verify the integrity of its data by executing a highly-efficient challenge-response protocol with the server.

Since their introduction in 2007 by Ateniese et al. (Computer and Communications Security, 2007) and Juels and Kaliski (Computer and Communications Security, 2007), PoS have received a lot of attention from the research community. This is due in large part to their potential practical applications (e.g., to the design of various kinds of secure cloud storage systems) but also due to their inherent theoretical properties and their connections to fundamental primitives like digital signatures, identification schemes, zero-knowledge proofs and error-correcting codes.

In this talk, I will survey the current state of PoS research. This will include the many variants of PoS that have been invented over the years, how to design them, the connections that have been established between PoS and other primitives and the many new applications PoS have enabled.

References

1. Ateniese, G., Burns, R., Curtmola, R., Herring, J., Kissner, L., Peterson, Z., Song, D.: Provable data possession at untrusted stores. In: ACM Conference on Computer and Communication Security (CCS 2007). ACM (2007)
2. Ateniese, G., Di Pietro, R., Mancini, L., Tsudik, G.: Scalable and efficient provable data possession. In: Conference on Security and Privacy in Communication Networks (SecureComm 2008), pp. 9:1–9:10 (2008)
3. Ateniese, G., Faonio, A., Kamara, S., Katz, J.: How to authenticate from a fully compromised system (under submission, 2013)
4. Ateniese, G., Kamara, S., Katz, J.: Proofs of storage from homomorphic identification protocols. In: Matsui, M. (ed.) ASIACRYPT 2009. LNCS, vol. 5912, pp. 319–333. Springer, Heidelberg (2009)
5. Benson, K., Dowsley, R., Shacham, H.: Do you know where your cloud files are? In: ACM Cloud Computing Security Workshop (CCSW 2011), pp. 73–82 (2011)
6. Bowers, K., Juels, A., Oprea, A.: Proofs of retrievability: Theory and implementation. Technical Report 2008/175. Cryptology ePrint Archive (2008)

T. Muntean, D. Poulakis, and R. Rolland (Eds.): CAI 2013, LNCS 8080, pp. 7–8, 2013.

7. Bowers, K., Juels, A., Oprea, A.: HAIL: a high-availability and integrity layer for cloud storage. In: ACM Conference on Computer and Communications Security (CCS 2009), pp. 187–198. ACM (2009)
8. Bowers, K., van Dijk, M., Juels, A., Oprea, A., Rivest, R.: How to tell if your cloud files are vulnerable to drive crashes. In: ACM Conference on Computer and Communications Security (CCS 2011), pp. 501–514. ACM (2011)
9. Cash, D., Küpçü, A., Wichs, D.: Dynamic proofs of retrievability via oblivious RAM. In: Johansson, T., Nguyen, P.Q. (eds.) EUROCRYPT 2013. LNCS, vol. 7881, pp. 279–295. Springer, Heidelberg (2013)
10. Dodis, Y., Vadhan, S., Wichs, D.: Proofs of retrievability via hardness amplification. In: Reingold, O. (ed.) TCC 2009. LNCS, vol. 5444, pp. 109–127. Springer, Heidelberg (2009)
11. Erway, C., Küpçü, A., Papamanthou, C., Tamassia, R.: Dynamic provable data possession. In: ACM Conference on Computer and Communications Security (CCS 2009), pp. 213–222. ACM, New York (2009)
12. Gondree, M., Peterson, Z.: Geolocation of data in the cloud. In: ACM Conference on Data and Application Security and Privacy (CODASPY 2013), pp. 25–36. ACM (2013)
13. Juels, A., Kaliski, B.: PORs: Proofs of retrievability for large files. In: Ning, P., De Capitani di Vimercati, S., Syverson, P. (eds.) ACM Conference on Computer and Communication Security (CCS 2007). ACM (2007)
14. Juels, A., Oprea, A.: New approaches to security and availability for cloud data. Communications of the ACM 56(2), 64–73 (2013)
15. Naor, M., Rothblum, G.: The complexity of online memory checking. In: IEEE Symposium on Foundations of Computer Science (FOCS 2005), pp. 573–584. IEEE Computer Society (2005)
16. Naor, M., Rothblum, G.: The complexity of online memory checking. Journal of the ACM 56(1), 2:1–2:46 (2009)
17. Peterson, A., Gondree, M., Beverly, R.: A position paper on data sovereignty: The importance of geolocating data in the cloud. In: USENIX Workshop on Hot Topics in Cloud Computing, HotCloud 2011 (2011)
18. Shacham, H., Waters, B.: Compact proofs of retrievability. In: Pieprzyk, J. (ed.) ASIACRYPT 2008. LNCS, vol. 5350, pp. 90–107. Springer, Heidelberg (2008)
19. Stefanov, E., van Dijk, M., Juels, A., Oprea, A.: Iris: a scalable cloud file system with efficient integrity checks. In: Annual Computer Security Applications Conference (ACSAC 2012), pp. 229–238. ACM, New York (2012)
20. van Dijk, M., Juels, A., Oprea, A., Rivest, R., Stefanov, E., Triandopoulos, N.: Hourglass schemes: how to prove that cloud files are encrypted. In: ACM Conference on Computer and Communications Security (CCS 2012), pp. 265–280. ACM (2012)
21. Wang, C., Chow, S., Wang, Q., Ren, K., Lou, W.: Privacy-preserving public auditing for secure cloud storage. Cryptology ePrint Archive, Report 2009/579 (2009), `http://eprint.iacr.org/2009/579`
22. Watson, G., Safavi-Naini, R., Alimomeni, M., Locasto, M., Narayan, S.: LoSt: location based storage. In: ACM Cloud Computing Security Workshop (CCSW 2012), pp. 59–70. ACM (2012)

Code Based Cryptography and Steganography

Pascal Véron

IMATH, Université du Sud Toulon-Var,
B.P. 20132, F-83957 La Garde Cedex, France
veron@univ-tln.fr

Abstract. For a long time, coding theory was only concerned by message integrity (how to protect against errors a message sent via some noisely channel). Nowadays, coding theory plays an important role in the area of cryptography and steganography. The aim of this paper is to show how algebraic coding theory offers ways to define secure cryptographic primitives and efficient steganographic schemes.

Cryptography

1 Introduction

Cryptography addresses the following problem: how to scramble a message before sending it in order to make it unintelligible to any outsider. In symmetric cryptography (or private key cryptography), the message is enciphered with a function e and deciphered using a function d. These two functions depend on a parameter k called the secret-key such that for all messages m, $d(e(m, k), k) = m$. As a consequence, this key must be shared by the sender and the recipient. In practice, this may be very difficult to achieve, especially if the key has to be sent via some channel. In 1976, W. Diffie and M.E. Hellman [37] laid the foundation for public key cryptography (or asymmetric cryptography) asking the following question: is it possible to use a pair of keys (k, ℓ) such that only k be necessary for encryption, while ℓ would be necessary for decryption ? For such a protocol, d and e must satisfy for all messages m, $d(e(m, k), \ell) = m$. A cryptosystem devised in this way is called a *public key cryptosystem* since k can be made public to all users. Obviously, it should be computationally infeasible to determine ℓ from k.

The security of all conventional public key cryptosystems actually deployed in practice depends on the hardness of two mathematical problems coming from number theory: integer factoring and discrete logarithm. At this time no one knows an efficient algorithm in order to solve them in a reasonable time although numerous researchers make good progress in this area. If the security of the schemes based on this two problems is well defined, one drawback is that they rely on arithmetic operations over large numbers. Moerover, Shor's quantum algorithm [96] published in 1994 poses a serious threat to the security of these conventional cryptosystems. Indeed, quantum computers (of an appropriate size) can potentially break them in polynomial time. Although such quantum computers still do not exist, there is a strong need to develop and study alternative public key cryptosystems that would be secured in a post quantum world.

T. Muntean, D. Poulakis, and R. Rolland (Eds.): CAI 2013, LNCS 8080, pp. 9–46, 2013.

Algebraic coding theory offers an alternative supposed to resist to quantum attackers. Remember that the aim of algebraic coding theory is to restore a message m sent via a channel disrupted by some natural perturbation and that the goal of cryptography is to intentionally scramble a message m before sending it, so that it becomes unintelligible except for its recipient. Obviously there are some links between these two fields. Security of code based cryptographic primitives depends on a problem which in its general form is a well known NP-complete problem: *the syndrome decoding problem.* Generally these protocols are easier to implement, use only basic operations over the two element field and provides fast encryption and decryption algorithms.

2 Minimal Background in Coding Theory

In this section, we recall few notions on coding theory in order to understand the sequel of this paper. For a more complete overview on this topic, the reader is addressed to [74].

Definition 1 (Linear code). *A linear code $\mathcal{C}$ is a k-dimensional subspace of an n-dimensional vector space over a finite field $\mathbb{F}_q$, where k and n are positive integers with $k \leqslant n$, and q a prime power. The error-correcting capability of such a code is the maximum number t of errors that the code is able to decode.*

Definition 2 (Hamming weight). *The (Hamming) weight of a vector x is the number of non-zero entries. We use $\omega(x)$ to represent the Hamming weight of x.*

Definition 3 (Generator and Parity Check Matrix). *Let $\mathcal{C}$ be a linear code over $\mathbb{F}_q$. A generator matrix G of $\mathcal{C}$ is a matrix whose rows form a basis of $\mathcal{C}$:*

$$\mathcal{C} = \{xG : x \in \mathbb{F}_q^k\}\,.$$

A parity check matrix H of $\mathcal{C}$ is is an $(n-k) \times n$ matrix whose rows form a basis of the orthogonal complement of the vector subspace $\mathcal{C}$, i.e. it holds that,

$$\mathcal{C} = \{x \in \mathbb{F}_q^n : H^t x = 0\}\,.$$

For the sequel, we will focus our attention on the decoding problem for binary linear codes (i.e. $q = 2$). First we recall two important results.

First Result. A binary linear code $\mathcal{C}$ of length n can correct t errors if for any $x, y \in \mathcal{C}$ $(x \neq y)$, $B(x,t) \cap B(y,t) = \emptyset$ where $B(x,t) = \{y \in \{0,1\}^n \mid d(x,y) \leqslant t\}$ and $d(x,y)$ denotes the Hamming distance.

Second Result. A binary linear code $\mathcal{C}(n,k)$ whose minimal distance is d can correct $\lfloor (d-1)/2 \rfloor$ errors.

Let $\mathcal{C}$ be a binary $[n, k, d]$ code. Let us consider a word c' such that $c' = c_0 + e$ where $c_0 \in \mathcal{C}$ and e is what is called an error vector. Let H be a parity check matrix of $\mathcal{C}$ and let s be the syndrome of c', i.e. $s = H^t c'$. Notice that the 2^k solutions x which satisfy the equation

$$H^t x = s\,, \tag{1}$$

are given by the set $\{u + e, u \in \mathcal{C}\}$ (remember that $\forall u \in \mathcal{C}, H^t u = 0$). If the Hamming weight of e (i.e. the number of non-zero bits of e) satisfies

$$\forall u \in \mathcal{C} \setminus \{0\},\ w(e) < w(u + e)\,, \tag{2}$$

then the error e is the minimum weight solution of (1).

Remark 1. If $w(e) \leqslant \lfloor (d-1)/2 \rfloor$, then e satisfies eq. (2).

Hence, without any extra information on the code, to decode c' one has to solve an optimization problem. Notice that searching for the minimum weight word which satisfies eq. (1) is equivalent to search for the closest codeword from c'. Indeed, it is easy to see that eq. (2) is equivalent to:

$$\forall u \in \mathcal{C} \setminus \{c_0\},\quad d(c_0, c') < d(u, c')\,. \tag{3}$$

One goal of coding theory is to find codes for which the minimum weight solution of (1) can be computed in polynomial time without constraints on the size of H. Such a problem can be stated in a more general setting as it will be developped in the next section.

3 The Syndrome Decoding Problem

Except for the Mc Eliece's cryptosystem and the CFS signature scheme, the security of all the code based cryptographic schemes that we are going to detail is based on the difficulty of the Syndrome Decoding Problem. The SD problem is a decision problem which can be stated as follows:

Name : SD
Input : $H(r, n)$ a binary matrix , s a binary column vector with r coordinates, p an integer.
Question : Is there a binary vector e of length n such that $H^t e = s$ and $w(e) \leq p$?

In the context of coding theory, if H is a parity check matrix, this means that the problem to decide wether there exists or not a word of given weight and syndrome is NP-complete.

This decision problem is linked to the optimization problem induced by maximum likelihood decoding. Indeed, searching for the closest codeword of a received word x is equivalent to find the minimum weight solution e of the equation

$H^t e = H^t x$. Now, let (H, s, p) be an instance of the SD problem, the vector e exists if and only if the minimum weight solution of $H^t x = s$ is less or equal than p. On the other hand, if one knows a polynomial time algorithm to solve SD, then it can be turned into a polynomial time algorithm to compute the minimal weight of a solution of the system $H^t x = s$. In 1978, E.R. Berlekamp, R.J. McEliece and H.C.A. Van Tilborg [13] proved that this problem is NP-complete reducing it to the THREE-DIMENSIONAL MATCHING problem [56].

Remark 2. The problem still remains NP-complete if:

- the matrix H is full rank (as it is the case for a parity check matrix),
- we ask for an s with exactly p 1's.

The SD problem can be stated in terms of the generator matrix since one can go from the parity-check matrix to the generator matrix (or vice versa) in polynomial time:

Name : G-SD
Input : $G(k, n)$ a generator matrix of a binary (n, k) code $\mathcal{C}$, $x \in \{0,1\}^n$ and $p > 0$ an integer.
Question : Is there a vector e of length n and weight p such that $x + e \in \mathcal{C}$?

While the SD problem is NP-complete, there exists weak matrices for which an efficient algorithm can be developed. Hence, one can alternatively define algebraic coding theory as the science whose one goal is to build easy instances of the SD problem, in order to set up polynomial time algorithms for decoding. However for a random matrix H, it is necessary to know for which parameters (n, r, p) the problem seems to be difficult to solve.

4 Algorithms for the SD Problem

Nowadays, there exists eight probabilistic algorithms to compute a solution to the SD problem: Lee and Brickell's algorithm [70], Leon's algorithm [71], Stern's algorithm [99], the toolbox of A. Canteaut and F. Chabaud [25], Johansson and Jönsonn's algorithm [69], the "ball-collision" decoding algorithm [18], the MMT algorithm [75] and the "1+1=0" decoding algorithm [10]. All these algorithms are devoted to search a word of small weight in a random code.

Proposition 1. *SD problem is equivalent to the following problem:*
Input : *$H(k, n)$ a binary matrix of rank k, $p > 0$ an integer.*
Question : *Is there a vector $x \in \{0,1\}^n$ such that $H^t x = 0$, $w(x) \leq p$ and $x_n = 1$?*

All these algorithms are based on the notion of information set decoding (ISD) introduced by Prange [87].

Definition 4. *Let G be a generator matrix of an $[n, k]$ code and $c = mG$ be a codeword. Let us denote by G_i the ith column of G and let $I = \{i_1, \ldots, i_k\}$ such that $G_I = (G_{i_1}, \ldots, G_{i_k})$ be a $k \times k$ invertible submatrix. Then these k coordinates uniquely determine the vector m, since $m = (c_{i_1}, \ldots, c_{i_k})G_I^{-1}$. The set I is called an information set.*

Now suppose that a received word $x = (c + e)$ is such that no errors occur in the information set I. The error pattern e can be recovered by computing $(x_{i_1}, \dots, x_{i_k})G_I^{-1} + x$. Hence, the main idea used in all the algorithms is to select random information sets from the generator matrix (or the parity check matrix for Stern's scheme) until the support of the error does not meet the selected set which leads to a probability of success of:

$$\frac{\binom{n-p}{k}}{\binom{n}{k}} \tag{4}$$

Using the usual binomial approximation this gives the following probability of success:

$$P_{\text{succ}} = \mathcal{O}(1).2^{-nH_2(p/n)-(1-k)H_2(p/(n-k))} \tag{5}$$

where $H_2(x)$ is the classical entropy function. Hence, the work factor (number of operations) needed to compute a solution for the SD problem can be roughly estimated by:

$$\frac{\text{Inv}(k)}{P_{\text{succ}}} \tag{6}$$

where $\text{Inv}(k)$ is the cost for inverting a $k \times k$ matrix. Usually this operation needs k^3 binary operations (notice that in order to be more precised, we should have take into account the probability for a random $k \times k$ matrix to be invertible). The algorithms of Lee and Brickell, Leon and Stern use some heuristic in order to minimize the call to the inverse procedure by:

1. taking into account information set which contains a small part (say w bits) of the support of the error pattern,
2. using a size-ℓ window of zeroes outside of I in order to constrain the possible locations for the error.

Canteaut and Chabaud combine these heuristics with a trick (proposed by J. Van Tilburg [104] and latter by H. Chabanne and B. Courteau [31]) in order to reduce the cost of the inverse procedure. Let I be the current information set for which the algorithm did not succeed, instead of randomly select k new columns, they exchange one column whose index is in I with a column whose index is in $\{1, \dots, n\} \setminus I$ which decreases the cost of the Gaussian elimination. Interested readers can find a complete description and analysis of the first four algorithms in [24,25]. It follows from the study of [25] that the modified version of Stern's algorithm is the best one to solve the SD problem.

"Ball-collision", MMT and "1+1=0" algorithms are improvement of modified Stern's scheme where the major contribution comes in that some positions of the error vector are also fixed in two subsets Z_1 and Z_2 outside I. Moreover, the "1+1=0" algorithm adds a further improvement in the initial search step. Here is a graphical representation (from [19]) which illustrates how the word e is searched for a given information set I.

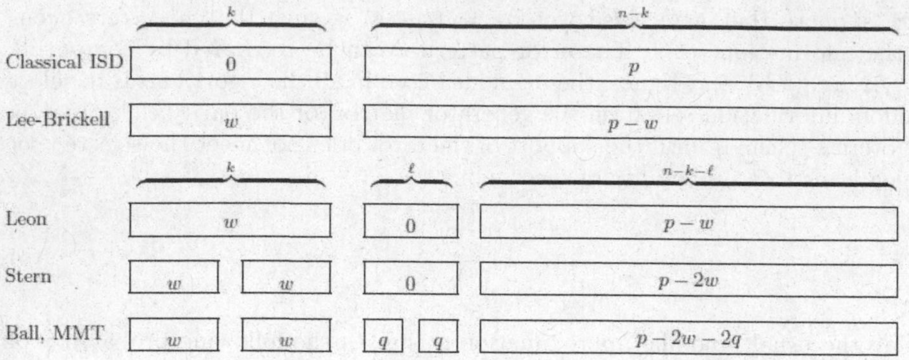

Nowadays, the "1+1=0" algorithm is the best one to solve the SD problem.

Another important result is that hard instances of the SD problem are obtained when the weight of the vector e is near from the theoretical minimal distance d of the code which is given by the Gilbert-Varshamov bound:

$$H_2(d/n) \simeq 1 - k/n. \tag{7}$$

Since random binary linear codes attain with overwhelming probability a rate $R(= k/n)$ (which is close to the Gilbert-Varshamov bound) the running time of the decoding algorithms (for random binary linear codes) can be expressed as a function of n and R only, namely $T(n, R)$. Let $T(n, R) = \mathcal{O}(2^{\theta n})$, where $\Theta = \lim_{n\to\infty} \frac{\log(T(n,R))}{n}$, table 1 gives the value of θ when R is close to the Gilbert Varshamov bound. In this table, half decoding means that we are searching for a word of weight $\lfloor (d-1)/2 \rfloor$ where d is the theoritical minimum distance of the code, while full decoding means that we are searching for the closest codeword from an arbitrary vector $x \in \mathbb{F}_2^n$ (see eq. 3). The algorithm of Johansson and Jönsson is slightly different from the other one. The input is a list of received words and the goal is to try to decode one of them. Since the algorithm works with information set, all the tricks used in the other algorithms can be used in order to optimize it. The probability of success grows with the size of the initial list. When this list is to small, the performances are not better than those of the other algorithms (see table 2).

4.1 The q-SD Problem

The SD problem can be considered over an arbitrary finite field.

Table 1. Complexity of the decoding algorithms for SD for random codes

	θ(half dec.)	θ(full dec.)
Lee − Brickell(1988)	0.05751	0.1208
Stern(1989)	0.05563	0.1167
Ball − collision(2011)	0.05558	0.1164
MMT(2011)	0.05364	0.1116
$1 + 1 = 0$(2012)	0.0497	0.1019

Table 2. Workfactor of Johansson and Jönsson algorithm

size of list	$n = 1024, k = 524, p = 50$	$n = 512, k = 256, p = 56$
1	$2^{68.1}$	$2^{72.2}$
2^5	$2^{63.7}$	$2^{68.9}$
2^{10}	$2^{59.5}$	$2^{65.9}$
2^{15}	$2^{56.2}$	$2^{64.1}$
2^{30}	$2^{50.2}$	2^{60}

Definition 5 (q-ary Syndrome Decoding (qSD) problem).
Input : *$H(r, n)$ a matrix over $\mathbb{F}_q$, s a vector with r coordinates over $\mathbb{F}_q$, an integer $p > 0$.*
Question : *Is there a q-ary vector e of length n such that $H^t e = s$ and $w(e) \leq p$?*

In 1994, A. Barg proved that this last problem remains NP-complete [8, in russian]. In [86], C. Peters generalizes all the ISD algorithms to the case of codes over $\mathbb{F}_q$ with $q > 2$. As an example, to reach a complexity of 2^{128}, it is enough to choose a $[961, 771]$ code over $\mathbb{F}_{31}$ and a word of weight 48. For the same complexity, in the binary case, we have to choose a $[2960, 2988]$ code and a word of weight 57. If the matrix is a public key, the matrix over $\mathbb{F}_{31}$ can be stored usion $90Kb$ while the one over $\mathbb{F}_2$ needs $188Kb$.

4.2 Quantum Computers and the SD Problem

The SD problem cannot be polynomially solved using quantum computers. However, the Grover's quantum algorithm [63,64] for computing roots of a function can be used in order to speedup the probabilistic algorithms against SD. In [15], the author shows that the quantum version of the information set decoding algorithms takes time only $c^{(1/2+o(1))n/\log_2 n}$ to break a length n and rate R code (with $c = 1/(1-R)^{1-R}$) where as the non quantum version takes time $c^{(1+o(1))n/\log_2 n}$. As a consequence, protecting against these quantum attacks requires essentially quadrupling the key size.

5 The SD Identification Scheme

5.1 Introduction

The SD Identification scheme is the first cryptographic protocol whose security relies on the difficulty of the SD problem. An identification scheme is a cryptographic protocol which enables party A (called the "prover") to prove his identity (by means of an on-line communication) polynomially many times to party B (called the "verifier") without enabling B to misrepresent himself as A to someone else. In 1985, S. Goldwasser, S. Micali and C. Rackoff described a very nice solution to this problem with zero-knowledge proofs [61], where a user convinces with a non-negligible probability an entity that he knows the solution

s of a public instance of a "difficult" problem without giving any information on s (see [88] for a nice introduction to zero-knowledge). In 1986, A. Fiat and A. Shamir proved the practical significance of zero-knowledge proofs for public-key identification [42]. Their scheme relies on the difficulty of factoring. Notice that, from a practical point of view, the prover may be identified to a smart card, hence it is supposed that he has reduced computational power and a small amount of memory. Since 1988, there were several attempts to build identification schemes which did not rely on number theory and use only very simple operations so as to minimize computing load. The idea to use error-correcting codes for identification is due to S. Harari [65] , unfortunately his scheme was not zero-knowledge and not really practical due to its heavy communication load. Moreover, the scheme has been proved to be insecure in [105]. Another scheme proposed by M. Girault [59] has been cryptanalysed in [93].

5.2 Stern's Scheme

The first truly practical scheme using error-correcting codes is due to J. Stern [100]. The scheme uses a fixed binary (k, n) parity check matrix H which is common to all users. In 1995, a dual version of Stern's scheme has been defined: the G-SD identification scheme [106]. This version improves the communication complexity (number of bits exchanged during the protocol) for exactly the same level of security as those of Stern's scheme.
Table 3 lists the secret and public data used in the SD protocol. The pair (i, p) is the public identification of the prover. His data can be computed by a certification center having the confidence of all users or the prover can choose his secret keys and the center certifies the corresponding public keys. The principle of the

Table 3. Public and secret data in the G-SD identification scheme

Common public data	: $H(k, n)$ a full rank binary matrix , a hash function denoted by $\langle . \rangle$.
Prover's secret data	: $s \in \{0,1\}^n$.
Prover's public data	: $i = Hs$ and $p = \omega(s)$.

protocol is the following: the prover (Alice) knows the secret vector s which satisfies $Hs = i$ and $p = \omega(s)$. Bob (the verifier) asks Alice a series of questions. If Alice really knows s, she can answer all the questions correctly. If she does not, she has a probability q of answering correctly. After r successful iterations of the protocol, Bob will be convinced that Alice knows s with probability $1 - q^r$.
The identification scheme relies on the notion of commitment. Commitment is a protocol between Alice and Bob which operates in 3 stages:

- Stage 1: Alice hides a sequence u of bits and sends it to Bob. The hidden function is public and hard to invert.
- Stage 2: Alice and Bob execute some protocol,

- Stage 3: Alice reveals u, Bob checks the validity of the hidden value received during stage 1.

From a practical point of view, u is hidden via a cryptographic public hash function. Hence Alice sends to Bob the image $\langle u \rangle$ of u. The hash function must be collision-free (i.e. it should be "infeasible" to compute $u' \neq u$ such that $\langle u' \rangle = \langle u \rangle$). Discussion on the length of the hash value $\langle u \rangle$ can be found in [60]. Let us denote by $x.y$ the concatenation of the binary strings x and y and by $y\sigma$ the image of $y \in \{0,1\}^n$ under the permutation σ of $\{1,\ldots,n\}$, the SD scheme includes r rounds each of these being performed as described in table 4.

Table 4. A round of the SD scheme

- A randomly computes :
  - $y \in \{0,1\}^n$,
  - σ a permutation of $\{1,\ldots,n\}$.

  and send to B three commitments:

$$c_1 = \langle \sigma, Hy \rangle,\ c_2 = \langle (y+s)\sigma \rangle, c_3 = \langle y\sigma \rangle$$

- B sends a random element $b \in \{0,1,2\}$ (challenge).
- if $b = 0$,
  - A reveals y and σ,
  - B checks the value of c_1 and c_3.
- if $b = 1$,
  - A reveals $y+s$ and σ,
  - B checks the value of c_1 and c_2.
- if $b = 2$,
  - A reveals $y\sigma$ and $s\sigma$,
  - B checks the value of c_2 and c_3 and verifies that $w(s\sigma) = p$.

5.3 Security and Performances

It can be proved that:

- the scheme is zero-knowledge i.e., informally speaking, during the protocol the transactions contain no information on s (more formally one can construct a polynomial time machine S which outputs a communication tape having the same probability distribution as a real communication).
- a cheater can bypass the protocol with a probability bounded by $(2/3)^r$, otherwise one can construct a polynomial-time probabilistic machine which either outputs a valid secret s or finds collision for the public hash function.

Practical security of the scheme is linked to the parameters n, k, p and r. Let H be the parity check matrix used in the scheme. In order to impersonate A, an intruder has to be able to compute a word s of weight p whose image under H is i (this is the SD problem). If p is chosen slightly below the value of the theoretical minimum distance of $\mathcal{C}$ then the probability that there exists a word $s' \neq s$ of weight p such that $Hs = Hs'$ is very low. Hence by choosing

$$n = 700, k = 350, p = 75,$$

searching the vector e with the probabilistic algorithms described in section 4 needs around 2^{70} operations. Moreover taking $r = 35$, the probability of success of a cheater is bounded by 10^{-6}.

If we envisage the prover as a smart card, essentially three parameters are to be taken into account: the communication complexity (number of bits exchanged during the protocol), the complexity of the computations done by the prover and the storage capacity needed by the prover. The SD identification scheme uses only very simple operations over the two element field (i.e . over bits) and can be implemented in hardware in a quite efficient way. One drawback is the size of the matrix H which must be stored by the prover. Another one is the communication complexity since at least 35 rounds are needed in order to achieve a reasonable level of security while for the same level (from a dishonest prover point of view) identification schemes based on number theory can be performed in only few rounds (4 rounds for Fiat-Shamir's scheme) . Table 5 sums up the performances of Stern's scheme, G-SD scheme and Fiat-Shamir's scheme (1024 bits version) giving for each one: the number of rounds needed to achieve a probability of success of 10^{-6} for a dishonest prover, the total communication complexity, the size of the ROM (number of bits stored by the prover), the total prover's computation complexity (number of binary operations performed by the prover during the whole protocol).

6 The McEliece's Public-Key Cryptosystem

Despite Mc Eliece's cryptosystem be the first code based cryptosystem, we decide to not describe it first because its security does not directly rely on the SD problem.

Soon after Diffie-Helmman's paper on public key cryptography , R.L. Rivest, A. Shamir and L. Adleman exhibited such a system: the well known RSA cryptosystem based on the factorization of integers [89]. Merkle and Hellman [78] proposed another cryptosystem based on the difficulty of the integer packing "knapsack" problem. There were several variants around this latter but the development of the LLL algorithm made most of them insecure. In 1978, R.J.

Table 5. SD schemes versus Fiat-Shamir scheme

	SD	G-SD	Fiat-Shamir
Rounds	35	35	4
ROM	123550	124250	5120
Computation complexity	$2^{23.04}$	$2^{23.04}$	$2^{25.4}$
Communication complexity	52523	44357	4628

McEliece defined the first public key cryptosystem using algebraic coding theory [76]. The basic idea is quite simple: use as a secret key a code $\mathcal{C}$ which belongs to a family of codes for which a polynomial time decoding algorithm exists and give as a public key an equivalent code $\mathcal{C}'$ which masks the algebraic structure of $\mathcal{C}$, so that $\mathcal{C}'$ looks like a random binary linear code. Table 6 describes the general protocol. Of course, one important parameter of this protocol is the code $\mathcal{C}$ to use:

- For n, k and d fixed, $\mathcal{C}$ must belong to a large family of codes so that it is impossible to find it via an exhaustive search. Notice that is is enough to find an equivalent code to the public one using an algorithm due to N. Sendrier [93] which can determine if two generator matrices define equivalent codes and can find back the permutation,
- a polynomial-time decoding algorithm must exist for $\mathcal{C}$,
- no information about the code $\mathcal{C}$ can be obtained from the generator matrix G'.

The third condition eliminates some classes of well known "decodable" codes such as generalized Reed-Solomon codes (as shown by V.M. Sidelnikov and S.O. Shestakov [98]), and concatenated codes (as shown by N. Sendrier [92]). The class of binary Goppa codes [62] as suggested by McEliece seems to satisfy these 3 conditions.

Definition 6. *Let $g(z) \in \mathbb{F}_{2^m}[z]$, $L = \{\alpha_1, \ldots, \alpha_n\} \subset \mathbb{F}_{2^m}$ such that $\forall i, g(\alpha_i) \neq 0$. The Goppa code $\Gamma(L, g)$, of length n over $\mathbb{F}_2$, is the set of codewords, i.e. $n-$tuples $(c_1, \ldots, c_n) \in \mathbb{F}_2^n$, satisfying*

$$\sum_{i=1}^{n} \frac{c_i}{z - \alpha_i} \equiv 0 \pmod{g(z)}.$$

Table 6. A code based public key cryptosystem

Secret Key:

- G a generator matrix of a binary linear $[n, k, d]$ code $\mathcal{C}$ for which a polynomial time decoding algorithm $\mathcal{A}$ is known,
- S a non-singular random $k \times k$ binary matrix,
- P a random binary $n \times n$ permutation matrix.

Public Key: $G' = SGP$ and $t = \lfloor (d-1)/2 \rfloor$.

Encryption :

. Message : $m \in \{0,1\}^k$,
. Cryptogram : $c = mG' + e$ where $e \in \{0,1\}^n$ satisfies $w(e) = t$.

Decryption : Since $w(eP^{-1}) = w(e)$, successively compute :

. $mS = \mathcal{A}(cP^{-1}) = \mathcal{A}((mS)G + eP^{-1})$,
. $m = (mS)S^{-1}$.

Proposition 2. *The dimension k of $\Gamma(L,g)$ and its minimal distance d satisfy*

$$k \geqslant n - m \deg g(z)$$
$$d \geqslant \deg \bar{g}(z) + 1.$$

where $\bar{g}(z)$ is the lowest degree perfect square which is divisible by $g(z)$.

Remark 3. For irreducible Goppa codes (i.e. codes for which $g(z)$ is irreducible), we deduce that the minimum distance satisfies $d \geqslant 2 \deg g(z) + 1$.

6.1 Cryptanalysis

McEliece recommended using an irreducible binary Goppa code of length 1024 with $L = \mathbb{F}_{2^{10}}$ and $g(z)$ an irreducible polynomial of degree 50. Since the number of monic irreducible polynomials of degree 50 over $\mathbb{F}_{2^{10}}$ is given by $(\sum_{d|50} \mu(d) 2^{500/d})/50$ (where μ is the Möbius function), this gives about 2^{500} candidates which clearly prevents any exhaustive search. However, two other kind of attacks can be envisaged against McEliece's cryptosystem:

- a structural attack,
- a generic attack.

A Structural Attack. A structural attack against McEliece's cryptosystem consists in studying the algebraic structure of the public code $\mathcal{C}$ in order to build a decoder (or at least to find some parameters of the hidden code). Remember that L and $g(z)$ are the two essential parameters for the decoding algorithm. Until know, there does not exist any algorithm which takes as input a generator matrix of a Goppa code and which outputs these two data. However, as pointed out by J.K. Gibson, if a generator matrix G of a binary Goppa code and L are known, it is then possible to find back the polynomial $g(z)$ [58] . Hence one can devise a cryptanalysis in three steps:

1. fix a permutation of $\mathbb{F}_{2^m}$ say $\bar{L} = \{\beta_1, \ldots, \beta_{2^m}\}$,
2. search for a permutation π of the columns of G' which transforms the public matrix into the generator matrix $\bar{G}$ of a $\Gamma(\bar{L}, \bar{g})$ Goppa code,
3. compute $\bar{g}$ from $\bar{G}$ and $\bar{L}$ and use the decoder of $\Gamma(\bar{L}, \bar{g})$ to decode the public code $\mathcal{C}$.

In [1], C.M. Adams and H. Meijer claim that there is no more than one permutation which satisfies step 2 of the cryptanalysis. This is not true, as proved by J.K. Gibson [58], who showed that there exists at least $m2^m(2^m - 1)$ such permutations. Unfortunately for $m = 10$, this represents less than $2^{-8713}\%$ of all the permutations !Nevertheless, P. Loidreau and N. Sendrier developed a nice attack when the polynomial $g(z)$ has only binary coefficients [73]. They use the support splitting algorithm (SSA) [93] which is able to decide if two linear codes are equivalent and outputs the permutation. Their structural attack uses the fact that Goppa codes defined from a binary polynomial have a non-trivial automorphism group (and so the automorphism group of the corresponding public code is also non-trivial). This cryptanalysis brings out weak keys in McEliece's cryptosystem even if their number is negligible as compared to the number of possible keys. A "real" structural attack to date necessitates a proper classification of Goppa codes.

A Generic Attack. Without the knowledge of L and g, it seems that it is computationally hard to make the difference between a random matrix and the generator matrix of a Goppa code. This is the Goppa code distinguishing problem (see section 6.4):

Name : GD
Input : $G(k,n)$ a binary matrix,
Question : Does there exists $m \in \mathbb{N}$, $L \subset \mathbb{F}_{q^m}$ and $g(z) \in \mathbb{F}_{q^m}[z]$ such that G be a generator matrix of the $\Gamma(L,g)$ code ?

Since there does not exist any suitable algorithm which uses the underlying Goppa code structure of McEliece's cryptosystem, cryptanalysis of the system boils down to the general problem of the decoding of a random binary linear code (the G-SD problem). In fact, cryptanalysis of McEliece's cryptosystem relies on a variant of the G-SD problem. Indeed, the weight t of the error

is linked to the parameters of the code. Let $n = 2^m$, it seems that for irreducible Goppa codes the dimension k always satisfies $k = n - mt$, hence $t = (n-k)/\log_2(n)$. The underlying problem to solve is then the following:

Name : GPBD (Goppa Parametrized Bounded Decoding)
Input : G a fullrank binary matrix $k \times n$, $y \in \{0,1\}^n$
Question : Does there exists $e \in \{0,1\}^n$ such that $y+e$ be a linear combination of rows from G and $w(e) \leqslant (n-k)/\log_2 n$?

This problem is NP-complete [43].

McEliece's cryptosystem with its original parameters can be cryptanalysed in $2^{64.2}$ binary operations using the algorithms to solve the SD problem [26]. Johansonn and Jönsson algorithm can output a cleartext from a list of 1024 cryptogram in $2^{59.5}$ operations. In order to obtain a security level of 2^{80} the parameters to use are [17]:

$$m = 11, n = 2048, k = 1685, t = 33\,.$$

For a security level of 2^{128}, a set of possible parameters is [46]:

$$m = 12, n = 4096, k = 3604, t = 41\,.$$

In [19], the authors proposed a bound on "future improvements" in attacks against the McEliece's cryptosystem, and suggested that designers use this bound to "choose durable parameters".

Remark 4. In its original form, the cryptosystem is vulnerable to active attacks where an intruder modifies the cryptogram and uses as an oracle a deciphering machine. The protocol is also vulnerable to message replay. That is to say that an intruder is able to distinguish the fact that two cryptogram come from the same plaintext and in this context he can devise an attack which can recover the message in less than 8 iterations for the original parameters.

6.2 Niederreiter's Variant

In 1986, Niederreiter [83] defined the dual version of McEliece's cryptosystem using the parity check matrix of the code instead of the generator matrix (see table 7). From a security point of view Niederreiter's cryptosystem and McEliece's cryptosystem are equivalent (if used with exactly the same parameters [72]). However they differ from a practical point of view. Unlike McEliece's cryptosystem, it is not necessary to use a pseudo-random generator for encryption process. Notice, however that the plaintext is a n-binary word of weight t, hence we need a practical algorithm which maps the integers between 1 and $\binom{n}{t}$ to the set of words of weight t and length n and vice-versa. Such algorithms can be found in [47,91].

Niederreiter's cryptosystem allows to reduce by a factor of 2 the size of the public key. Indeed, the matrix H can be expressed as $H = (I_{n-k} \mid M)$, hence it is enough to store the $(n-k) \times n$ matrix M. Such a trick is impossible in McEliece's cryptosystem since if $G' = (I_k \mid M)$ and the original message is not random, the cryptogram $c = mG' + e$ would reveal a part of the plaintext.

Table 7. Niederreiter's cryptosystem

Secret key :

- A binary linear code $\mathcal{C}[n,k,d]$ for which there exists a polynomial algorithm $\mathcal{A}$ able to correct $t \leqslant \lfloor (d-1)/2 \rfloor$ errors,
- $S(n-k, n-k)$ an invertible matrix,
- $P(n,n)$ a permutation matrix.

Public key : $(H' = SHP,\ t)$ where H is a parity check matrix of $\mathcal{C}$.

Encryption :

. Message : $m \in \{0,1\}^n$ of weight t,
. Cryptogram : $c = H'\,{}^t\!m$.

Decryption :

. Compute $S^{-1}c = HP\,{}^t\!m$,
. Since $w(P\,{}^t\!m) \leqslant t$, apply $\mathcal{A}$ to find back $P\,{}^t\!m$,
. Compute $m = {}^t(P^{-1}P\,{}^t\!m)$.

Since the public key in Niederreiter's cryptosystem is smaller and the plaintext is a word of small weight, this implies that the number of operations involved during the encryption process is less than what is done in McEliece's cryptosystem. Finally, depending on the parameters, the transmission rate (number of information symbols/ number of transmitted symbols) which is equal to $\log_2 \binom{n}{t}/(n-k)$ can be better or worst that those of McEliece (k/n).

Table 8 sums up these differences and makes a comparison with the RSA cryptosystem when used with a 2048 modulus and a public exponent e equal to $2^{16}+1$ as in `openssl` toolbox (the complexity is given as the number of binary operations to perform per information bit):

Table 8. A comparison between McEliece, Niederreiter and RSA cryptosystems

	McEliece $(2048, 1718, t=30)$	Niederreiter $(2048, 1718, t=30)$	RSA-2048 $e=2^{16}+1$
Public-key size (Kbytes)	429.5	69.2	0.5
Transmission rate	83.9%	67.3%	100%
Encryption complexity	1025	46.63	40555
Decryption complexity	2311	8450	6557176, 5

Remark 5. Notice that in his original paper, Niederreiter suggested using either a binary $[104, 24, 32]$ code (obtained by concatenation of other binary codes) or a $[30, 12, 19]$ Reed-Solomon code over $\mathbb{F}_{31}$. These two codes were verified as insecure by Brickell and Odlyzko [23] using the LLL algorithm.

6.3 Hardware and Software Implementations

To assess the performances of Mc Eliece's cryptosystem, several implementation have been realized. In [22], on a 32 bits processor, for a security level of 2^{128}, the software implementation of Mc Eliece's cryptosystem gains an order of magnitude for both encryption and decryption compared to RSA-2048 (CPU cycles are divided by 5 per byte to encrypt and by 100 per byte to decrypt).

An 8-bit version for AVR microprocessors and for FPGA is described in [39]. Once again, results show that Mc Eliece's cryptosystem gives better results compared to RSA but not compared to elliptic cryptosystems.. A smart card implementation (16 bits processor) is described in [101], ciphering and deciphering is done in less than 2 seconds for a 2048 code length.

Hardware implementations of Mc Eliece's cryptosystem gave rise to several side channel attacks [102,97,66,27,80].

6.4 The Goppa Distinguishing Problem

This problem has been stated in [32] by N. Courtois, M. Finiasz and N. Sendrier. It has been widely believed for ten years that this problem was computationally hard. As a consequence, this hardness assumption has been used in numerous proofs of security of code based cryptosystems [22,84,38,29,35]. However, notice that even if a security proof cannot be stated for a cryptosystem, it does not mean that there exists an efficient cryptanalysis against this scheme. Unformally speaking, it just means that it cannot be formally proved that an efficient algorithm breaking the scheme can be turned into an efficient algorithm being able to solve a well known difficult problem.

In 2010, J.C Faugère, A. Otmani, L. Perret and J.-P. Tillich proposed the first algorithm which can decide if a binay (k, n) matrix is a random one or generates a Goppa code [41]. The distinguisher is highly discriminant for high rate code (i.e. when k is near from n).

The main idea is to compute the rank of a linear system deduced from the generator matrix G. Goppa codes are a subset of alternant codes whose parity check matrix is:

$$V_r(x,y) = \begin{pmatrix} y_1 & y_n \\ y_1 x_1 & y_n x_n \\ \vdots & \vdots \\ y_1 x_1^{r-1} & y_n x_n^{r-1} \end{pmatrix}$$

where $x_i, y_i \in \mathbb{F}_{q^m}$. The corresponding alternant code (whose dimension is greater or equal than $n - mr$) is Ker $V_r(x, y) \cap \mathbb{F}_q^n$. Using this matrix, one can build a polynomial decoder which can correct up to $\lfloor r/2 \rfloor$ errors.

By defintion of the public encryption matrix G, we have $V_r(x, y)G^t = 0$, where the elements x_i and y_i are the solution of the system:

$$\{g_{i,1} y_1 x_1^j + \cdots + g_{i,n} y_n x_n^j = 0 \mid i \in \{1, \ldots, k\},\ j \in \{0, \ldots, r-1\}\}\,. \quad (8)$$

For the parameters used in Mc Eliece's cryptosystem, such a system cannot be solved. Moreover, if we recover the x_i's and the y_i's only $r/2$ errors can be decoded instead of r. However, a distinguisher can be designed from this system. Using a linearization process, this system can be transformed in another one with k equations and $\binom{mr}{2}$ unknowns. For high-rate Goppa codes, the rank of this system is (with high probability):

$$mr((2\ell + 1)r - 2^\ell - 1)\,,$$

where $\ell = \lfloor \log_2 r \rfloor + 1$. It holds that the rank of the same system, obtained from a random binary matrix G, will be 0 or $\binom{mr}{2} - k$ depending wether $k \geqslant \binom{mr}{2}$ or not. Table 9 gives, for codes of length 2^m, the smallest r for which the distinguisher does not work. This result has to be seriously taken into account for the parameters to use in a code based cryptosystem whose security proof relies on the hardness of GD assumption.

Table 9. Smallest order r of a binary Goppa code of length $n = 2^m$ for which the distinguisher does not work

m	8	9	10	11	12	13	14	15	16	17	18	19	20	21	22	23
r_{min}	5	8	8	11	16	20	26	34	47	62	85	114	157	213	290	400

7 Some other Code Based Cryptosystems

During the past 20 years, a lot of code based cryptosystems have been designed. Here are few comments about these schemes and a list of bibliographical notes for further reading. Some of the protocols listed below are described in [107].

A Pseudo Random Generator. In 1996, B. Fischer and J. Stern [47] defined a set of strongly one-way functions related to the SD problem. Using this set, they described an efficient pseudo random generator which can output 3500 bits/sec as compared to an RSA based generator (512 bits modulus) which outputs 1800 bits/sec. Their scheme has been improved in 2007 [54] using regular words and circulants codes (see table 10 section 8.3).

A Signature Scheme. A signature scheme is a protocol where the recipient of a message M can check its integrity, the sender's identity and such that the sender cannot refute that he sent M. Usually the sender does not sign the message M itself but a hash value $h(M)$ of M. Every public key cryptosystem can be used to sign a message, using the deciphering algorithm with $h(M)$ as input, and the output of this algorithm as the signature s. The recipient uses s and the public key of the sender as inputs of the enciphering algorithm and checks that the output is equal to $h(M)$. Using Niederreiter's cryptosystem, N. Courtois, M. Finiasz and N. Sendrier proposed in 2001 [32] a signature scheme which ouputs very short signatures. The main problem is that hash values lie in the set of syndromes and must match the syndrome of an error of weight t in order to apply the deciphering function. For Goppa codes, the probability for a syndrome to be a "decodable" syndrome is roughly $1/t!$. Hence instead of directly compute the value $h(m)$ the idea is to successively compute $h(m||i)$ where i is a counter which is increased by 1 until a decodable syndrome be obtained ($||$ is the concatenating operator). The proof of security of the scheme [34] relies on the hardness of GPBD problem (see sec. 6.1) and GD problem (see sec. 6.4). Notice however that, since t must be small (at mots $t!$ attempts are needed to find a decodable syndrome), the scheme must use very large Goppa codes to resist against the various ISD algorithms. Since the dimension of a Goppa code is $n-mt$, it means that CFS uses high rate Goppa code and thus the GD problem falls in the area of parameters where it can be easily solved ! To reach a security level of 2^{80}, the scheme uses a code of length 2^{21} and codimension 210, and produces 211 bits signature. The size of the public matrix H is $52.5Mb$. This size can be reduced using parallel CFS [44]. CFS has been implemented in hardware on a FPGA (Field Programmable Gate Array) giving a signature time of 0.86 second [21] for a security level of 2^{63}.

A Hash Function. At Mcrypt 2005, a provably collision resistant family of hash functions have been proposed by D. Augot, N. Finiasz and N. Sendrier [6]. The Fast Syndrome Based Hash function is based on the Merkle-Damgård design [36] which consists in iterating a compression function $\mathcal{F}$. This function takes as input a word of s bits, maps it to a word of length n and weight t and computes its syndrome from a given $r \times n$ parity check matrix (with $r < s$). The mapping is done using regular words in order to speed up the process.

Definition 7. *Let consider a binary word of size n as n/t consecutive blocks of size t. A (n, t) regular word is a word which has exactly one non-zero coordinate in each block.*

From an algorithmic point of view, the generation of (n, t) regular worlds is obviously easiest than the one of constant weight words. The security of the hash function relies on two new NP-complete problems [43] linked to the original SD problem: RSD and 2-RNSD.

Name : RSD (Regular Syndrome Decoding);
Input : H a fullrank $r \times n$ binary matrix , an integer t and a syndrome y,
question : Does there exists a (n, t) regular word $e \in \{0, 1\}^n$ such that $H^t e = y$?
Name : 2-RNSD (2-Regular Null Syndrome Decoding)
Input : H a full rank $r \times n$ binary matrix r, p an integer,
Question : Does there exists a 2-regular (n, p) word e such that $H^t e = 0$?

Remark 6. A 2-regular (n, p) word is a word of length n such that each of the p consecutive blocks of size n/p contains either zero or two one.

Depending on the value of n, r and t, the hash function can be cryptanalysed using ISD algorithms or Wagner's generalized birthday technique [108]. Taking into account this two kind of attacks, the size of the output functions must be of at least 5ℓ bits for a security level of 2^ℓ. The proposed scheme has two main drawbacks:

- the size of the matrix H is large (around 1Mbytes for the parameters suggested in [6]). Paradoxically, the speed of the compression function can be improved with larger n while keeping a constant security level of 2^{80},
- usually the security of a hash function must be half its output size.

In 2007, an improvement of this scheme has been proposed by N. Finiasz, P. Gaborit and N. Sendrier [45]. Unfortunately the proposed parameters lead to two kind of cryptanalysis [90,48]. Taking into account these two attacks, a new version has been proposed for the SHA-3 challenge [5], but the function was quite slow and was not selected for the second round of competition. Later, an optimization (RFSB) has been proposed in [14]. The RFSB hash function runs at 13.62 cycles/byte while SHA-256 runs at 15 cycles/byte.

An Identity Based Identification Scheme. The main problem in "real life" public key cryptography is to establish a link between a public key and its owner's identity. In 1984, Shamir introduced the notion of identity based public key cryptography [95]. The concept make use of a trusted third party: the KGC (Key Generation Center). This one has a master public key and a master secret key. From an identity i and the master public key, any one can derive the public key linked to i. In 2004, Bellare, Neven and Namprempre described a generic method to derive an identity base identification scheme from a standard authentication scheme [11]. As usual this concept has only been applied to number theory schemes. In 2007 [29], P.-L. Cayrel, P. Gaborit and M. Girault considered the combination of two code based schemes (CFS signature scheme and Stern's identification scheme) in order to produce the first identity based identification scheme using error correcting codes. The generation of Alice's parameters is obtained from an execution of the CFS signature's scheme. Hence in order to prevent an intruder to be able to compute Alice's secret key from her identity, one has to consider the parameters that guarantee the security of the CFS scheme. The drawback is that the CFS scheme uses very long Goppa codes while Stern's scheme uses shorter ones. Since the same matrix has to be used by the KGC and by the identification process, this will overload the communication complexity.

A Ring Signature Scheme. A t-out-of-N threshold ring signature scheme is a protocol which enable any t participating users belonging to a set of N users to produce a signature in such a way that the verifier cannot determine the identity of the t actual signers. Classical t-out-of-N threshold ring signature schemes based on number theory have complexity $\mathcal{O}(tN)$. Using Stern's three-pass identification scheme, Aguilar et al. [2] defined the first t-out-of-N threshold ring signature scheme whose complexity is $\mathcal{O}(N)$. Performances of the scheme has been improved in [28] and a security proof is given in [40].

8 Improving Code Based Cryptosystems

There are essentially two drawbacks in code based cryptography. First, some protocols needs the generation of constant weight. This is a problem which involves computation which slow down the whole process. Next, all the schemes depend on a public matrix whose size is greater than the usual public data used in number theory based cryptography. An issue to the first problem is to used regular words (see preceeding section) instead of constant weight words. For the second problem, numerous research have been done in order to find codes with a "compact" representation. At this stage, it is important to distinguish protocols which use Goppa codes (like Mc Eliece or CFS) from those which use random codes.

8.1 List Decoding Algorithms, Specific Polynomials

In Mc Eliece's cryptosystem, for a given keysize, the security level will be increased by adding extra errors. Symmetrically, adding extra errors makes it possible to use shorter keys while keeping a similar security level, but it also requires the receiver to decode the additional errors. Let t be the error capacity of the code, in [4] authors described a "list decoding algorithm" which can correct up to $(n - \sqrt{n(n-4t-2)})/2) \geqslant t+1$ errors which is an improvement of a first algorithm decribed in [16]. Since we add extra errors, encrypting distinct codewords can lead to the same cryptogram. A list decoding algorithm outputs a list of candidates. Hence, if the initial message m has been first formatted, before computing mG, it should be easy to find back the correct codeword. Adding only one extra error, a security level of 2^{80} can be obtained using a $(1632, 1269, 33)$ Goppa code instead of $(2048, 1751, 27)$ Goppa code. The size of the public matrix will be 460647 bits instead of 520047 bits, i.e. 12% smaller [16]. Using a list decoding algorithm leads to shorter keys at the expense of a moderately increased decryption time.

Another idea to reduce the size of the public key, is to use Goppa codes over $\mathbb{F}_q$ built on polynomials of the form g^{q-1} where $g \in \mathbb{F}_q^m$ is an irreducible polynomial of degree t over $\mathbb{F}_q^m$. From [103], these codes have a better error-correction capacity: they can correct up to $\lfloor qt/2 \rfloor$ errors. Combining this trick with the preceeding one, a $[1633, 1297, 49]$ code over $\mathbb{F}_7$ with 2 extra errors achieves a security level of 2^{128} and leads to a public matrix of 1223423 bits [20]. For the same security level, over $\mathbb{F}_2$, the size of the matrix will be 1537536 bits using a $[2960, 2288, 57]$ code with one extra error.

8.2 Quasi Cyclic and Dyadic Codes

Another way to reduce the size of the public key in Mc Eliece's cryptosystem is to used some structured codes which admit a "compact" representation. This issue has been first addressed in 2005 by P. Gaborit [52] by using set of s quasi-cyclic subcodes of a given BCH code. The particularity of quasi-cyclic codes is that the whole generator matrix can be derived from the knowledge of few rows. Hence it is enough to publish these few rows (a kind of compressed version of the public matrix) instead of the whole matrix. In 2007, M. Baldi and F. Chiaraluce proposed to use quasi-cyclic LDPC codes [7]. LDPC codes are defined by a very sparse parity-check matrix and can be represented in a compact form. These two propositions have been cryptanalyzed in [85].

In 2009, two new modifications have been proposed using alternant quasi-cyclic codes and quasi dyadic codes [12,79] and cryptanalyzed in [41,57]. The generator matrix of these two families can be derived from the knowledge of one row. Only the binary version of quasi dyadic codes has not been cryptanalyzed. With these codes, the size of the public key of Mc Eliece's cryptosystem and CFS signature scheme can be highly reduced [9] (see table 10).

8.3 Circulant Codes

For protocols using random codes, a particular class of quasi-cyclic codes can be used, those whose generator matrix is obtained by concatenation of circulant matrix.

Definition 8. *A $r \times r$ circulant matrix is such that the $r - 1$ latest rows are obtained by cyclic shifts of the first row.*

It was shown in [55] that, if one admits a small constraint on the size n of the code then such codes behave like purely random codes (in particular they satisfy the Gilbert-Varshamov bound). Hence they are well suited to be used in code base schemes for which a random matrix is needed. Although all classical algorithms used to find a word of given weight in a code do not give better results when applied to quasi cyclic codes, nowadays it is not known if the decoding of a random quasi cyclic code is an NP-complete problem.

In 2007, a modification of Stern's identification scheme has been proposed using as public matrix H, the concatenation of two $k \times k$ circulant matrices (the identity matrix and a random one) [53]. This way, the public matrix can only be described from the first line of the random matrix which in particular decreases the size of the data which must be stored by the prover. The underlying difficult problem upon which the security of the scheme is linked can be stated as follows:

Name : Syndrome Decoding of Double Circulant Linear Codes
Input : $H(k, 2k)$ a double binary circulant matrix , s a binary column vector with r coordinates, p an integer.
Question : Is there a binary vector e of length n such that $H^t e = s$ and $w(e) \leq p$?

Nowadays, it is not known if this problem is NP-complete.

Using this same trick and the regular words C. Laudauroux, P Gaborit, and N. Sendrier have defined in 2007 a modified version of Fischer-Stern's algorithm in order to speed the output of the generator : the SYND pseudo random generator [54]. They obtain this way a pseudo random generator as fast as AES in counter mode[67] with few memory requirement (around 1Kbytes). Moreover, the scheme has a formal proof of security.

8.4 Codes Over $\mathbb{F}_q$

Stern's identification scheme has two major drawbacks:

1. since the probability of a successful impersonation is 2/3 for Stern's construction instead of 1/2 as in the case of Fiat-Shamir's protocol based on integer factorization, Stern's scheme uses more rounds to achieve the same security, typically 28 rounds for an impersonation resistance of 2^{16},
2. there is a common data shared by all users (from which the public identication is derived) which is very large, typically 66 Kbits. In Fiat Shamir's scheme, this common data is 1024 bits long.

In [30], using the q-SD problem, the authors proposed a 5-pass identification scheme for which the success probability of a cheater is 1/2, reducing this way the number of rounds needed for an identification process. Using quasi dyadic codes, they also reduce the size of the public data.

We sum up in table 10 the characteristics of this different improvements when applied to various code based schemes.

Remark 7. For the modified version of Stern's identification scheme there exists a variant in which the secret key is embedded in the public one. This allows to reduce again the size of the public and private data but increases the complexity computation and the global transmission rate (see [53] for more details).

9 Secret Sharing Schemes

A (k, n) secret sharing scheme is a protocol where a secret S is split into n pieces, each one being distributed to n users. If strictly fewer than k users meet together, they must not be able to compute S. Any assembly of k (or more) users can retrieve S. This problem was first considered by A. Shamir and he gives a solution using interpolation of polynomials over $\mathbb{Z}_p$, the secret being the constant term of a polynomial f of degree $k-1$. Each participant owns a pair $(i, f(i))$ $(i \in \mathbb{Z}_p^\times)$ and using Lagrange's formulas, any k users can compute f and deduce its constant term [94]. R.J. McEliece and D.V. Sarwate show that this scheme can be generalized using Reed-Solomon codes [77] for which a polynomial time decoding algorithm is known. Let $\{\alpha_1, \ldots, \alpha_n\}$ be the non-zero elements of the field $\mathbb{F}$ and $\mathcal{C}$ an $[n, k]$ RS code over $\mathbb{F}$, then each word $(m_0, \ldots, m_{k-1})$ can be encoded into the codeword $c = (c_1, \ldots, c_n)$ such that $c_i = m(\alpha_i)$ where $m(x) = \sum_{j=0}^{k-1} m_j x^j$ (Shamir's scheme corresponds to the case where $n+1$ is prime and $\alpha_i = i$). The secret to be shared is the information symbol m_0. Table 11 describes the protocol. When r users meet together, they know r symbols (and their positions) of the whole codeword c. The remaining $n - r$ symbols are called *erasures*: simply replace them with 0 and they become special errors whose positions are known.

Remark 8. Notice that since the protocol is used over $\mathbb{F}_q$ we have $n = q - 1$.

Proposition 3. *Reed-Solomon codes can polynomially decode n_e errors and n_ε erasures provided that $2n_e + n_\varepsilon < n - k + 1$.*

Table 10. Some characteristics of the improved schemes

McEliece[79]	
(n, k)	$(4096, 2048)$
t	128
Pub.key	4 Ko
Security level	2^{128}

QD version

Pseudo Random Generator[54]	
(n, k)	$(8192, 256)$
t	32
Data	1.03 Ko
Trans. rate	1Gbits/sec
Security level	2^{152}

QC version + Regular words

CFS[9]	
(m, t)	(21,10)
Signature cost	$2^{22.8}$
Pub. key	24.49 Mo
Security level	$2^{81.5}$

QD Version

SD identification scheme[53]	
(n, k)	$(634, 317)$
t	69
Public data	634 bits
Private data	951 bits
Trans. rate	40096 bits
Security level	2^{85}

QC Version

q-SD identification scheme ($\mathbb{F}_{2^8}$)[30]	
(n, k)	$(134, 67)$
t	49
Public data	1072 bits
Private data	1072 bits
Trans. rate	33040 bits
Security level	2^{87}

QD Version

Table 11. A code based secret sharing scheme

Secret : $m_0 \in \mathbb{F}_q$

Secret sharing :

. Compute the codeword $c = (c_1, \ldots, c_n)$ from the information symbols $(m_0, \ldots, m_{k-1})$, $(m_1, \ldots, m_{k-1})$ being randomly generated.
. Each user receives a pair (i, c_i).

Secret recovering :

. From $r(\geqslant k)$ pairs $(i_1, c_{i_1}), \ldots, (i_r, c_{i_r})$, build an n bits word c' such that $c'_i = c_i$ if $i \in \{i_1, \ldots, i_r\}$, $c'_i = 0$ otherwise.
. Use the erasure decoding algorithm to compute c and then m_0.

In our case, we have $n_e = 0$ and $n_\varepsilon = n - r$, thus if $r \geqslant k$, every assembly of r users can compute the whole codeword c using the decoding algorithm of RS codes and deduce m_0.

Remark 9. Notice that $m_0 = -\sum_{i=1}^{n} c_i$. Moreover the encoding of RS code can be done in an efficient way without the generator matrix of the code. Hence in this protocol, there is no need to store this matrix.

This protocol has a non-negligible advantage as compared to Shamir's scheme. Suppose that a dishonest party want to denied access to the secret to legitimate users by tampering some of the pieces c_i (or being less paranoiac, just envisage that some c_i's have been tampered with some "natural" phenomena). Let t be the number of invalid c_i. Suppose r users meet together and t of them have corrupted pieces, the whole codeword c can be computed if $2t+n-r < n-k+1$, i.e. $r \geqslant k + 2t$. Hence, if some pieces are damaged, it is still possible to retrieve the secret. On the other hand, since there are n users, the opponent has to alter more than $\lfloor (n-k)/2 \rfloor$ pieces to ensure that the secret be inaccessible.

A more general situation is to specify some users who have greater privileges of access to the secret than to others. An access structure consists of all subsets of participants that should be able to compute the secret but that contains no proper subset that also could determine it. J.L. Massey proposed to treat this problem using linear codes and the notion of minimal codewords [68],[3].

10 Conclusion

While code based cryptosystems use only elementary operations over the two elements field, they were not really considered by cryptographic community because of the size of the public data. Since these last years, numerous works have been developed in order to enhance the performance of code based cryptography leading to realistic alternatives to number based theory schemes even in constrained environments such as smart cards or RFID tags. Nowadays code based cryptography has to be considered as a real alternative to number theory based cryptography especially since:

. despite several speedups and improvements, best cryptanalysis against the Syndrome Decoding problem is still exponential whereas it is subexponential for factoring,
. there does not exist a quantum algorithm which can polynomially solve the SD problem while Shor's algorithm can factor an integer N in $\mathcal{O}((\log N)^3)$ operations on a quantum computer.

Steganography

1 Introduction

Steganography (from greek *steganos*, or "covered", and *graphie*, or "writing") is the art and sciencc of hidding a secret message within an ordinary message (the *cover-medium*) in such a way that no one, apart from the sender and intended recipient, even realizes there is a hidden message. While cryptography intends to make a message unreadable from a third party without hiding the secret communication, the aim of steganography is covert communication to hide the message from a third party. As an increasing amount of data is stored on computers and transmitted over networks, multimedia objects like image, audio and video files are today's most common cover-media.

Usually, the sender extracts from the cover-medium some of its components to construct a *cover-data* vector (for example the least significant bit of each byte of the cover medium). Then, the message is embedded into the cover-data to produce the *stego-data*. Finally, the cover-data is replaced by the stego-data in the cover-medium, which gives the *stego-medium* communicated to the recipient. From the *stego-medium*, the recipient uses a recovering algorithm in order to extract the embedded message. The embedding and recovering algorithms form the *steganographic scheme* (or *stegoscheme*).

Only the sender and the receiver should be able to tell if the stego-medium carries an hidden message or not. This means that the stego-medium should be statistically indistinguishable from the cover-medium. Especially, it is of importance to embed the message while modifying as less components of the cover-data as possible.

2 Definitions, Properties

Definition 9 (Stegoscheme). *Let $\mathcal{A}$ a finite alphabet, $r, n \in \mathbb{N}$ such that $r < n$, $\mathbf{x} \in \mathcal{A}^n$ denote the* cover-data, *$\mathbf{m} \in \mathcal{A}^r$ denote the message to embed, and T*

be a strictly positive integer. A stegoscheme *is defined by a pair of functions Ext and Emb such that:*

$$\begin{array}{ll} Emb : \mathbb{F}_2^n \times \mathbb{F}_2^r \longrightarrow \mathbb{F}_2^n & Ext(Emb(\mathbf{x},\mathbf{m})) = \mathbf{m} \\ Ext\ : \mathbb{F}_2^n \longrightarrow \mathbb{F}_2^r & d(\mathbf{x}, Emb(\mathbf{x},\mathbf{m})) \leq T \end{array}$$

where $d(.,.)$ *denotes the Hamming distance over* $\mathcal{A}^n$.

We focus in this paper on binary stegoscheme, i.e. $\mathcal{A} = \mathbb{F}_2$. The efficiency of a stegoscheme is usually evaluated through two quantities: the embedding efficiency and the relative payload.

Definition 10 (Embedding efficiency). *The* average embedding efficiency *of a stegoscheme, is usually defined by the ratio of the number of message symbols we can embed by the average number of symbols changed. We denote it by* e.

Definition 11 (Relative payload). *The* relative payload *of a stegoscheme, denoted by* α, *is the ratio of the number of message symbols we can embed by the number of (modifiable) symbols of covered data.*

3 LSB Embedding

The simplest and most common steganographic algorithm uses LSB (Least Significant Bit) embedding. Let us assume that the cover-medium is an image composed of n pixels. The cover-data is the sequence $x_1, \ldots, x_n$ where x_i is the LSB of the ith pixel of the image. The message to embed is composed of n bits $m_1, \ldots, m_n$. The functions Ext and Emb are defined as:

$$\begin{array}{lcl} Emb : \mathbb{F}_2^n \times \mathbb{F}_2^n & \longrightarrow & \mathbb{F}_2^n \\ \quad ((x_1, \ldots, x_n), (m_1, \ldots, m_n)) & \longmapsto & (m_1, \ldots, m_n) \\ Ext\ : \mathbb{F}_2^n & \longrightarrow & \mathbb{F}_2^n \\ \quad (y_1, \ldots, y_n) & \longmapsto & (y_1, \ldots, y_n) \end{array}$$

Hence, each bit of the cover-data conveys one bit of the message, and if the bits of the message are uniformally distributed (which should be the case if it has been encrypted before) then on average one bit over 2 is not modified in the cover-data, i.e. on average we modify only one bit to insert two bits of the message. Hence for this system: $\alpha = 1$ and $e = 2$.

Unfortunately, one can easily detect the presence of a secret message by looking at the image histogram. Let us consider each pixel as an integer, and denote by:

- $h[j]$ the number of pixels whose value is j in the cover-medium,
- $h_s[j]$ the number of pixels whose value is j in the stego-medium.

Notice that if a pixel is equal to $2i$ in the cover-medium and if the bits of the message to hide are uniformally distributed, then in the stego-medium, this same pixel is equal to $2i$ (with probability $1/2$) or $2i+1$ (with probability $1/2$), hence:

$$E(h_s[2i]) = \frac{h[2i] + h[2i+1]}{2}.$$

Similarly, a pixel whose value is $2i+1$ gives rise to a pixel equal to $2i+1$ (with probability $1/2$) or $2i$ (with probability $1/2$), hence:

$$E(h_s[2i+1]) = \frac{h[2i]+h[2i+1]}{2} = E(h_s[2i]) .$$

Such a result shows that LSB embedding has a tendency to even out the histogram within each pair of bin representing a pair $(2i, 2i+1)$. This is the starting point of several powerful attacks against this scheme.

Another drawback of this scheme comes from the embedding efficiency. Let us suppose that the message we aim to hide contains $2n/3$ bits. It is obvious that the size of the message has no impact on the embedding efficiency for the LSB scheme, we will always (on average) modify 1 bit of the cover-data to insert two bits of the message.

Let us now consider the cover-data as a vector composed of $n/3$ blocks (x_0, x_1, x_2) and the message as vector of $n/2$ blocks (m_0, m_1). For each block, apply the following algorithm to compute the stego-data:

If $x_0 \oplus x_2 \neq m_0$ **and** $x_1 \oplus x_2 \neq m_1$ **then** flip x_2
elsif $x_0 \oplus x_2 \neq m_0$ **then** flip x_0
elsif $x_1 \oplus x_2 \neq m_1$ **then** flip x_1

In each block of the cover-data, the probability that one bit is changed is:

$$1 - \Pr(x_0 \oplus x_2 = m_0 \text{ and } x_1 \oplus x_2 = m_1) = \frac{3}{4} .$$

Thus the embedding efficiency of this scheme is:

$$e = \frac{2n/3}{(3/4)(n/3)} = \frac{8}{3} > 2 .$$

Hence, less bits are modified in this scheme to insert the message as compared to the LSB scheme.

4 From LSB Embedding to Matrix Embedding and Coding Theory

In the preceeding scheme, in order to extract the message m, consider the stego-data as $n/3$ blocks of three bits (y_0, y_1, y_2) and compute for each block:

$$m_0 = y_0 \oplus y_2, \quad m_1 = y_1 \oplus y_2 .$$

Let $y = \begin{pmatrix} y_0 \\ y_1 \\ y_2 \end{pmatrix}$ and $m = \begin{pmatrix} m_0 \\ m_1 \end{pmatrix}$, then for each block (y_0, y_1, y_2), the extraction algorithm computes $m = Hy$ where

$$H = \begin{pmatrix} 1\,0\,1 \\ 0\,1\,1 \end{pmatrix} ,$$

is a matrix over $\mathbb{F}_2$. This method named matrix embedding has been proposed in 1998 by Crandall [33]. Notice that H is the parity check matrix of the $[3,1]$ binary Hamming code.

Definition 12. *The binary Hamming code is a linear code whose columns of the parity check matrix are all the non zero vectors of $\mathbb{F}_2^n$. Code length is $2^n - 1$, dimension is $2^n - 1 - n$ and minimal distance is 3.*

To embed any message m in a fixed cover-data x we have to solve the equation $Hy = m$ and we substitute x by y. Since there exists $e \in \mathbb{F}_2^3$ such that $y = x + e$, the embedding process is equivalent to find e such that $He = m - Hx$.

With this method, we can embed 2 bits in 3 pixels and at most one bit is modified. Remember that a good steganographic scheme has to embed as much information as possible in the cover with as few changes as possible. Suppose now that we want to embed any sequence of p bits into a set of s fixed pixels allowing one change at most. What is the minimum value of s ? Since there are 2^p sequences of p bits and since changing at most 1 bit in s gives $s+1$ new pixels, then we must have $s + 1 \geqslant 2^p$.

Theorem 1. *Let H be the parity check matrix of the $[2^p-1, 2^p-1-p]$ Hamming code, let $x \in \mathbb{F}_2^{2^p-1}$, the system*

$$He = m - Hx\,,$$
$$\omega(e) \leqslant 1\,.$$

where $\omega(e)$ denotes the Hamming weight of e, always admits, for any $m \in \mathbb{F}_2^p$, a solution $e \in \mathbb{F}_2^{2^p-1}$.

Proof. Since H contains all the non zero vectors of $\mathbb{F}_2^p$, if $m - Hx \neq 0$, then $m - Hx$ is one of the column of H.

From this, we can deduce the following theorem:

Theorem 2. *Let H be the parity check matrix of the $[2^p-1, 2^p-1-p]$ Hamming code, the corresponding stegoscheme verifies:*

$$\alpha = \frac{p}{2^p - 1}, \; e = \frac{p}{1 - 2^{-p}}\,.$$

Proof. H can be used to embed p bits in $2^p - 1$ pixels, hence the relative payload is $p/(2^p - 1)$. During the embedding process, the $2^p - 1$ bits are not modified with probability $1/2^p$, and exactly one bit is modified with probability $1 - 1/2^p$. The average number of bits modified is thus $0 \times 1/2^p + 1 \times (1 - 1/2^p)$.

From this theorem, we can see that embedding efficiency increases with p while relative payload decreases (see tab. 12). Hamming codes are well suited when the size of the message to embed is a small fraction of the cover-data since many bits can be embedded with a single change. For example, when the size of the message is 18% of the size of the cover-data, 9 bits of information are embedded

Table 12. Relative payload α_p and embedding efficiency e_p for stegoscheme defined from the $[2^p-1, 2^p-1-p]$ Hamming code

p	α_p	e_p
1	1	2
2	0.667	2.667
3	0.429	3.429
4	0.267	4.267
5	0.161	5.161
6	0.093	6.093
7	0.055	7.055
8	0.031	8.031
9	0.018	9.018

with a single bit modification. Notice that when $p = 1$, matrix embedding leads to classical LSB embedding. Moreover, for any relative payload α, since one has to choose the largest α_p such that $\alpha_p \geqslant \alpha$ to embed a message using Hamming codes, this method boils down to LSB embedding when $\alpha > 2/3$.

Let us now consider a random binary linear $[n, k]$ code $\mathcal{C}$ and let $x \in \mathbb{F}_2^n$ be a stego-data. To build a stegoscheme from $\mathcal{C}$, we have to solve the following system:

$\forall m \in \mathbb{F}_2^{n-k}$, find $e \in \mathbb{F}_2^n$, such that:

$$\begin{aligned} He &= m - Hx\,, \\ \omega(e) &\leqslant T\,. \end{aligned} \tag{9}$$

where H is a $(n-k, n)$ parity check matrix of $\mathcal{C}$, and T must be as "small" as possible in order to minimize the number of changes in x. Hence, for any message m, we have to solve an instance of the well-known SD problem which is NP-complete. In other words, for general linear codes, computing the vector e is a problem whose complexity will exponentially increase with n.

Now for any code $\mathcal{C}$, we have to answer to the following questions:

1. What is the maximum number of changes needed to embed a message m ?
2. What is the relative payload ?
3. What is the embedding efficiency ?

As we are going to show, all these values are well determined by the parameters of the code. The first problem is to determine for a given code $\mathcal{C}$, what is the maximal number of changes needed to embed any message m. In other words, we need an upperbound on T. Let us denote by R the covering radius of $\mathcal{C}$ which is determined by the most distant point y from the code, i.e.:

$$R = \max_{y \in \mathbb{F}_2^n} d(y, \mathcal{C})\,.$$

For any $s \in \mathbb{F}_2^{n-k}$, let $C(s) = \{e \in \mathbb{F}_2^n,\ He = s\}$. This set has 2^{n-k} members.

Definition 13. *A coset leader e_s for s is a member of $C(s)$ with the smallest Hamming weight.*

Proposition 4. *The Hamming weight of any coset leader is at most R.*

Proof. Let $z \in \mathbb{F}_2^n$, and $s = Hz$. From elementary linear algebra, $C(s) = \{x \in \mathbb{F}_2^n \mid x = z - c, c \in \mathcal{C}\}$. Let e_s be a coset leader,

$$R = \max_{y \in \mathbb{F}_2^n} d(y, \mathcal{C}) \geqslant d(z, \mathcal{C}) = \min_{c \in \mathcal{C}} \omega(z - c) = \omega(e_s).$$

The minimum number of changes in the stego-data is obtained when the solution e of the problem (9) is a coset leader of $C(m - Hx)$. Hence, this problem always admits a solution $e \in \mathbb{F}_2^n$ such that $\omega(e) \leqslant R$.

Theorem 3 (Matrix embedding theorem). *A stegoscheme defined from an $[n, k]$ binary code $\mathcal{C}$ whose covering radius is R can embed $n - k$ bits in n pixels by making at most R changes. The relative payload is $(n - k)/n$ and the embedding efficiency is $(n - k)/R_{\mathcal{C}}$ where:*

$$R_{\mathcal{C}} = \frac{1}{2^n} \sum_{x \in \mathbb{F}_2^n} d(x, \mathcal{C}),$$

is the average distance to the code.

Proof. As already mentionned, the bound on the number of changes comes from the property that the Hamming weight of any coset leader is bounded by the covering radius R of the code. Next, by definition, we have $\alpha = (n - k)/n$. Now, let us suppose that the messages to embed are uniformally distributed, so that $m - Hx$ is uniformally ditributed in $\mathbb{F}_2^{n-k}$, to find the average number of changes, we thus have to compute the expected weight of a coset leader:

$$\frac{1}{2^{n-k}} \sum_{s \in \mathbb{F}_2^{n-k}} \omega(e_s) = \frac{1}{2^n} \sum_{s \in \mathbb{F}_2^{n-k}} 2^k \omega(e_s).$$

Let $s \in \mathbb{F}_2^{n-k}$, from proposition 4, for any $x \in C(s)$, $d(x, \mathcal{C}) = \omega(e_s)$, hence:

$$\frac{1}{2^n} \sum_{s \in \mathbb{F}_2^{n-k}} 2^k \omega(e_s) = \frac{1}{2^n} \sum_{s \in \mathbb{F}_2^{n-k}} \sum_{x \in C(s)} d(x, \mathcal{C}) = \frac{1}{2^n} \sum_{x \in \mathbb{F}_2^n} d(x, \mathcal{C}).$$

since $\cup_{s \in \mathbb{F}_2^{n-k}} C(s) = \mathbb{F}_2^n$.

To end this section we will give (without proofs, see [51]) asymptotic bounds on optimal matrix embedding schemes when embedding into cover-medium containing n pixels:

Proposition 5. *Let $\mathcal{H}_2(x)$ be the binary entropy function defined by:*

$$\mathcal{H}_2(x) = -x \log_2(x) - (1 - x) \log_2(1 - x),$$

and $\mathcal{H}_2^{-1}()$ be its inverse function, then:

1. *The maximal number of bits which can be embedded making at most R changes is $n\mathcal{H}_2(R/n)$.*
2. *The average number of embedding changes to embed m bits is $n\mathcal{H}_2^{-1}(m/n)$.*
3. *The maximal embedding efficiency to embed m bits is $\frac{m/n}{\mathcal{H}_2^{-1}(m/n)}$.*

Last property can be generalized to obtain:

Proposition 6 (Sphere-covering bound [49]). *For any binary stegoscheme,*

$$e \leq \frac{\alpha}{\mathcal{H}_2^{-1}(\alpha)},$$

where α is the relative payload associated to the stegoscheme.

Remark 10. These bounds are still valid for q-ary codes using the q-entropy function:

$$H_q(x) = x\log_2(q-1) - x\log_2(x) - (1-x)\log_2(1-x)\,.$$

5 Wet Paper Codes

Usually, the sender does not use all pixels of the image to embed a message m. He may select part of the image where embedding changes will be more difficult to detect. The set of pixels which can be modified is called the *selection channel.* Most of the time, the selection channel is unknown to the receiver, he may even not know the selection rules used by the sender, we then call it a *non-shared selection channel.*

Wet paper codes, introduced in [50], have been designed to tackle the non-shared selection channel context. The idea is to consider that the cover-medium has been altered (like a sheet of paper) by rain. Hence a subset $\mathcal{W}$ of the components are "wet" and cannot be changed. Only a subset $\mathcal{D}$ of components (the "dry" components) can be modified to embed the message. During the transmission, the cover medium dries out and the receiver cannot determine $\mathcal{D}$ and $\mathcal{W}$.

Let $\mathcal{C}$ be an $[n, k]$ linear binary code, $\mathcal{D} \subset \{1, \ldots, n\}$, $\mathcal{W} = \{1, \ldots, n\} \setminus \mathcal{D}$, to build a stegoscheme for the non-shared selection channel defined by $\mathcal{D}$, we have to solve the following problem:

Let $x \in \mathbb{F}_2^n$, $\forall m \in \mathbb{F}_2^{n-k}$, find $e \in \mathbb{F}_2^n$, such that:

$$\begin{aligned} He &= m - Hx\,, \\ e_i &= 0\,,\ \forall i \in \mathcal{W}\,. \end{aligned} \tag{10}$$

Notice that in this context, we do not seek for a word of minimum weight, but for a word e whose support is contained in $\mathcal{D}$. Let $H^{\mathcal{D}}$ denote the matrix composed of the columns of H whose index is in $\mathcal{D}$, then (10) is equivalent to:

Let $x \in \mathbb{F}_2^n$, $\forall m \in \mathbb{F}_2^{n-k}$, find $\tilde{e} \in \mathbb{F}_2^{\#\mathcal{D}}$, such that:

$$H^{\mathcal{D}}\tilde{e} = m - Hx. \tag{11}$$

The problem is that $H^{\mathcal{D}}$ depends on $\mathcal{D}$, that in turn depends on the cover object, hence even if H comes from some structure code for which the computation of a coset leader is easy, the sender cannot always deduce nice properties on $H^{\mathcal{D}}$. In particular, this means that trying to choose $\tilde{e}$ as a coset leader will constitute a much harder task than computing an arbitrary coset member.

Proposition 7 ([81]). *Problem (10) has a solution if and only the matrix $G_{\mathcal{W}}$ is of full rank, where $G_{\mathcal{W}}$ is the projection over $\mathcal{W}$ of the columns of a generator matrix G of the code $\mathcal{C}$.*

Proof. Let us denote by $\pi_{\mathcal{W}}$ the projection over the set $\mathcal{W}$. Let $x \in \mathbb{F}_2^n$, notice that (10) has a solution, if and only if for any $m \in \mathbb{F}_2^{n-k}$, $\pi_{\mathcal{W}}(x) \in \pi_{\mathcal{W}}(C(m))$, where $C(m) = \{z \in \mathbb{F}_2^n \mid Hz = m\}$. Now, for any m,

$$\#\pi_{\mathcal{W}}(C(m)) = \#\pi_{\mathcal{W}}(\mathcal{C}) = 2^{\mathrm{rank}(G_{\mathcal{W}})},$$

since $C(m) = z + \mathcal{C}$, where z satisfies $Hz = m$. For any x, we must have $\pi_{\mathcal{W}}(x) \in \pi_{\mathcal{W}}(C(m))$, it means that $\pi_{\mathcal{W}}(\mathbb{F}_2^n) \subset \pi_{\mathcal{W}}(C(m))$, hence $\mathrm{rank}(G_{\mathcal{W}}) = \#\mathcal{W}$ (notice that $\#\mathcal{W} \leqslant k$ since we need to embed $n-k$ symbols in $\#\mathcal{D}$ dry symbols).

Proposition 8 ([81]). *$G_{\mathcal{W}}$ is full rank iff there is no word in $\mathcal{C}^{\perp}$ with support contained in $\mathcal{W}$.*

Proof. Can be easily deduced from the fact that there exists a word of weight δ in $\mathcal{C}^{\perp}$ iff there are δ linear dependent columns in G.

Proposition 9 ([81]). *Problem (10) has a solution for any $\mathcal{W}$ iff $\#\mathcal{W} < d_{\min}(\mathcal{C}^{\perp})$ and in this case the number of solutions is exactly $q^{k-\#\mathcal{W}}$.*

Proof. If $\#\mathcal{W} < d_{\min}(\mathcal{C}^{\perp})$ then no codeword of $\mathcal{C}^{\perp}$ has its support contained in $\mathcal{W}$ hence, from proposition 7 and 8, problem (10) has a solution. Conversely, suppose that problem (10) has a solution for any $\mathcal{W}$ and that $\#\mathcal{W} \geqslant d_{\min}(\mathcal{C}^{\perp})$. Choose a set $\mathcal{W}$ and a word c of $C^{\perp}$ such that its support be contained in $\mathcal{W}$ then, from propostion 8, $\mathrm{rank}(G_{\mathcal{W}}) < \#\mathcal{W}$ which is a contradiction with proposition 7. Last, when $\mathrm{rank}(G_{\mathcal{W}}) = \#\mathcal{W}$, the number of solutions is $\#\mathcal{C}/\#\pi_{\mathcal{W}}(\mathcal{C}) = q^{k-\#\mathcal{W}}$.

From these propositions, we deduce that for a general $[n,k]$ code $\mathcal{C}$, $n-k$ symbols can be embed in a cover medium if there are strictly less than $d^{\perp}$ wet positions. As an example, using the binary Hamming code, p bits can be embed in $2^p - 1$ bits, if there are at most $2^{p-1} - 1$ wet positions.

Remark 11. A more general result states that, for n large enough, the number of dry symbols needed on average to transmit k informations symbols is roughly equal to k [81].

6 The $\varepsilon + 1$ Matrix Embedding Scheme

In this section we describe how to use wet paper codes to transform an optimal binary matrix embedding scheme into an optimal ternary matrix embedding scheme. Let us suppose that we have a binary code $\mathcal{C}$ with embedding efficiency equal to ε, i.e. k bits can be embed in n bits by making on average k/ε changes. Let $(x_1,\ldots,x_n)$ be the cover-data obtained by taking the LSB of the n pixels of the image. Let us denote by $\mathcal{D} \subset \{1,\ldots,n\}$ the indices of the modified pixels during the embedding process, and let $\mathcal{W} = \{1,\ldots,n\} \setminus \mathcal{D}$. When the sender flips the last bit of the pixel p_i, $i \in \mathcal{D}$, he also adjusts the second LSB of p_i to insert one more bit of information. Here is the description of the embedding process:

1. Let m a message of length k, x the cover-data, find e such that $H(x+e) = m$. Let t be the Hamming weight of e (on average $t \simeq k/\varepsilon$).
2. Let $\tilde{m}$ a message of lenght t and $\tilde{x}$ the cover-data computed from the second LSB of the cover-medium. Find $\tilde{e}$ such that $\tilde{H}(\tilde{x} + \tilde{e}) = \tilde{m}$ and $\tilde{e}_i = 0$ for $i \in \mathcal{W}$ (where $\tilde{H}$ is obtained from the t first rows of H).

The value t must be communicated to the receiver, a small portion of the cover image can be used to embed this value.

Notice that instead of flipping a bit (or adding 1 if the bit is even and -1 is the bit is odd) , we now modify a pixel by adding +1 or -1 regardless its parity. On average, $k + k/\epsilon$ bits are embedded making k/ε modifications, the embedding efficiency is then:

$$\frac{k + k/\varepsilon}{k/\varepsilon} = \varepsilon + 1\,.$$

From proposition 5, if the binary stegoscheme is optimal than the maximal number of bits which can be embed making at most R changes is $n\mathcal{H}_2(R/n)$. Using this scheme with wet paper trick, we can embed at most $n\mathcal{H}_2(R/n) + R$ bits. Now,

$$n\mathcal{H}_2(R/n) + R = n(\mathcal{H}_2(R/n) + R/n)$$

$$= n(\mathcal{H}_2(R/n) + R/n \log_2(3-1)) = nH_3(R/n)$$

which is the maximal number of bits that can be embed using an optimal ternary stegoscheme (see remark 10).

The first practical steganographic scheme which incorporates the matrix embedding mechanism is the F5 algorithm [109]. A good starting point on Steganography and matrix embedding is [51]. In [82], steganography is described from a coding theory point of view and numerous bibliographical notes are given about the study of some well known codes in this context (Hamming, Golay, BCH, Reed-Solomon, $\mathbb{Z}_4$ linear codes).

References

1. Adams, C., Meijer, H.: Security-related comments regarding McEliece's public-key cryptosystem. IEEE Trans. Inform. Theory 35, 454–455 (1989)
2. Aguilar Melchor, C., Cayrel, P.-L., Gaborit, P.: A new efficient threshold ring signature scheme based on coding theory. In: Buchmann, J., Ding, J. (eds.) PQCrypto 2008. LNCS, vol. 5299, pp. 1–16. Springer, Heidelberg (2008)
3. Ashikmin, A.E., Barg, A.: Minimal vectors in linear codes. IEEE Transaction on Information Theory 44(5) (1998)
4. Augot, D., Barbier, M., Couvreur, A.: List-decoding of binary goppa codes up to the binary johnson bound. In: IEEE, ITW 2011, pp. 229–233 (October 2011)
5. Augot, D., Finiasz, M., Gaborit, P., Manuel, S., Sendrier, N.: Sha-3 proposal: Fsb. Submission to NIST (2008)
6. Augot, D., Finiasz, M., Sendrier, N.: A family of fast syndrome based cryptographic hash functions. In: Dawson, E., Vaudenay, S. (eds.) Mycrypt 2005. LNCS, vol. 3715, pp. 64–83. Springer, Heidelberg (2005)
7. Baldi, M., Chiaraluce, F.: Cryptanalysis of a new instance of McEliece cryptosystem based on qc-ldpc codes. In: IEEE International Symposium on Information Theory, ISIT 2007, pp. 2591–2595 (June 2007)
8. Barg, S.: Some new NP-complete coding problems. Probl. Peredachi Inf. 30, 23–28 (1994)
9. Barreto, P.S.L.M., Cayrel, P.-L., Misoczki, R., Niebuhr, R.: Quasi-dyadic CFS signatures. In: Lai, X., Yung, M., Lin, D. (eds.) Inscrypt 2010. LNCS, vol. 6584, pp. 336–349. Springer, Heidelberg (2011)
10. Becker, A., Joux, A., May, A., Meurer, A.: Decoding random binary linear codes in $2^{n/20}$: How $1 + 1 = 0$ improves information set decoding. In: Pointcheval, D., Johansson, T. (eds.) EUROCRYPT 2012. LNCS, vol. 7237, pp. 520–536. Springer, Heidelberg (2012)
11. Bellare, M., Namprempre, C., Neven, G.: Security proofs for identity-based identification and signature schemes. In: Cachin, C., Camenisch, J.L. (eds.) EUROCRYPT 2004. LNCS, vol. 3027, pp. 268–286. Springer, Heidelberg (2004)
12. Berger, T.P., Cayrel, P.-L., Gaborit, P., Otmani, A.: Reducing key length of the McEliece cryptosystem. In: Preneel, B. (ed.) AFRICACRYPT 2009. LNCS, vol. 5580, pp. 77–97. Springer, Heidelberg (2009)
13. Berlekamp, E.R., McEliece, R.J., van Tilborg, H.C.A.: On the intractability of certain coding problems. IEEE Transactions on Information Theory 24(3), 384–386 (1978)
14. Bernstein, D.J., Lange, T., Peters, C., Schwabe, P.: Really fast syndrome-based hashing. In: Nitaj, A., Pointcheval, D. (eds.) AFRICACRYPT 2011. LNCS, vol. 6737, pp. 134–152. Springer, Heidelberg (2011)
15. Bernstein, D.J.: Grover vs. McEliece (2008), `http://cr.yp.to/papers.html`
16. Bernstein, D.J.: List decoding for binary goppa codes (2008), `http://cr.yp.to/codes/goppalist-20081107.pdf`
17. Bernstein, D.J., Lange, T., Peters, C.: Attacking and defending the McEliece cryptosystem. In: Buchmann, J., Ding, J. (eds.) PQCrypto 2008. LNCS, vol. 5299, pp. 31–46. Springer, Heidelberg (2008)
18. Bernstein, D.J., Lange, T., Peters, C.: Smaller decoding exponents: Ball-collision decoding. In: Rogaway, P. (ed.) CRYPTO 2011. LNCS, vol. 6841, pp. 743–760. Springer, Heidelberg (2011)

19. Bernstein, D.J., Lange, T., Peters, C.: Smaller decoding exponents: Ball-collision decoding. In: Rogaway, P. (ed.) CRYPTO 2011. LNCS, vol. 6841, pp. 743–760. Springer, Heidelberg (2011)
20. Bernstein, D.J., Lange, T., Peters, C.: Wild McEliece. In: Biryukov, A., Gong, G., Stinson, D.R. (eds.) SAC 2010. LNCS, vol. 6544, pp. 143–158. Springer, Heidelberg (2011)
21. Beuchat, J.L., Sendrier, N., Tisserand, A., Villard, G.: Fpga implementation of a recently published signature scheme. Tech. Rep. 5158, Inria (March 2004)
22. Biswas, B., Sendrier, N.: McEliece cryptosystem implementation: Theory and practice. In: Buchmann, J., Ding, J. (eds.) PQCrypto 2008. LNCS, vol. 5299, pp. 47–62. Springer, Heidelberg (2008)
23. Brickell, E., Odlyzko, A.: Cryptanalysis: A survey of recent results. In: Comtemporary Cryptology - the Science of Information Integrity, pp. 501–540 (1992)
24. Canteaut, A.: Attaques de cryptosystèmes à mots de poids faible et construction de fonctions t-résilientes. PhD thesis, Université Paris VI (1996)
25. Canteaut, A., Chabaud, F.: A new algorithm for finding minimum-weight words in a linear code: Application to McEliece's cryptosystem and to narrow-sense bch codes of length 511. IEEE Transactions on Information Theory 44(1), 367–378 (1998)
26. Canteaut, A., Sendrier, N.: Cryptanalysis of the original McEliece cryptosystem. In: Ohta, K., Pei, D. (eds.) ASIACRYPT 1998. LNCS, vol. 1514, pp. 187–199. Springer, Heidelberg (1998)
27. Cayrel, P.-L., Dusart, P.: McEliece/niederreiter pkc: sensitivity to fault injection. In: International Workshop on Future Engineering, Applications and Services, FEAS (2010)
28. Cayrel, P.-L., El Yousfi Alaoui, S.M., Hoffmann, G., Véron, P.: An improved threshold ring signature scheme based on error correcting codes. In: Özbudak, F., Rodríguez-Henríquez, F. (eds.) WAIFI 2012. LNCS, vol. 7369, pp. 45–63. Springer, Heidelberg (2012)
29. Cayrel, P.-L., Gaborit, P., Girault, M.: Identity-based identification and signature schemes using correcting codes. In: Augot, D., Sendrier, N., Tillich, J.P. (eds.) WCC 2007. INRIA (2007)
30. Cayrel, P.-L., Véron, P., El Yousfi Alaoui, S.M.: A zero-knowledge identification scheme based on the q-ary syndrome decoding problem. In: Biryukov, A., Gong, G., Stinson, D.R. (eds.) SAC 2010. LNCS, vol. 6544, pp. 171–186. Springer, Heidelberg (2011)
31. Chabanne, H., Courteau, B.: Application de la méthode de décodage itérative d'omura à la cryptanalyse du système de mc eliece. Rapport de Recherche 122, Université de Sherbrooke (October 1993)
32. Courtois, N.T., Finiasz, M., Sendrier, N.: How to achieve a McEliece-based digital signature scheme. In: Boyd, C. (ed.) ASIACRYPT 2001. LNCS, vol. 2248, pp. 157–174. Springer, Heidelberg (2001)
33. Crandall, R.: Some notes on steganography (1998), Posted on the steganography mailing list
34. Dallot, L.: Towards a concrete security proof of courtois, finiasz and sendrier signature scheme. In: Lucks, S., Sadeghi, A.-R., Wolf, C. (eds.) WEWoRC 2007. LNCS, vol. 4945, pp. 65–77. Springer, Heidelberg (2008)
35. Dallot, L., Vergnaud, D.: Provably secure code-based threshold ring signatures. In: Parker, M.G. (ed.) Cryptography and Coding 2009. LNCS, vol. 5921, pp. 222–235. Springer, Heidelberg (2009)

36. Damgård, I.B.: A design principle for hash functions. In: Brassard, G. (ed.) CRYPTO 1989. LNCS, vol. 435, pp. 416–427. Springer, Heidelberg (1990)
37. Diffie, W., Hellman, M.E.: New directions in cryptography. IEEE Transactions on Information Theory IT-22(6), 644–654 (1976)
38. Dowsley, R., Müller-Quade, J., Nascimento, A.C.A.: A CCA2 secure public key encryption scheme based on the McEliece assumptions in the standard model. In: Fischlin, M. (ed.) CT-RSA 2009. LNCS, vol. 5473, pp. 240–251. Springer, Heidelberg (2009)
39. Eisenbarth, T., Güneysu, T., Heyse, S., Paar, C.: Microeliece: Mceliece for embedded devices. In: Clavier, C., Gaj, K. (eds.) CHES 2009. LNCS, vol. 5747, pp. 49–64. Springer, Heidelberg (2009)
40. El Yousfi Alaoui, S.M., Dagdelen, Ö., Véron, P., Galindo, D., Cayrel, P.-L.: Extended Security Arguments for Signature Schemes. In: Mitrokotsa, A., Vaudenay, S. (eds.) AFRICACRYPT 2012. LNCS, vol. 7374, pp. 19–34. Springer, Heidelberg (2012)
41. Faugère, J.C., Otmani, A., Perret, L., Tillich, J.P.: A distinguisher for high rate McEliece cryptosystems. IACR Eprint archive, 2010/331 (2010)
42. Fiat, A., Shamir, A.: How to prove yourself: Practical solutions to identification and signature problems. In: Odlyzko, A.M. (ed.) CRYPTO 1986. LNCS, vol. 263, pp. 186–194. Springer, Heidelberg (1987)
43. Finiasz, M.: Nouvelles constructions utilisant des codes correcteurs d'erreurs en cryptographie à clé publique. PhD thesis, Ecole Polytechnique (2004)
44. Finiasz, M.: Parallel-CFS: Strengthening the CFS McEliece-based signature scheme. In: Biryukov, A., Gong, G., Stinson, D.R. (eds.) SAC 2010. LNCS, vol. 6544, pp. 159–170. Springer, Heidelberg (2011)
45. Finiasz, M., Gaborit, P., Sendrier, N.: Improved fast syndrome based cryptographic hash function. In: ECRYPT Hash Workshop 2007 (2007)
46. Finiasz, M., Sendrier, N.: Security bounds for the design of code-based cryptosystems. In: Matsui, M. (ed.) ASIACRYPT 2009. LNCS, vol. 5912, pp. 88–105. Springer, Heidelberg (2009)
47. Fischer, J.-B., Stern, J.: An efficient pseudo-random generator provably as secure as syndrome decoding. In: Maurer, U.M. (ed.) EUROCRYPT 1996. LNCS, vol. 1070, pp. 245–255. Springer, Heidelberg (1996)
48. Fouque, P.-A., Leurent, G.: Cryptanalysis of a hash function based on quasi-cyclic codes. In: Malkin, T. (ed.) CT-RSA 2008. LNCS, vol. 4964, pp. 19–35. Springer, Heidelberg (2008)
49. Fridrich, J.: Asymptotic behavior of the ZZW embedding construction. IEEE Transactions on Information Forensics and Security 4(1), 151–153 (2009)
50. Fridrich, J., Goljan, M., Lisonek, P., Soukal, D.: Writing on wet paper. IEEE Trans. on Signal Processing 53(10), 3923–3935 (2005)
51. Fridrich, J.: Steganography in Digital Media: Principles, Algorithms, and Applications, 1st edn. Cambridge University Press, New York (2009)
52. Gaborit, P.: Shorter keys for code based cryptography. In: Proceeedings of WCC 2005, pp. 81–90 (2005)
53. Gaborit, P., Girault, M.: Lightweight code-based identification and signature. In: Proceeedings of ISIT 2007 (2007)
54. Gaborit, P., Laudauroux, C., Sendrier, N.: Synd: a fast code-based stream cipher with a security reduction. In: Proceeedings of ISIT 2007 (2007)
55. Gaborit, P., Zémor, G.: Asymptotic improvement of the gilbert-varshamov bound for linear codes. In: Proceeedings of ISIT 2006, pp. 287–291 (2006)
56. Garey, M.R., Johnson, D.S.: Computers and Intractability, A Guide to the Theory of NP-Completeness. W.H. Freeman and Company, New York (1979)

57. Gauthier Umana, V., Leander, G.: Practical key recovery attacks on two McEliece variants. IACR Eprint archive, 2009/509 (2009)
58. Gibson, J.K.: Equivalent goppa codes and trapdoors to McEliece's public key cryptosystem. In: Davies, D.W. (ed.) EUROCRYPT 1991. LNCS, vol. 547, pp. 517–521. Springer, Heidelberg (1991)
59. Girault, M.: A (non-practical) three-pass identification protocol using coding theory. In: Seberry, J., Pieprzyk, J.P. (eds.) AUSCRYPT 1990. LNCS, vol. 453, pp. 265–272. Springer, Heidelberg (1990)
60. Girault, M., Stern, J.: On the length of cryptographic hash-values used in identification schemes. In: Desmedt, Y.G. (ed.) CRYPTO 1994. LNCS, vol. 839, pp. 202–215. Springer, Heidelberg (1994)
61. Goldwasser, S., Micali, S., Rackoff, C.: The knowledge complexity of interactive proof systems. SIAM, Journal of Computing 18, 186–208 (1989)
62. Goppa, V.D.: A new class of linear error correcting codes. Probl. Pered. Inform., 24–30 (1970)
63. Grover, L.K.: A fast quantum mechanical algorithm for database search. In: Proceedings of the Twenty-Eighth Annual ACM Symposium on Theory of Computing, STOC 1996, pp. 212–219. ACM, New York (1996)
64. Grover, L.K.: Quantum mechanics helps in searching for a needle in a haystack. Phys. Rev. Lett. 79(2), 325–328 (1997)
65. Harari, S.: A new authentication algorithm. In: Wolfmann, J., Cohen, G. (eds.) Coding Theory 1988. LNCS, vol. 388, pp. 91–105. Springer, Heidelberg (1989)
66. Heyse, S., Moradi, A., Paar, C.: Practical power analysis attacks on software implementations of McEliece. In: Sendrier, N. (ed.) PQCrypto 2010. LNCS, vol. 6061, pp. 108–125. Springer, Heidelberg (2010)
67. Housley, R.: Using advanced encryption standard (aes) counter mode with ipsec encapsulating security payload (esp). RFC 3686, Network Working Group (January 2004)
68. Massey, J.L.: Minimal codewords and secret sharing. In: 6th Joint Swedish-Russian Workshop on Information Theory, pp. 276–279 (1993)
69. Johansson, T., Jönsson, F.: On the complexity of some cryptographic problems based on the general decoding problem. IEEE Transactions on Information
Theory 48(10), 2669–2678 (2002)
70. Lee, P.J., Brickell, E.F.: An observation on the security of McEliece's public-key cryptosystem. In: Günther, C.G. (ed.) EUROCRYPT 1988. LNCS, vol. 330, pp. 275–280. Springer, Heidelberg (1988)
71. Leon, J.S.: A probabilistic algorithm for computing minimum weights of large error-correcting codes. IEEE Transactions on Information Theory 34(5), 1354–1359 (1988)
72. Li, Y.X., Deng, R.H., Wang, X.M.: On the equivalence of McEliece's and niederreiter's public-key cryptosystems. IEEE Transactions on Information Theory 40(1), 271–273 (1994)
73. Loidreau, P., Sendrier, N.: Weak keys in the McEliece public-key cryptosystem. IEEE Transactions on Information Theory 47(3), 1207–1211 (2001)
74. MacWilliams, F.J., Sloane, N.J.A.: The Theory of Error-Correcting Code. North-Holland (1977)
75. May, A., Meurer, A., Thomae, E.: Decoding random linear codes in $\tilde{\mathcal{O}}(2^{0.054n})$. In: Lee, D.H., Wang, X. (eds.) ASIACRYPT 2011. LNCS, vol. 7073, pp. 107–124. Springer, Heidelberg (2011)

76. McEliece, R.J.: A public-key cryptosystem based on algebraic coding theory. JPL DSN Progress Report, pp. 114–116 (1978)
77. McEliece, R.J., Sarwate, D.V.: On sharing secrets and Reed-Solomon codes. Communications of the ACM 24(9), 583–584 (1981)
78. Merkle, R., Hellman, M.: Hiding information and signatures in trapdoor knapsacks. IEEE Trans. Inform. Theory 24, 525–530 (1978)
79. Misoczki, R., Barreto, P.S.L.M.: Compact McEliece keys from goppa codes. In: Jacobson Jr., M.J., Rijmen, V., Safavi-Naini, R. (eds.) SAC 2009. LNCS, vol. 5867, pp. 376–392. Springer, Heidelberg (2009)
80. Molter, H., Stöttinger, M., Shoufan, A., Strenzke, F.: A simple power analysis attack on a McEliece cryptoprocessor. Journal of Cryptographic Engineering 1, 29–36 (2011)
81. Munuera, C., Barbier, M.: Wet paper codes and the dual distance in steganography. Advances in Mathematics of Communications 6(3), 237–285 (2012)
82. Munuera, C.: Steganography from a coding theory point of view. Series on Coding Theory and Cryptology, vol. 8. World Scientific Publishing Co. Pte. Ltd. (2013)
83. Niederreiter, H.: Knapsack-type cryptosystems and algebraic coding theory. Problems Control Inform. Theory 15(2), 159–166 (1986)
84. Nojima, R., Imai, H., Kobara, K., Morozov, K.: Semantic security for the McEliece cryptosystem without random oracles. Designs, Codes and Cryptography 49, 289–305 (2008), doi:10.1007/s10623-008-9175-9
85. Otmani, A., Tillich, J.P., Dallot, L.: Cryptanalysis of two McEliece cryptosystems based on quasi-cyclic codes. Mathematics in Computer Science 3, 129–140 (2010)
86. Peters, C.: Information-set decoding for linear codes over $\mathbf{F}_q$. In: Sendrier, N. (ed.) PQCrypto 2010. LNCS, vol. 6061, pp. 81–94. Springer, Heidelberg (2010)
87. Prange, E.: The use of information sets in decoding cyclic codes. IRE Trans. IT-8, 85–89 (1962)
88. Quisquater, J.-J., Guillou, L.C., Berson, T.: How to explain zero-knowledge protocols to your children. In: Brassard, G. (ed.) CRYPTO 1989. LNCS, vol. 435, pp. 628–631. Springer, Heidelberg (1990)
89. Rivest, R.L., Shamir, A., Adleman, L.: A method for obtaining digital signatures and public-key cryptosystems. Commun. ACM 26(1), 96–99 (1983)
90. Saarinen, M.-J.O.: Linearization attacks against syndrome based hashes. In: Srinathan, K., Rangan, C.P., Yung, M. (eds.) INDOCRYPT 2007. LNCS, vol. 4859, pp. 1–9. Springer, Heidelberg (2007)
91. Sendrier, N.: Efficient generation of binary words of given weight. In: Boyd, C. (ed.) Cryptography and Coding 1995. LNCS, vol. 1025, pp. 184–187. Springer, Heidelberg (1995)
92. Sendrier, N.: On the structure of a randomly permuted concateneted code. In: EUROCODE 1994, 169–173. Inria (1994)
93. Sendrier, N.: Finding the permutation between equivalent linear codes: The support splitting algorithm. IEEE Transactions on Information Theory 46(4), 1193–1203 (2000)
94. Shamir, A.: How to Share a Secret. Communications of the ACM 22(11), 612–613 (1979)
95. Shamir, A.: Identity-based cryptosystems and signature schemes. In: Blakely, G.R., Chaum, D. (eds.) CRYPTO 1984. LNCS, vol. 196, pp. 47–53. Springer, Heidelberg (1985)
96. Shor, P.W.: Polynomial-time algorithms for prime factorization and discrete logarithms on a quantum computer. In: Proceedings of the 35th Annual Symposium on Foundations of Computer Science, pp. 20–22 (1994)

97. Shoufan, A., Strenzke, F., Molter, H.G., Stöttinger, M.: A Timing Attack against Patterson Algorithm in the McEliece PKC. In: Lee, D., Hong, S. (eds.) ICISC 2009. LNCS, vol. 5984, pp. 161–175. Springer, Heidelberg (2010)
98. Sidelnikov, V., Shestakov, S.: On cryptosystems based on generalized reed-solomon codes. Diskretnaya Math. 4, 57–63 (1992)
99. Stern, J.: A method for finding codewords of small weight. In: Wolfmann, J., Cohen, G. (eds.) Coding Theory 1988. LNCS, vol. 388, pp. 106–113. Springer, Heidelberg (1989)
100. Stern, J.: A new identification scheme based on syndrome decoding. In: Stinson, D.R. (ed.) CRYPTO 1993. LNCS, vol. 773, pp. 13–21. Springer, Heidelberg (1994)
101. Strenzke, F.: A smart card implementation of the McEliece PKC. In: Samarati, P., Tunstall, M., Posegga, J., Markantonakis, K., Sauveron, D. (eds.) WISTP 2010. LNCS, vol. 6033, pp. 47–59. Springer, Heidelberg (2010)
102. Strenzke, F., Tews, E., Molter, H.G., Overbeck, R., Shoufan, A.: Side channels in the McEliece PKC. In: Buchmann, J., Ding, J. (eds.) PQCrypto 2008. LNCS, vol. 5299, pp. 216–229. Springer, Heidelberg (2008)
103. Sugiyama, Y., Kasahara, M., Hirasawa, S., Namekawa, T.: Further results on goppa codes and their applications to constructing efficient binary codes. IEEE Transactions on Information Theory 22, 518–526 (1976)
104. van Tilburg, J.: On the McEliece public-key cryptosystem. In: Goldwasser, S. (ed.) CRYPTO 1988. LNCS, vol. 403, pp. 119–131. Springer, Heidelberg (1990)
105. Véron, P.: Cryptanalysis of harari's identification scheme. In: Boyd, C. (ed.) Cryptography and Coding 1995. LNCS, vol. 1025, pp. 264–269. Springer, Heidelberg (1995)
106. Véron, P.: Improved identification schemes based on error-correcting codes. Appl. Algebra Eng. Commun. Comput. 8(1), 57–69 (1996)
107. Véron, P.: Public key cryptography and coding theory. In: Woungang, I., Misra, S., Misra, S. (eds.) Selected Topics in Information and Coding Theory, vol. 7. World Scientific Publications (March 2010)
108. Wagner, D.: A generalized birthday problem. In: Yung, M. (ed.) CRYPTO 2002. LNCS, vol. 2442, pp. 288–304. Springer, Heidelberg (2002)
109. Westfeld, A.: F5-A steganographic algorithm. In: Moskowitz, I.S. (ed.) IH 2001. LNCS, vol. 2137, pp. 289–302. Springer, Heidelberg (2001)

Strong Prefix Codes of Pictures*

Marcella Anselmo[1], Dora Giammarresi[2], and Maria Madonia[3]

[1] Dipartimento di Informatica, Università di Salerno
I-84084 Fisciano (SA) Italy
anselmo@dia.unisa.it

[2] Dipartimento di Matematica, Università di Roma
"Tor Vergata", via della Ricerca Scientifica, 00133 Roma, Italy
giammarr@mat.uniroma2.it

[3] Dipartimento di Matematica e Informatica, Università di Catania
Viale Andrea Doria 6/a, 95125 Catania, Italy
madonia@dmi.unict.it

Abstract. A set $X \subseteq \Sigma^{**}$ of pictures is a code if every picture over Σ is tilable in at most one way with pictures in X. The definition of *strong prefix code* is introduced and it is proved that the corresponding family of finite strong prefix codes is decidable and it has a polynomial time decoding algorithm. Maximality for finite strong prefix codes is also considered. Given a strong prefix code, it is proved that there exists a unique maximal strong prefix code that contains it and that has a minimal size. The notion of completeness is also investigated in relation to maximality.

1 Introduction

The notion of codes in two dimensions is an interesting subject for researchers both from theoretical and applicative side due to the important role that images have nowadays in human communications. The aim is to generalize to 2D the well established theory of string codes (see [7] for a complete reference).

In the last two decades, two dimensional codes were studied in different contexts and it were defined polyomino codes, picture codes, and brick codes. A set C of polyominoes is a code if every polyomino that is tilable with (copies of) elements of C, it is so in a unique way. Most of the results show that in the 2D context we loose important properties. A major result due to D. Beauquier and M. Nivat states that the problem whether a finite set of polyominoes is a code is undecidable, and the same result holds also for dominoes ([6]). Related particular cases were studied in [1]. In [12] codes of directed polyominoes equipped with catenation operations are considered, and some special decidable cases are detected. Codes of labeled polyominoes, called bricks, are studied in [13] and further undecidability results are proved.

* Partially supported by MIUR Project *"Aspetti matematici e applicazioni emergenti degli automi e dei linguaggi formali"*, by 60% Projects of University of Catania, Roma "Tor Vergata", Salerno.

T. Muntean, D. Poulakis, and R. Rolland (Eds.): CAI 2013, LNCS 8080, pp. 47–59, 2013.

As major observation, remark that all mentioned results consider 2D codes independently from a 2D language theory. The first attempt to connect this two sides was presented by S. Bozapalidis in [8]. The paper considers codes of pictures, i.e. rectangular arrays of symbols. Between pictures there can be defined two partial concatenation operations, sometimes referred to as horizontal and vertical concatenation: pictures to be concatenated need to have same number of rows or columns, respectively. Using these operations, doubly-ranked monoids are introduced and picture codes are studied in order to extend syntactic properties to two dimensions. Unfortunately many results are again negative and involve undecidability issues. Even the definition of prefix picture codes in [11] does not lead to any wide enough class.

Very recently, in [3], a new definition for picture codes was introduced in relation to the family REC of picture languages recognized by tiling systems (see [10]). Instead of referring to horizontal and vertical concatenation, it is considered the operation of *tiling star* as defined in [14]: the tiling star of a set X is the set X^{**} of all pictures that are tilable (in the polyominoes style) by elements of X. Then X is a code if any picture in X^{**} is tilable in one way. Remark that if $X \in$ REC then X^{**} is also in REC. By analogy to the string case, it holds that if X is a finite picture code then, starting from pictures in X we can easily construct an unambiguous tiling system for X^{**} [4]. Unfortunately, despite this nice connection to the word code theory, it is proved that it is still undecidable whether a given set of pictures is a code. This is actually coherent with the known result of undecidability for unambiguity inside REC.

Looking for decidable subclasses of picture codes, in [3] the definition of *prefix code* is proposed. Pictures are then considered with a preferred scanning direction: from top-left corner to the bottom-right one. Then a picture p is a prefix of a picture q, if p coincides with the "top-left portion" of q. Observe that it is not possible to define a set X to be prefix by merely imposing that its pictures are not mutually prefixes: this would not automatically imply that X is a code. The property that is maintained going from string to pictures is then the following: if X is a prefix code, when decoding a picture p starting from top-left corner, it should be univocally decided which element in X we can start with. The formal definition of prefix sets involves special kind of polyominoes: in fact "pieces" of pictures get in the intermediate steps of a decoding process are not in general pictures itself. And this is actually what makes the major difference when passing from string to pictures.

In [3] it is proved that it is decidable whether a finite set of picture is a prefix set and that, as in the 1D case, every prefix set of pictures is a code. Moreover a polynomial time decoding algorithm for finite prefix codes is presented. Prefix codes for pictures inherit several properties from the original family of prefix string codes and several non trivial examples can be exhibited. Nevertheless it is worth to say that the definition is sometimes difficult to manage, since the presence of a specific picture in the prefix set depends on a tiling combination of (possibly) many other pictures in the same set.

In this paper we take back the definition of prefix set for strings and generalize the notion to 2D in a different way. We introduce the notion of *horizontal prefix* and *vertical prefix* restricted to pairs of pictures with the same number of rows or columns, respectively. More specifically, picture x is horizontal prefix of p when x is prefix of p and moreover x and p have the same numbers of rows. Similarly for vertical prefix. Then a set of picture X will be said *strong prefix* if no picture in X is prefix of another one and moreover no two pictures p and q in X can have a common prefix x that is horizontal prefix of p and vertical prefix of q.

Strong prefix sets are again a decidable family of picture codes with a simple polynomial decoding algorithm. Then maximal strong prefix sets are considered and the maximality of a given finite strong prefix set is shown to be decidable. The embedding of a strong prefix set in a maximal one can be realized by a polynomial algorithm. Moreover it is proved that, given a strong prefix set X, there exists a unique maximal strong prefix set containing X that has minimal size. This result is quite surprising since a picture can be "grown" in several ways, following the horizontal or the vertical direction. Some results concerning completeness and its relations with maximality for strong prefix sets are also described.

2 Preliminaries

We introduce some definitions about pictures and two-dimensional languages (see [10] for a complete reference).

A *picture* over a finite alphabet Σ is a two-dimensional rectangular array of elements of Σ. Given a picture p, $|p|_{row}$ and $|p|_{col}$ denote the number of rows and columns, respectively; $|p| = (|p|_{row}, |p|_{col})$ denotes the picture *size*. Differently from the one-dimensional case, we can define an infinite number of empty pictures namely pictures of size $(m, 0)$ and of size $(0, n)$, for all $m, n \geq 0$, will be called *empty columns* and *empty rows*, and denote by $\lambda_{m,0}$ and $\lambda_{0,n}$ respectively.

The set of all pictures over Σ of fixed size (m, n) is denoted by $\Sigma^{m,n}$, while Σ^{m*} and Σ^{*n} denote the set of all pictures over Σ with m rows and n columns, respectively. The set of all pictures over Σ is denoted by Σ^{**}. A *two-dimensional language* (or *picture language*) over Σ is a subset of Σ^{**}.

The *domain* of a picture p is the set of coordinates

$$\mathrm{dom}(p) = \{1, 2, \ldots, |p|_{row}\} \times \{1, 2, \ldots, |p|_{col}\}.$$

We let $p(i, j)$ denote the symbol in p at coordinates (i, j). Positions in dom(p) are ordered following the lexicographic order: $(i, j) < (i', j')$ if either $i < i'$ or $i = i'$ and $j < j'$. Moreover, to easily detect border positions of pictures, we use initials of words "top", "bottom", "left" and "right": then, for example the *tl-corner* of p refers to position $(1, 1)$. A *subdomain* of dom(p) is a set d of the form $\{i, i+1, \ldots, i'\} \times \{j, j+1, \ldots, j'\}$, where $1 \leq i \leq i' \leq |p|_{row}$, $1 \leq j \leq j' \leq |p|_{col}$, also specified by the pair $[(i, j), (i', j')]$. The *subpicture of* p associated to $[(i, j), (i', j')]$ is the portion of p corresponding to positions in the subdomain and is denoted by $p[(i,j),(i',j')]$. Given pictures x, p, with $|x|_{row} \leq |p|_{row}$ and $|x|_{col} \leq |p|_{col}$, we say that x is a *prefix* of p if x is a subpicture of p corresponding to its top-left portion, i.e. if $x = p[(1,1),(|x|_{row},|x|_{col})]$.

Dealing with pictures, two "classical" concatenation products are defined. Let $p, q \in \Sigma^{**}$ pictures of size (m,n) and (m',n'), respectively, the *column concatenation* of p and q (denoted by $p \varobar q$) and the *row concatenation* of p and q (denoted by $p \ominus q$) are partial operations, defined only if $m = m'$ and if $n = n'$, respectively, as:

$$p \varobar q = \begin{array}{|c|c|} \hline p & q \\ \hline \end{array} \qquad\qquad p \ominus q = \begin{array}{|c|} \hline p \\ \hline q \\ \hline \end{array}.$$

These definitions can be extended to define two-dimensional languages row- and column- concatenations and *row-* and *column- stars*. If $X \subseteq \Sigma^{**}$ is a set of pictures then the row- and column- star of X will be denoted by $X^{\ominus *}$ and $X^{\varobar *}$, respectively. ([10]).

We also consider another interesting star operation for picture language introduced by D. Simplot in [14]. The idea is to compose pictures in a way to cover a rectangular area without the restriction that each single concatenation must be a $\ominus$ or $\varobar$ operation. For example, the following figure sketches a possible kind of composition that is not allowed applying only $\ominus$ or a $\varobar$ operations.

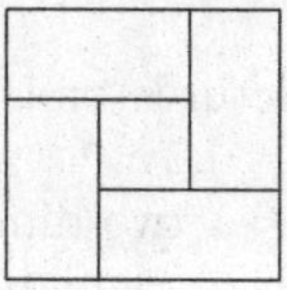

Definition 1. *The* tiling star *of X, denoted by X^{**}, is the set of pictures p whose domain can be partitioned in disjoint subdomains $\{d_1, d_2, \ldots, d_k\}$ such that any subpicture p_h of p associated with the subdomain d_h belongs to X, for all $h = 1, ..., k$.*

Language X^{**} is called the set of all tilings by X in [14]. In the sequel, if $p \in X^{**}$, the partition $t = \{d_1, d_2, \ldots, d_k\}$ of dom(p), together with the corresponding pictures $\{p_1, p_2, \ldots, p_k\}$, is called a *tiling decomposition* of p in X.

In this paper, while dealing with tiling star of a set X, we will need to manage also non-rectangular "portions" of pictures composed by elements of X: those are actually labeled polyominoes, that we will call polyominoes, for the sake of simplicity. We extend to polyominoes the notion of tiling decomposition in a set of pictures X. We also define a sort of tiling star that, applied to a set of pictures X, produces the set of all polyominoes that have a tiling decomposition in X. If a polyomino p belongs to the polyomino star of X, we say that p is *tilable* in X.

2.1 Two-Dimensional Codes

In this paper we refer to the definition of code given in [3] where two-dimensional codes are introduced in the setting of the theory of recognizable two-dimensional languages and coherently to the notion of language unambiguity as in [2,4].

Definition 2. *Let Σ be a finite alphabet. $X \subseteq \Sigma^{**}$ is a* code *iff any $p \in \Sigma^{**}$ has at most one tiling decomposition in X.*

We show some simple examples. Let $\Sigma = \{a, b\}$ be the alphabet.

Example 1. Let $X = \left\{ \boxed{a\,b}, \begin{array}{|c|}\hline a \\ \hline b \\ \hline \end{array}, \begin{array}{|cc|}\hline a & a \\ a & a \\ \hline \end{array} \right\}$. It is easy to see that X is a code. Any picture $p \in X^{**}$ can be decomposed starting at tl-corner and checking the size $(2,2)$ subpicture $p[(1,1),(2,2)]$: it can be univocally decomposed in X. Then, proceed similarly for the next contiguous size $(2,2)$ subpictures.

Example 2. Let $X = \left\{ \boxed{a\,b}, \boxed{b\,a}, \begin{array}{|c|}\hline a \\ a \\ \hline \end{array} \right\}$. Notice that no picture in X is prefix of another picture in X; nevertheless X is not a code. Indeed picture $\begin{array}{|ccc|}\hline a & b & a \\ a & b & a \\ \hline \end{array}$ has the two following different tiling decompositions in X: $t_1 = \begin{array}{|c|c|}\hline a\,b & a \\ \cline{1-1} a\,b & a \\ \hline \end{array}$ and $t_2 = \begin{array}{|c|c|}\hline a & b\,a \\ \cline{2-2} a & b\,a \\ \hline \end{array}$.

For the rest of the section we summarize the main results in [3].

First the problem whether a given set of pictures is a code is in general undecidable.

With the aim of defining a subclass of codes that is decidable two-dimensional prefix codes are then introduced as a generalization to two dimensions of the family of string prefix codes.

The basic idea in defining a *prefix code* is to prevent the possibility to start decoding a picture in two different ways (as it is for the prefix string codes). One major difference going from 1D to 2D case is that, while any initial part of a decomposition of a string is still a string, the initial part of a decomposition of a picture has not necessarily a rectangular shape: it is in general a (labeled) polyomino. Hence a notion related to tiling and referred to as *covering* is introduced. Informally a picture p is *covered* by (pictures in a set) X, if p can be tiled with pictures that possibly "exit" p throughout the bottom and the right border. For example, in the figure below, the picture with thick borders is *(properly) covered* by the others. Refer to [3] for the formal definition.

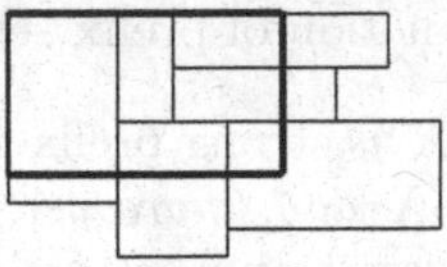

Then the definition of prefix set given in [3] is equivalent to the following one.

Definition 3. *A set X is prefix if and only if every $x \in X$ cannot be properly covered by pictures in X.*

It is easy to verify that the set X of Example 1 is prefix. On the contrary, the set X of Example 2 is not prefix: picture $\begin{array}{|c|}\hline a \\ a \\ \hline \end{array}$, can be covered by two copies of $\boxed{a\,b}$.

Definition 3 is a good generalization of prefix set of strings in fact it is proved that a prefix set is a code referred to as *prefix code*. Contrarily to the case of all other known classes of 2D codes, the family of finite prefix codes has the important property to be decidable. Furthermore a polynomial decoding algorithm for a finite prefix picture code is given.

Maximality is a central notion in theory of (word) codes: the subset of any code is a code, and then the investigation may restrict to maximal codes. In 1D, the notion of maximality of (prefix) codes is related to the one of (right) completeness. The notions of maximality and completeness are extended to picture codes. In 2D, a prefix code $X \subseteq \Sigma^{**}$ is said *maximal prefix* over Σ if it is not properly contained in any other prefix code over Σ. In 1D maximality coincides with completeness (for thin codes), while in 2D, complete prefix codes are a proper subset of maximal prefix codes, that is fully characterized. Maximality of finite prefix codes is decidable.

3 Strong Prefix Codes

We now introduce an alternative definition of prefix set for pictures; it can be viewed as a more direct generalization from the notion of prefix sets for strings. We will refer to it as *strong prefix*. Despite the notion will correspond to a smaller family of codes, such family has many remarkable properties that generalize the theory of codes from one to two dimensions.

We first specialize the definition of "picture p prefix of picture q" in the particular cases when p and q have the same number of rows (columns resp.): in this case p will correspond to a left (top resp.) portion of q. Here below there is the formal definition.

Definition 4. *Let $p, q \in \Sigma^{**}$. Picture p is a* horizontal prefix *of q, denoted by $p \leq_h q$, if there exists $x \in \Sigma^{**}$ such that $q = p \oplus x$. Picture p is a* vertical prefix *of q, denoted by $p \leq_v q$, if there exists $y \in \Sigma^{**}$ such that $q = p \ominus y$. If x (y resp.) is empty the horizontal (vertical, resp.) prefix is* proper.

Then, combining the previous two definitions of horizontal and vertical prefix for pictures, we generalize the notion of prefix set from strings to pictures.

Definition 5. *Let $X \subseteq \Sigma^{**}$. X is* strong prefix *if there is no picture in X that is prefix of another picture in X and, moreover, for any two different pictures p and q in X, there does not exist any non-empty picture $x \in \Sigma^{**}$ such that $x \leq_h p$ and $x \leq_v q$.*

Remark that the previous Definition 5 can be stated by referring directly to picture domains by imposing that any two distinct pictures in X differ in the common part of the domain, i.e. they do not overlap if we let their tl-corners coincide. For example, the following two pictures p and q cannot both belong to a strong prefix set:

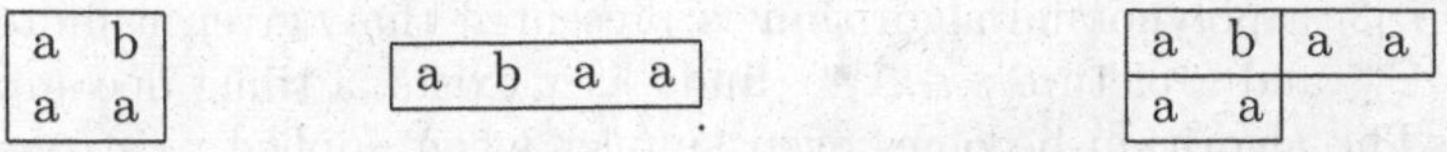

p q p and q "overlapped"

Let us give some examples.

Example 3. Let $X = \left\{ \boxed{a\,b\,a}\,,\ \boxed{a\,b\,b}\,,\ \boxed{\begin{array}{c} b \\ b \end{array}}\,,\ \boxed{\begin{array}{cc} a & a \\ a & a \end{array}}\,,\ \boxed{\begin{array}{cc} a & a \\ a & b \end{array}}\,,\ \boxed{\begin{array}{cc} b & a \\ a & a \end{array}}\,,\ \boxed{\begin{array}{cc} b & a \\ a & b \end{array}}\,,\ \boxed{\begin{array}{cc} b & b \\ a & a \end{array}}\,,\ \boxed{\begin{array}{cc} b & b \\ a & b \end{array}} \right\}$.
Language X is strong prefix: no two pictures in X overlap on their tl-corner.

Example 4. Let $\Sigma = \{a, b\}$ and $X = \left\{ \boxed{\begin{array}{cc} a & a \\ b & b \\ b & b \end{array}}\,,\ \boxed{\begin{array}{ccc} a & b & a \\ a & b & a \\ b & b & b \end{array}}\,,\ \boxed{\begin{array}{cccc} a & b & a & b \\ a & b & a & a \\ b & b & a & b \end{array}}\,,\ \boxed{\begin{array}{cccc} b & a & b & b \\ a & a & b & b \\ b & a & a & b \end{array}} \right\}$. Language X is strong prefix. Note that $X \subseteq \Sigma^{3*}$.

Definition 5 seems a valid generalization from the 1D case; in fact it can be easily proved the following result.

Proposition 1. *If $X \subseteq \Sigma^{**}$ is strong prefix then X is a code.*

Proof. Suppose by contradiction that there exists a picture $u \in \Sigma^{**}$ that admits two different tiling decompositions in X, say t_1 and t_2. Now, let (i_0, j_0) the smallest position (in lexicographic order) of u, where t_1 and t_2 differ. Position (i_0, j_0) corresponds in t_1 to position $(1, 1)$ of some $x_1 \in X$, and in t_2 to position $(1, 1)$ of some $x_2 \in X$, with $x_1 \neq x_2$. Consider now the size of x_1 and x_2: if $|x_1|_{row} = |x_2|_{row}$, then one of them is a horizontal prefix of the other one. If, instead, $|x_1|_{row} \neq |x_2|_{row}$, suppose without loss of generality that $|x_1|_{row} \geq |x_2|_{row}$. This implies that either x_2 is a prefix of x_1, (in the case $|x_1|_{col} \geq |x_2|_{col}$) or there exists a non-empty picture $x \in \Sigma^{**}$ such that $x \leq_h x_1$ and $x \leq_h x_2$ (in the case $|x_1|_{col} < |x_2|_{col}$), against X strong prefix. □

Applying directly the definition it can be shown that, given a set of pictures X, one can decide whether X is strong prefix in time polynomial with respect to the total area of pictures in X (just compare every pair of pictures). Hence strong prefix sets are a decidable family of picture codes.

Remark that strong prefix sets are in particular prefix sets (in the sense introduced in [3]). They are a proper subclass of prefix sets (see Example 1) that is simpler to handle with. The definitions of "prefix set" and of "strong prefix set" both reduce to the definition of "prefix set" of strings, when restricted to one-row pictures (identifiable with strings). Moreover observe that they also coincide on a more general kind of languages: languages $X \subseteq \Sigma^{m*}$ and $X \subseteq \Sigma^{*n}$ (see Example 4). Such languages can be viewed as "one-dimensional" languages, over the alphabet $\Sigma^{m,1}$, or $\Sigma^{1,n}$. Their properties will be addressed to in the sequel.

Strong prefix codes inherit some properties from the prefix codes family. For example, in [3] a polynomial algorithm is presented that, given a finite prefix code $X \subseteq \Sigma^{**}$ and a picture $p \in \Sigma^{**}$, finds, if it exists, a tiling decomposition of p in X. The algorithm becomes even simpler when applied to strong prefix codes.

4 Maximal Strong Prefix Codes

In this section we present the main results concerning maximality of codes introduced in the previous section.

Definition 6. *A strong prefix set $X \subseteq \Sigma^{**}$ is* maximal strong prefix *over Σ if it is not properly contained in any other strong prefix set over Σ; that is, $X \subseteq Y \subseteq \Sigma^{**}$ and Y strong prefix imply $X = Y$.*

The following lemma gives a general tool to decide whether a finite strong prefix set is maximal strong prefix. It shows that if a strong prefix set is not maximal strong prefix, there is always a "small" picture that witnesses it. As a consequence, one can check whether a strong prefix set is maximal strong prefix by restricting the test to a finite number of pictures.

Lemma 1. *Let X be a strong prefix finite set, $r_X = max\{|x|_{row}, x \in X\}$, and $c_X = max\{|x|_{col}, x \in X\}$. If X is not maximal strong prefix, then there exists $p' \in \Sigma^{**}$, $p' \notin X$, such that $X \cup \{p'\}$ is still strong prefix and $|p'|_{row} \leq r_X$, $|p'|_{col} \leq c_X$.*

Proof. Let $p \in \Sigma^{**}, p \notin X$, such that $X \cup \{p\}$ is still strong prefix. If $|p|_{row} \leq r_X$, $|p|_{col} \leq c_X$, then let $p' = p$. Otherwise, let $h = \min\ \{|p|_{row}, r_X\}$, $k = \min\ \{|p|_{col}, c_X\}$ and let p' be the prefix of p of size (h, k). Let us show that $X \cup \{p'\}$ is strong prefix.

By contradiction, suppose that $X \cup \{p'\}$ is not strong prefix. Then, if $|p'| = (r_X, c_X)$, since X is strong prefix, there exists $x \in X$ prefix of p'. Therefore x is also prefix of p, against $X \cup \{p\}$ strong prefix. If instead, $|p'| \neq (r_X, c_X)$, let us suppose $|p'| = (|p|_{row}, c_X)$ (the case $|p'| = (r_X, |p|_{col})$ is analogous). Since X is strong prefix, two different cases can occur. In the first case, there exists $x \in X$ such that x is prefix of p' or p' is prefix of x. But, if $x \in X$ is prefix of p' then $x \in X$ is also prefix of p, against $X \cup \{p\}$ strong prefix. If instead p' is prefix of x, then it must be $p' \leq_v x$ and, since $p' \leq_h p$, this contradicts $X \cup \{p\}$ strong prefix. In the second case, there exist $x \in X$ and $y \in \Sigma^{**}$ such that $y \leq_h p'$ and $y \leq_v x$. Since $p' \leq_h p$, it implies $y \leq_h p$ against $X \cup \{p\}$ strong prefix. □

The following proposition is a direct consequence of Lemma 1.

Proposition 2. *It is decidable whether a finite strong prefix set X is maximal strong prefix.*

Let X be a strong prefix set. By applying the definition, one can show that if X is maximal prefix then it is maximal strong prefix. We let open the question whether the vice versa holds.

4.1 Embedding of Strong Prefix Codes

Lemma 1 can be also used to embed finite strong prefix codes into maximal finite ones.

Proposition 3. *Let $X \subseteq \Sigma^{**}$ be a finite strong prefix set. Then it is possible to construct a finite set $Y \subseteq \Sigma^{**}$ such that Y is maximal strong prefix and $X \subseteq Y$.*

Proof. Let $r_X = max\{|x|_{row}, x \in X\}$, $c_X = max\{|x|_{col}, x \in X\}$ and Z be the finite set of pictures of size (m, n) with $m \leq r_X$ and $n \leq c_X$. Language Y can be incrementally obtained starting from X, and adding one by one all pictures in Z that do not overlap any picture of the current Y. Let us show that Y is maximal strong prefix. By contradiction, suppose that Y is not maximal strong prefix. Then, from Lemma 1, there exists $p \in \Sigma^{**}, p \notin Y$, such that $Y \cup \{p\}$ is still strong prefix and $|p|_{row} \leq max\{|y|_{row}, y \in Y\} = r_X$, $|p|_{col} \leq max\{|y|_{col}, y \in Y\} = c_X$. But this is not possible since all pictures p, with $|p|_{row} \leq r_X$, $|p|_{col} \leq c_X$, have already been considered. □

The proof of the previous proposition shows the correctness of an algorithm that constructs a maximal strong prefix code containing a given strong prefix code X. The procedure can output different sets depending on the order in which it processes the candidate pictures to be added. See the next example.

Example 5. Let $X = \left\{ \boxed{a\ b\ a}\ , \boxed{a\ b\ b}\ , \boxed{\begin{matrix} b \\ b \end{matrix}} \right\}$. Following Proposition 3, we can construct the following two sets, that are both maximal strong prefix sets and contain X:

$$Y = X \cup \left\{ \boxed{\begin{matrix} a & a \\ a & a \end{matrix}}, \boxed{\begin{matrix} a & a \\ a & b \end{matrix}}, \boxed{\begin{matrix} a & a \\ b & a \end{matrix}}, \boxed{\begin{matrix} a & a \\ b & b \end{matrix}}, \boxed{\begin{matrix} b & a \\ a & a \end{matrix}}, \boxed{\begin{matrix} b & a \\ a & b \end{matrix}}, \boxed{\begin{matrix} b & b \\ a & a \end{matrix}}, \boxed{\begin{matrix} b & b \\ a & b \end{matrix}} \right\} \text{ and}$$

$$Y' = X \cup \left\{ \boxed{a\ a}\ , \boxed{\begin{matrix} b \\ a \end{matrix}} \right\}.$$

In 1D, given a finite prefix code, there exists a unique maximal finite code that contains it, and that is minimum both in cardinality and in the total length of its strings. We ask whether a similar situation holds in 2D. The setting looks like more involved, since pictures can "extend" both horizontally and vertically. Surprisingly, the following result holds.

Define the *area* of a picture of size (m, n) as $m \times n$, and the *size* of a finite picture language X, denoted $size(X)$, as the sum of the areas of its pictures. Then define the following order on pictures p and q with $size(p) = (m, n)$ and $size(q) = (m', n') : p \leq q$, if $m \times n < m' \times n'$; when p and q have the same area, $p \leq q$ when (m, n) is lexicographically smaller than (m', n'); and when p and q have same area and same size, $p \leq q$ when the string obtained reading p row by row is lexicographically smaller than the string obtained reading q row by row.

Proposition 4. *Let $X \subseteq \Sigma^{**}$ be a finite strong prefix set. There exists a unique finite maximal strong prefix set $Y \subseteq \Sigma^{**}$ that contains X and has minimum size. Moreover Y is minimum in cardinality too.*

Proof. Specialize the algorithm provided by the proof of Proposition 3, by choosing pictures from the finite set of pictures of size (m, n) with $m \leq r_X$ and $n \leq c_X$, following the order on pictures defined above. First show that this algorithm provides a solution of minimal size. Let $A = \{p_1, \cdots, p_h\}$ be the maximal strong prefix set returned by the execution of the algorithm on X, and $O = \{q_1, \cdots, q_k\}$ be a maximal strong prefix set of minimal size. Suppose that pictures in both A and O are in increasing order. The goal is to prove that $A = O$. Suppose by the contrary that $A \neq O$. Consider p_1. If $p_1 \in O$ then $p_1 = q_1$ (by minimality of p_1) and repeat the considerations for p_2. Suppose without loss of generality that $p_1 \notin O$. Since O is maximal strong prefix then p_1 "overlaps" with some $q_i \in O$. Let x be the "intersection" of p_1 with q_i. Then one could replace in O, picture q_i with x (x is "compatible" with all the other pictures in O). Then the minimality of O implies that $size(x) = size(q_i)$, that is $q_i \leq_h p_1$, or $q_i \leq_v p_1$. Now q_i cannot be a proper prefix of p_1 (for the minimality of p_1) then $q_i = p_1$ against $p_1 \notin O$. This proves that $A = O$. Then, given two maximal strong prefix sets of minimal size, say O_1 and O_2, they are both equal to A, and then $O_1 = O_2$.

Finally observe that the proof also holds when minimality with respect to the cardinality of sets is concerned. □

Example 6. Referring to the set X in Example 5, and using the algorithm provided by the proof of Proposition 4, one can prove that the set Y' is the maximal strong prefix set of minimum size that contains X.

Let us consider again prefix languages of pictures of fixed number of rows/columns, as in Example 4. They form a special family of strong prefix codes that warrants many properties. The following proposition regards the embedding of such languages. It states that among all possible embedding there is always one (not necessarily the minimal one), preserving the fixed number of rows/columns. Despite such languages are somehow "one-dimensional" languages, the result is not straightforward, since they have to be compared with pictures of any number of rows/columns.

Proposition 5. *Let $X \subseteq \Sigma^{m*}$ ($X \subseteq \Sigma^{*n}$, resp.) be a finite (strong) prefix set. Then it is possible to construct a finite set $Y \subseteq \Sigma^{m*}$ ($Y \subseteq \Sigma^{*n}$, resp.) such that Y is maximal strong prefix and $X \subseteq Y$.*

Proof. Setting $\Gamma = \Sigma^{m,1}$, X can be considered as a set of strings over Γ and, in particular, X is a prefix set of strings. From classical theory of codes (see e.g. [7]), we know that there exists $Y \subseteq \Gamma^*$ such that $X \subseteq Y$, Y finite maximal prefix (set of strings). Moreover, Y can be chosen so that the maximal length of a string in X is equal to the maximal length of a string in Y. Remark that $Y \subseteq \Sigma^{m*}$ and that Y, viewed as a set of pictures, is strong prefix. Let us show that Y is maximal strong prefix. By contradiction, suppose that Y is not maximal strong prefix that is there exists $p \in \Sigma^{**}$, $p \notin Y$, such that $Y \cup \{p\}$ is still strong prefix. Clearly $|p|_{row} \neq m$. If $|p|_{row} > m$, let $c_Y = max\{|y|_{col}, y \in Y\}$ and consider the prefix p' of p of size (m, n'), where $n' = min\{|p|_{col}, c_Y\}$. Then there exists $y \in Y$ such that either $p' \leq_h y$ and $p' \leq_v p$ or p is a prefix of y and this contradicts $Y \cup \{p\}$ strong prefix. If $|p|_{row} < m$, consider any picture $q \in \Sigma^{**}$ such that $p' = p \ominus q \in \Sigma^{m*}$. Picture p' can be considered as a string over Γ and, since Y considered as a set of strings is right-complete (recall that in 1D a set is maximal prefix if and only if it is right-complete [7]), there exist $y_1, \ldots, y_k \in Y$ and $r \in \Sigma^{m*}$ such that $p' \oplus r = y_1 \oplus \ldots \oplus y_k$. But this implies that either p prefix of y_1 or $z \leq_h p$ and $z \leq_v y'$ for some $z \in \Sigma^{**}$ and, again, this contradicts $Y \cup \{p\}$ strong prefix. □

4.2 Maximality and Completeness

In 1D the notion of maximality coincides with that of (right-) completeness for thin (prefix) codes. Let us compare the two notions for strong prefix two-dimensional codes. The definition below was first given in [3], and refers to the notion of covering recalled in Section 2.1.

Definition 7. *A set $X \subseteq \Sigma^{**}$ is* br-complete *if every $p \in \Sigma^{**}$ can be covered by (pictures in) X.*

Proposition 6. *Let X be a strong prefix code. If X is br-complete then it is strong prefix maximal.*

Proof. By contradiction, suppose that there exists $p \in \Sigma^{**}$, $p \notin X$ such that $X \cup \{p\}$ is still strong prefix. Since X is br-complete, p can be properly covered by pictures in X. We will show that this contradicts $X \cup \{p\}$ strong prefix. Indeed, let $x \in X$ be the picture that covers position $(1,1)$ of p. If $|x|_{row} = |p|_{row}$ and $|x|_{col} < |p|_{col}$ ($|x|_{col} > |p|_{col}$, resp.) then x (p, resp.) is a prefix of p (x, resp.). If $|x|_{row} > |p|_{row}$ and $|x|_{col} \geq |p|_{col}$, then p is a prefix of x. If $|x|_{row} > |p|_{row}$ and $|x|_{col} < |p|_{col}$, then there exists $y \in \Sigma^{**}$ such that $y \leq_v x$ and $y \leq_h p$. The cases with $|x|_{row} < |p|_{row}$ are analogous. □

The vice versa does not hold, as shown by the following example.

Example 7. Let $Y \subseteq \Sigma^{**}$ as in Example 5: Y is a maximal strong prefix set. Let us show that Y is not br-complete. Consider the picture $p = \begin{array}{|cccc|} b & b & a & b \\ b & b & b & x \\ a & b & y & z \end{array}$ with $x, y, z \in \Sigma$: p cannot be covered with pictures in Y. Indeed, from a careful analysis of possible compositions of pictures in Y, it follows that the symbol b in position $(2,3)$ of p cannot be tiled by pictures in Y.

Maximality and completeness, that do not coincide in general for strong prefix codes, are in fact equivalent for the family of languages of pictures with fixed number of rows/columns.

Proposition 7. *Let $X \subseteq \Sigma^{m*}$ or $X \subseteq \Sigma^{*n}$ be a finite (strong) prefix set. X is maximal strong prefix set iff X is br-complete.*

Proof. Suppose that $X \subseteq \Sigma^{m*}$ is maximal strong prefix. Setting $\Gamma = \Sigma^{m,1}$, X can be considered as a set of strings over Γ and, in particular, X is a maximal prefix set of strings over Γ. Then X is right-complete. Let us show that X is br-complete. Let $p \in \Sigma^{**}$. If $|p|_{row} = m$ then, obviously, p can be covered by pictures of X, since X is right-complete. If $|p|_{row} < m$ then consider a picture p', $|p'|_{row} = m$ obtained by adding some rows to p: $p' \in \Gamma^*$ and, since X is right-complete, it can be "covered" by strings in X. But this implies that picture p can be covered by pictures of X. If $|p|_{row} > m$ then p can be considered as the row concatenation of some pictures in Σ^{m*}, and a picture with a number of rows less than or equal to m. Applying previous considerations, each of these pictures can be "covered" by pictures of X and, therefore p too can be covered by pictures of X. The case $X \subseteq \Sigma^{*n}$ is analogous.

The fact that X br-complete implies X maximal strong prefix follows from Proposition 6. □

5 Conclusions

We introduced the definitions of strong prefix code that generalizes in two dimensions the definition of prefix string code, and inherits many of its properties. In particular, given a strong prefix code, we proved the existence and unicity of a maximal strong prefix code that contains it, and that has a minimal size. Then we showed that the br-completeness of a picture language implies the strong prefix maximality, but, differently from the 1D case, the converse does not hold. The two notions are equivalent on the restricted family $\mathcal{S}$ of strong prefix languages containing pictures of fixed number of rows/columns.

The following hierarchy summarizes the results on families of 2D finite codes obtained in this paper and in our previous work [3]:

$$\begin{aligned} \text{Prefix maximal codes in } \mathcal{S} &= \text{Prefix (maximal and) br-complete codes} \\ &\subsetneq \text{Prefix maximal codes} \subseteq \text{Strong prefix maximal codes} \\ &\subsetneq \text{Strong prefix codes} \subsetneq \text{Prefix codes} \\ &\subsetneq \text{Codes.} \end{aligned}$$

We let open the problem whether (finite) prefix maximal codes equals (finite) strong prefix maximal codes or not. As future research, we will try to remove the finiteness hypothesis and consider prefix sets belonging to particular sub-families in REC, such as deterministic ones ([2,5]).

References

1. Aigrain, P., Beauquier, D.: Polyomino tilings, cellular automata and codicity. Theoretical Computer Science 147, 165–180 (1995)
2. Anselmo, M., Giammarresi, D., Madonia, M.: Deterministic and unambiguous families within recognizable two-dimensional languages. Fund. Inform. 98(2-3), 143–166 (2010)

3. Anselmo, M., Giammarresi, D., Madonia, M.: Two dimensional prefix codes of pictures. In: Béal, M.-P., Carton, O. (eds.) DLT 2013. LNCS, vol. 7907, pp. 46–57. Springer, Heidelberg (2013), `http://www.di.unisa.it/professori/anselmo/DLT2013-preliminary.pdf`
4. Anselmo, M., Giammarresi, D., Madonia, M., Restivo, A.: Unambiguous Recognizable Two-dimensional Languages. RAIRO: Theoretical Informatics and Applications 40(2), 227–294 (2006)
5. Anselmo, M., Madonia, M.: Deterministic and unambiguous two-dimensional languages over one-letter alphabet. Theoretical Computer Science 410(16), 1477–1485 (2009)
6. Beauquier, D., Nivat, M.: A codicity undecidable problem in the plane. Theoretical Computer Science 303, 417–430 (2003)
7. Berstel, J., Perrin, D., Reutenauer, C.: Codes and Automata. Cambridge University Press (2009)
8. Bozapalidis, S., Grammatikopoulou, A.: Picture codes. ITA 40(4), 537–550 (2006)
9. Giammarresi, D., Restivo, A.: Recognizable picture languages. Int. Journal Pattern Recognition and Artificial Intelligence 6(2-3), 241–256 (1992)
10. Giammarresi, D., Restivo, A.: Two-dimensional languages. In: Rozenberg, G., et al. (eds.) Handbook of Formal Languages, vol. III, pp. 215–268. Springer (1997)
11. Grammatikopoulou, A.: Prefix Picture Sets and Picture Codes. In: Procs. CAI 2005, pp. 255–268 (2005)
12. Kolarz, M., Moczurad, W.: Multiset, Set and Numerically Decipherable Codes over Directed Figures. In: Smyth, B. (ed.) IWOCA 2012. LNCS, vol. 7643, pp. 224–235. Springer, Heidelberg (2012)
13. Moczurad, M., Moczurad, W.: Some Open Problems in Decidability of Brick (Labelled Polyomino) Codes. In: Chwa, K.-Y., Munro, J.I. (eds.) COCOON 2004. LNCS, vol. 3106, pp. 72–81. Springer, Heidelberg (2004)
14. Simplot, D.: A Characterization of Recognizable Picture Languages by Tilings by Finite Sets. Theoretical Computer Science 218(2), 297–323 (1991)

The Algebraic Theory of Parikh Automata

Michaël Cadilhac[1], Andreas Krebs[2], and Pierre McKenzie[1,*]

[1] DIRO at U. de Montréal and Chaire Digiteo ENS Cachan-École Polytechnique
{cadilhac,mckenzie}@iro.umontreal.ca
[2] WSI at U. Tübingen
mail@krebs.net

Abstract. The Parikh automaton model equips a finite automaton with integer registers and imposes a semilinear constraint on the set of their final settings. Here the theory of typed monoids is used to characterize the language classes that arise algebraically. Complexity bounds are derived, such as containment of the unambiguous Parikh automata languages in NC^1. Noting that DetAPA languages are positive supports of rational $\mathbb{Z}$-series, DetAPA are further shown stronger than Parikh automata on unary langages. This suggests unary DetAPA languages as candidates for separating the two better known variants of uniform NC^1.

Introduction

The Parikh automaton model was introduced in [19]. It amounts to a nondeterministic finite automaton equipped with registers tallying up the number of occurrences of each transition along an accepting run. Such a run is then deemed successful iff the tuple of final register settings falls within a fixed semilinear set. An *affine* variant of the model in which transitions further induce an affine transformation on the registers was considered in [10]. An *unambiguous* variant of the model was considered in [11]. Tree Parikh automata and other variants were considered in [18].

Recall the tight connection between AC^0, ACC^0 and NC^1 and aperiodic monoids, solvable monoids and nonsolvable monoids respectively [2,3]. This connection was refined and studied in depth (see [24] for a lovely account), but the class $TC^0 \subseteq NC^1$ was left out of the picture because the MAJ gate in circuits could not be translated into the operation of a finite algebraic structure. Typed monoids were introduced in [20] as a means of capturing TC^0 meaningfully in the algebraic framework.

In both the classical and the typed monoid framework, a compelling notion of a natural class of monoids is that of a variety. In both frameworks, different monoid varieties capture different classes of languages as inverse homomorphic images of an accepting subset of the monoid [13,6]. The internal structure of NC^1 hinges on whether different monoid varieties still capture different classes of languages when the classical notion of a homomorphism is appropriately generalized to capture as above complexity classes such as ACC^0, TC^0 and NC^1.

* Supported by NSERC of Canada and Digiteo.

T. Muntean, D. Poulakis, and R. Rolland (Eds.): CAI 2013, LNCS 8080, pp. 60–73, 2013.

Our contribution is an algebraic characterization of the language classes defined by the deterministic and unambiguous variants of the Parikh automaton (called CA, for "constrained automaton") and the affine Parikh automaton. We show:

- the class $\mathcal{L}_{\text{DetCA}}$ of languages accepted by deterministic CA is the set of languages recognized by typed monoids from $\mathbf{Z}^+ \wr \mathbf{M}$, i.e., by wreath products of the monoid of integers with some finite monoid; the least typed monoid variety generated by $\mathbf{Z}^+ \wr \mathbf{M}$ also captures $\mathcal{L}_{\text{DetCA}}$
- the class $\mathcal{L}_{\text{UnCA}}$ of languages accepted by unambiguous CA is the set of languages recognized by typed monoids from $\mathbf{Z}^+ \Box \mathbf{M}$, i.e., by block products of the monoid of integers with some finite monoid; the least typed monoid variety generated by $\mathbf{Z}^+ \Box \mathbf{M}$ also captures $\mathcal{L}_{\text{UnCA}}$
- the classes $\mathcal{L}_{\text{DetAPA}}$ and $\mathcal{L}_{\text{UnAPA}}$, of languages accepted by deterministic and by unambiguous affine Parikh automata respectively (where an affine Parikh automaton generalizes the constrained automaton by allowing each transition to perform an affine transformation on the automaton registers), are the Boolean closure of the positive supports of rational series over the integers.

The first two characterizations above add legitimacy to the theory of typed monoids, and they suggest further relevance of that theory to our understanding of NC^1. It follows from the characterization of $\mathcal{L}_{\text{UnCA}}$ that $\mathcal{L}_{\text{UnCA}} \subseteq \text{NC}^1$, a fact which is not immediately obvious from the operation of an unambiguous constrained automaton.

The Boolean closure of the class of positive supports of rational series over the integers, hence $\mathcal{L}_{\text{DetAPA}} = \mathcal{L}_{\text{UnAPA}}$, can be viewed as a very tightly uniform version of the (DLOGTIME-uniform) class PNC^1, introduced in [12] as the log depth analog of the poly time and log space classes PP and PL [16]. Fulfilling $\text{NC}^1 \subseteq \text{PNC}^1 \subseteq \text{L}$, PNC^1 is robust, pointedly characterized using iterated products of constant dimension integer matrices, but also characterized using paths in bounded width graphs, proof trees in log depth circuits, accepting paths in nondeterministic finite automata or evaluation of a log depth $\{+, \times\}$-formula [12]. An elaborate structural complexity evolved around PNC^1 with the work of [21]. We note that using formal power series as a tool to investigate counting classes below L was already suggested in [1], but with emphasis there on the complexity of performing operations such as inversion and root extraction on such series.

1 Preliminaries

Monoids, integers, vectors. A monoid is a set M with an associative operation, usually denoted multiplicatively $(x, y) \mapsto xy$, and an identity element denoted 1. For $S \subseteq M$, we write S^* for the monoid generated by S, i.e., the smallest submonoid of M containing S. A (monoid) *morphism* from M to N is a map preserving product and identity. Moreover, if $M = \Sigma^*$ for some alphabet Σ (i.e., Σ is a finite set of symbols), then h need only be defined on the elements of Σ.

We write $\mathbb{N}$, $\mathbb{Z}$, $\mathbb{Z}^+$, $\mathbb{Z}_0^-$ for the sets of nonnegative integers, integers, positive integers, and nonpositive integers respectively. Vectors in $\mathbb{N}^d$ are noted in bold, e.g., $\mathbf{v}$ whose elements are $v_1, v_2, \ldots, v_d$. We write $\mathbf{e}_i \in \{0,1\}^d$ for the vector having a 1 only in position i, and $\mathbf{0}$ for the all-zero vector. We view $\mathbb{N}^d$ as the additive monoid $(\mathbb{N}^d, +)$, with $+$ the component-wise addition and $\mathbf{0}$ the identity element. We let $\mathcal{M}_{\mathbb{Z}}(k)$, for $k \geq 1$, be the monoid of square matrices of dimension $k \times k$ with values in $\mathbb{Z}$ and with the operation mapping (M_1, M_2) to M_2M_1. In particular, a morphism $h\colon \{a,b\}^* \to \mathcal{M}_{\mathbb{Z}}(k)$ is such that $h(ab) = h(b).h(a)$ with . the usual matrix multiplication. We write Ψ_i for the projection on the i-th component, $\Psi_i(a_1, a_2, \ldots, a_i, \ldots) = a_i$.

Semilinear sets, Parikh image. A subset C of $\mathbb{N}^d$ is *linear* if there exist $\mathbf{c} \in \mathbb{N}^d$ and a finite $P \subseteq \mathbb{N}^d$ such that $C = \mathbf{c} + P^*$. The subset C is said to be *semilinear* if it is equal to a finite union of linear sets: $\{4n + 56 \mid n > 0\}$ is semilinear while $\{2^n \mid n > 0\}$ is not. We will often use the fact that the semilinear sets are those sets of natural numbers definable in first-order logic with addition [17]. Let $\Sigma = \{a_1, a_2, \ldots, a_n\}$ be an (ordered) alphabet and 1 be the empty word. The *Parikh image* is the morphism $\mathsf{Pkh}\colon \Sigma^* \to \mathbb{N}^n$ defined by $\mathsf{Pkh}(a_i) = \mathbf{e}_i$, for $1 \leq i \leq n$ — in particular, we have that $\mathsf{Pkh}(1) = \mathbf{0}$. For $w \in \Sigma^*$, with $\mathsf{Pkh}(w) = \mathbf{x}$ and $a_i \in \Sigma$, we write $|w|_{a_i}$ for x_i. The Parikh image of a language L is defined as $\mathsf{Pkh}(L) = \{\mathsf{Pkh}(w) \mid w \in L\}$. The name of this morphism stems from Parikh's theorem [22], stating that for L context-free, $\mathsf{Pkh}(L)$ is semilinear; outside language theory, it is also referred to as the *commutative image*.

Affine functions. A function $f\colon \mathbb{N}^d \to \mathbb{N}^d$ is a (total and positive) *affine function* of dimension d if there exist a matrix $M \in \mathbb{N}^{d\times d}$ and $\mathbf{v} \in \mathbb{N}^d$ such that for any $\mathbf{x} \in \mathbb{N}^d$, $f(\mathbf{x}) = M\mathbf{x} + \mathbf{v}$. We abusively write $f = (M, \mathbf{v})$. We let $\mathcal{F}_d$ be the monoid of such functions under the operation $\diamond$ defined by $(f \diamond g)(\mathbf{x}) = g(f(\mathbf{x}))$, where the identity element is the identity function, i.e., $(Id, \mathbf{0})$ with Id the identity matrix of dimension d.

Automata. An automaton is a quintuple $A = (Q, \Sigma, \delta, q_0, F)$ where Q is a finite set of states, Σ is an alphabet, $\delta \subseteq Q \times \Sigma \times Q$ is a set of transitions, $q_0 \in Q$ is the initial state, and $F \subseteq Q$ is a set of final states. For a transition $t = (q, a, q') \in \delta$, define $\mathsf{From}(t) = q$ and $\mathsf{To}(t) = q'$. We define $\mathsf{Label}_A\colon \delta^* \to \Sigma^*$ as the morphism given by $\mathsf{Label}_A(t) = a$, with, in particular, $\mathsf{Label}_A(1) = 1$, and write Label when A is clear from the context. The set of accepting paths of A, i.e., the set of words over δ describing paths starting from q_0 and ending in F, is written $\mathsf{Run}(A)$. The language of the automaton is $L(A) = \mathsf{Label}_A(\mathsf{Run}(A))$. An automaton is *unambiguous* if for all $w \in L(A)$ there is a unique $\pi \in \mathsf{Run}(A)$ with $\mathsf{Label}(\pi) = w$.

A *constrained automaton (CA)* [10] is a pair (A, C) where A is an automaton with d transitions and $C \subseteq \mathbb{N}^d$ is semilinear. Its language is $L(A, C) = \mathsf{Label}_A(\{\pi \in \mathsf{Run}(A) \mid \mathsf{Pkh}(\pi) \in C\})$. The CA is said to be *deterministic (DetCA)* if A is deterministic, and *unambiguous (UnCA)* if A is unambiguous. We write $\mathcal{L}_{\mathrm{CA}}$, $\mathcal{L}_{\mathrm{DetCA}}$, and $\mathcal{L}_{\mathrm{UnCA}}$ for the classes of languages recognized by CA, DetCA, and UnCA, respectively.

An *affine Parikh automaton (APA)* [10] of dimension d is a triple (A, U, C) where A is an automaton with transition set δ, $U : \delta^* \to \mathcal{F}_d$ is a morphism, and $C \subseteq \mathbb{N}^d$ is semilinear. Its language is $L(A, U, C) = \mathsf{Label}_A(\{\pi \in \mathsf{Run}(A) \mid [U(\pi)](\mathbf{0}) \in C\})$. The APA is said to be *deterministic (DetAPA)* if A is deterministic, and *unambiguous (UnAPA)* if A is unambiguous. We write $\mathcal{L}_{\text{DetAPA}}$ and $\mathcal{L}_{\text{UnAPA}}$ for the classes of languages recognized by DetAPA and UnAPA, respectively.

Transition monoid. Let $A = (Q, \Sigma, \delta, q_0, F)$ be a complete deterministic automaton. For $a \in \Sigma$, define $f_a : Q \to Q$ by $f_a(q) = q'$ iff $(q, a, q') \in \delta$. The *transition monoid* M of A is the closure under composition of the set $\{f_a \mid a \in \Sigma\}$. The monoid M acts on Q naturally by $q.m = m(q)$, $m \in M, q \in Q$. Write $\eta : \Sigma^* \to M$ for the canonical surjective morphism associated, that is, the morphism defined by $\eta(a) = f_a$, $a \in \Sigma$. Then $q.\eta(w)$ is the state reached by reading $w \in \Sigma^*$ from the state $q \in Q$.

2 Normal Forms of CA and APA

We present several technical lemmata on CA and APA that will help us in devising concise proofs for the algebraic characterizations that follow. Their main purpose is to simplify the constraint set, so that only sign checks on linear combinations of variables are performed.

Recall (e.g., [14]) that for any semilinear set $C \subseteq \mathbb{Z}^d$, there is a Boolean combination of expressions of the form: $\sum_{i=1}^{d} \alpha_i x_i > c$ and $\sum_{i=1}^{d} \alpha_i x_i \equiv_p c$, with $\alpha_i, c \in \mathbb{Z}$ and $p > 1$, which is true iff $(x_1, x_2, \ldots, x_d) \in C$. Note that the α_i may be zero. We define two notions which refine this point of view:

Definition 1. *We say that a semilinear set C is* modulo-free *if it can be expressed as a Boolean combination of expressions of the form $\sum_i \alpha_i x_i > c$, for $\alpha_i \in \mathbb{Z}$. We say that C is* basic *if it can further be expressed as a* positive *Boolean combination of expressions of the form $\sum_i \alpha_i x_i > 0$.*

The first normal form concerns DetCA and UnCA:

Lemma 1. *Every DetCA (resp. UnCA) has the same language $L \subseteq \Sigma^+$ as another DetCA (resp. UnCA) (A, C) with $L(A) = \Sigma^*$ and C a basic set.*

We also note the following simple fact:

Lemma 2. *For $(A, C_1 \cap C_2)$ a DetCA or an UnCA it holds that:*

$$L(A, C_1 \cap C_2) = L(A, C_1) \cap L(A, C_2) \ .$$

The same holds for $\cup$.

We show more in the context of APA to allow the forthcoming proofs of characterization to translate smoothly from CA to APA. In the following, we consider that a matrix $M \in \mathcal{M}_{\mathbb{Z}}(k)$ is in a set $C \subseteq \mathbb{Z}^{k^2}$ if the vector consisting of the columns of M is in C.

Lemma 3. *Let $L \subseteq \Sigma^+$ be in $\mathcal{L}_{\text{DetAPA}}$. There is a morphism $h\colon \Sigma^* \to \mathcal{M}_{\mathbb{Z}}(k)$, for some k, and a set $\mathcal{Z} \subseteq \mathbb{Z}^{k^2}$ expressible as a Boolean combination of expressions $x_i > 0$, such that $L = h^{-1}(\mathcal{Z})$.*

Similarly, let $L \subseteq \Sigma^+$ be in $\mathcal{L}_{\text{UnAPA}}$. There is an unambiguous automaton A with transition set δ, a morphism $h\colon \delta^* \to \mathcal{M}_{\mathbb{Z}}(k)$, for some k, and a set $\mathcal{Z} \subseteq \mathbb{Z}^{k^2}$ expressible as a Boolean combination of expressions $x_i > 0$, such that $L = \mathsf{Label}_A(h^{-1}(\mathcal{Z}) \cap \mathsf{Run}(A))$.

3 Capturing Parikh Automata Classes Algebraically

In this section we characterize DetCA, UnCA, DetAPA, and UnAPA using the theory of (finitely) typed monoids [20].

3.1 Typed Monoids

In the following, we will use such notions as language recognition by a finite monoid, varieties of languages, and pseudovarieties of finite monoids (see, e.g., [13]). In algebraic language theory, central tools for finite monoid composition include the *block* and *wreath* products, the definitions of which we recall here, before giving similar definitions in the theory of typed monoids.

Let M and N be finite monoids. To distinguish the operation of M and N, we denote the operation of M as $+$ and its identity element as 0 (although this operation is not necessarily commutative) and the operation of N implicitly and its identity element as 1. A left action of N on M is a function mapping pairs $(n, m) \in N \times M$ to $nm \in M$ and satisfying $n(m_1+m_2) = nm_1+nm_2$, $n_1(n_2m) = (n_1n_2)m$, $n0 = 0$ and $1m = m$. Right actions are defined symmetrically. If we have both a right and a left action of N on M that further satisfy $n_1(mn_2) = (n_1m)n_2$, we define the bilateral semidirect product $M ** N$ as the monoid with elements in $M \times N$ and multiplication defined as $(m_1, n_1)(m_2, n_2) = (m_1n_2 + n_1m_2, n_1n_2)$. This operation is associative and $(0, 1)$ acts as an identity for it. Given only a left action, the unilateral semidirect product $M * N$ is the bilateral semidirect product $M ** N$ where the right action on M is trivial ($mn = m$).

Let M, N be two monoids. The *wreath product* of M and N, written $M \wr N$, is defined as the unilateral semidirect product of M^N and N, where the left action of N on M^N is given by $(n \cdot f)(n') = f(n'n)$, for $f\colon N \to M$ and $n, n' \in N$. The *block product* of M and N, written $M \Box N$, is defined as the bilateral semidirect product of $M^{N\times N}$ and N, where the right (resp. left) action of N on $M^{N\times N}$ is given by $(f \cdot n)(n_1, n_2) = f(n_1, nn_2)$ (resp. $(n \cdot f)(n_1, n_2) = f(n_1n, n_2)$), for $f\colon N \times N \to M$ and $n, n_1, n_2 \in N$.

We now turn to the theory of typed monoids.

Definition 2 (Typed monoid [20]). *A* typed monoid *is a pair $(S, \mathfrak{S})$ where S is a finitely generated monoid and $\mathfrak{S}$ is a finite Boolean algebra of subsets of S whose elements are called* types. *We write $(S, \{\mathcal{S}_1, \mathcal{S}_2, \ldots, \mathcal{S}_n\})$ for the typed monoid $(S, \mathfrak{S})$ where $\mathfrak{S}$ is generated by the $\mathcal{S}_i$'s. If $n = 1$, we simply write*

$(S, \mathcal{S}_1)$. *For two typed monoids* $(M, \mathfrak{M})$, $(N, \mathfrak{N})$, *their direct product* $(S, \mathfrak{S}) = (M, \mathfrak{M}) \times (N, \mathfrak{N})$ *is defined by* $S = M \times N$, *and* $\mathfrak{S}$ *is the Boolean algebra generated by* $\{\mathcal{M} \times \mathcal{N} \mid \mathcal{M} \in \mathfrak{M}$ *and* $\mathcal{N} \in \mathfrak{N}\}$. *A typed monoid* $(S, \mathfrak{S})$ recognizes *a language* L *if there are a morphism* $h \colon \Sigma^* \to S$ *and a type* $\mathcal{S} \in \mathfrak{S}$ *such that* $L = h^{-1}(\mathcal{S})$. *We write* $\mathcal{L}((S, \mathfrak{S}))$ *for the class of languages, over any alphabet, recognized by* $(S, \mathfrak{S})$ *and extend this notation naturally to classes of typed monoids.*

We view a finite monoid M as the typed monoid $(M, 2^M)$, and write $\mathbf{M}$ for the class of typed finite monoids; note that the usual notion of language recognition then coincides with the one given here.

The usual wreath product (resp. block product) of M and N, i.e., the unilateral (resp. bilateral) semidirect product of M^N (resp. $M^{N \times N}$) and N, results, in the infinite monoid case, in monoids with uncountably many elements, failing to fall within the definition of typed monoid. Thus the block product was restricted, in [20], to *type-respecting* functions, that is, functions that only depend on the type of their arguments (multiplied by some constants). Here, we are not concerned with this technicality as all our monoids N will be finite. Hence we define:

Definition 3 (Typed block [20] and wreath products). *Let* $(M, \mathfrak{M})$ *and* $(N, \mathfrak{N})$ *be typed monoids. The* block product *(resp.* wreath product*) of* $(M, \mathfrak{M})$ *and* $(N, \mathfrak{N})$, *written* $(M, \mathfrak{M}) \Box (N, \mathfrak{N})$ *(resp.* $(M, \mathfrak{M}) \wr (N, \mathfrak{N})$*), is* $(M \Box N, \mathfrak{S})$ *(resp.* $(M \wr N, \mathfrak{S})$*) with* $\mathfrak{S} = \{\mathcal{S}_{\mathcal{M}} \mid \mathcal{M} \in \mathfrak{M}\}$ *where:*

$$\mathcal{S}_{\mathcal{M}} = \{(f, n) \in S \mid f(1,1) \in \mathcal{M}\} \ (\textit{resp. } \mathcal{S}_{\mathcal{M}} = \{(f, n) \in S \mid f(1) \in \mathcal{M}\}) \ .$$

The appropriateness of typed monoids in the study of the algebraic properties of nonregular languages is witnessed by the following Eilenberg-like theorem of Behle, Krebs, and Reifferscheid:

Theorem 1 ([6]). *Varieties of typed monoids and varieties of languages are in a one-to-one correspondence, i.e., (1) Let* $\mathcal{V}$ *be a variety of languages and* $\mathbf{V}$ *the smallest variety of typed monoids that recognizes all languages in* $\mathcal{V}$, *then* $\mathcal{L}(\mathbf{V}) = \mathcal{V}$; *(2) Let* $\mathbf{V}$ *be a variety of typed monoids and* $\mathbf{W}$ *be the smallest variety that recognizes all languages of* $\mathcal{L}(\mathbf{V})$, *then* $\mathbf{V} = \mathbf{W}$.

Similar to the untyped algebraic theory of languages, if a typed monoid recognizes a language, it also recognizes its complement. This implies that $\mathcal{L}_{\mathrm{CA}}$, which is not closed under complement, does not accept a typed monoid characterization. We will thus focus on characterizing the deterministic and unambiguous classes. Note that we will frequently focus on languages which do not contain the empty word. This is a technical simplification which introduces no loss of generality, as all our typed monoid classes recognize $\{1\}$ and are closed under union.

3.2 Capturing DetCA and UnCA

Let $\mathbf{Z}^+$ be the set of typed monoids $\{(\mathbb{Z}, \mathbb{Z}^+)^k \mid k \geq 1\}$.

Theorem 2. $\mathcal{L}(\mathbf{Z}^+ \wr \mathbf{M}) = \mathcal{L}_{\text{DetCA}}$.

Proof. ($\mathcal{L}_{\text{DetCA}} \subseteq \mathcal{L}(\mathbf{Z}^+ \wr \mathbf{M})$) We first show that $\mathcal{L}(\mathbf{Z}^+ \wr \mathbf{M})$ is closed under union and intersection. Let $L_1, L_2 \in \mathcal{L}(\mathbf{Z}^+ \wr \mathbf{M})$ be two languages over Σ, that is, for $i = 1, 2$, there exist a finite monoid M_i, an integer k_i, a morphism $h_i\colon \Sigma^* \to \mathbb{Z}^{k_i} \wr M_i$, and a type $\mathcal{T}_i$ of $(\mathbb{Z}, \mathbb{Z}^+)^{k_i}$ such that $L_i = h_i^{-1}(\mathcal{T}_i)$.

Consider the typed monoid $(\mathbb{Z}, \mathbb{Z}^+)^{k_1+k_2} \wr (M_1 \times M_2) \in \mathbf{Z}^+ \wr \mathbf{M}$. This monoid recognizes both the intersection and union of L_1 and L_2 as follows. Define $h\colon \Sigma^* \to \mathbb{Z}^{k_1+k_2} \wr (M_1 \times M_2)$ by $h(a) = (f_a, (\Psi_2(h_1(a)), \Psi_2(h_2(a))))$ where $a \in \Sigma$ and $f_a((m_1, m_2)) = ([\Psi_1(h_1(a))](m_1), [\Psi_1(h_2(a))](m_2)) \in \mathbb{Z}^{k_1+k_2}$. Now let $\oslash \in \{\cup, \cap\}$. We define $\mathcal{T}_\oslash = (\mathcal{T}_1 \times \mathbb{Z}^{k_2}) \oslash (\mathbb{Z}^{k_1} \times \mathcal{T}_2)$, and thus $h^{-1}(\mathcal{T}_\oslash) = L_1 \oslash L_2$.

Now let (A, C) be a DetCA with $A = (Q, \Sigma, \delta, q_0, F)$, and suppose (by Lemma 1) that $F = Q$ and that the constraint set is expressed by a positive Boolean combination of clauses of the form $\sum_{t\in\delta} \alpha_t x_t > 0$. Closure of $\mathcal{L}(\mathbf{Z}^+ \wr \mathbf{M})$ under $\cup$ and $\cap$ together with Lemma 2 imply that it is enough to argue the case in which C is defined by a single such clause.

Let M be the transition monoid of A, $\eta\colon \Sigma^* \to M$ the canonical morphism associated. We now define $h\colon \Sigma^* \to \mathbb{Z} \wr M$ as follows. Let $\tau\colon M \times \Sigma \to \delta$ be defined by $\tau(m, a) = (q_0.m, a, q_0.m\eta(a))$. Then:

$$h(a) = (f_a, \eta(a)), \text{ where } f_a(m) = \alpha_{\tau(m,a)} \ .$$

Now let $w = w_1 w_2 \cdots w_n \in \Sigma^*$ and $\pi = \pi_1 \pi_2 \cdots \pi_n$ where $w_i \in \Sigma$ and $\pi_i \in \delta$ for every $1 \le i \le n$, such that π is the unique accepting path in A from q_0 labeled w. We have:

$$h(w) = (f_{w_1} + \eta(w_1) \cdot f_{w_2} + \cdots + \eta(w_1 w_2 \cdots w_{n-1}) \cdot f_{w_n}, \eta(w))$$

$$[\Psi_1(h(w))](\eta(1)) = \alpha_{\tau(\eta(1), w_1)} + \sum_{i=2}^{n} \alpha_{\tau(\eta(w_1 \cdots w_{i-1}), w_i)} \ ,$$

note that $q_0.\eta(w_1 \cdots w_{i-1})$ is $\mathsf{From}(\pi_i)$ and thus $\tau(\eta(w_1 \cdots w_{i-1}), w_i) = \pi_i$, hence:

$$[\Psi_1(h(w))](\eta(1)) = \sum_{i=1}^{n} \alpha_{\pi_i} = \sum_{t\in\delta} |\pi|_t \times \alpha_t \ .$$

Thus, $\mathsf{Pkh}(\pi) \in C$ iff $[\Psi_1(h(w))](\eta(1)) > 0$. Hence with the type $\mathcal{T} = \{(f, m) \in (\mathbb{Z}, \mathbb{Z}^+) \wr M \mid f(\eta(1)) > 0\}$, which is indeed a type of $(\mathbb{Z}, \mathbb{Z}^+) \wr M$, we have that $h^{-1}(\mathcal{T}) = L(A, C)$.

($\mathcal{L}(\mathbf{Z}^+ \wr \mathbf{M}) \subseteq \mathcal{L}_{\text{DetCA}}$) Let $L \subseteq \Sigma^*$ be recognized by $(\mathbb{Z}, \mathbb{Z}^+)^k \wr M$ using a type $\mathcal{T}$ and a morphism $h\colon \Sigma^* \to (\mathbb{Z}^k)^M \times M$, and write for convenience $h_i(w) = \Psi_i(h(w))$, $i = 1, 2$. Let A be the automaton $(M, \Sigma, \delta, 1, M)$, where:

$$\delta = \{(m, a, m') \mid m \in M, a \in \Sigma \text{ and } m' = m.h_2(a)\} \ .$$

Now as $\mathcal{L}_{\mathrm{DetCA}}$ is closed under union and intersection, we may suppose that the type $\mathcal{T}$ is of the following form:

$$\mathcal{T} = \prod_{i=1}^{k} \{(f, m) \mid f(1) \in \mathcal{T}_i\} ,$$

where each $\mathcal{T}_i \in \{\emptyset, \mathbb{Z}_0^-, \mathbb{Z}^+, \mathbb{Z}\}$. Define $T = \mathcal{T}_1 \times \mathcal{T}_2 \times \cdots \times \mathcal{T}_k$, and the semilinear set C consisting of elements:

$$(x_{t_1}, x_{t_2}, \ldots, x_{t_{|\delta|}}) \text{ s.t. } \sum_{t \in \delta} x_t \times [h_1(\mathsf{Label}(t))](\mathsf{From}(t)) \in T .$$

We claim that the language of the DetCA (A, C) is L. Let $w = w_1 w_2 \cdots w_n \in \Sigma^*$. There is an (accepting) path in A labeled w going through the states $1 = h_2(1)$, $h_2(w_1)$, $h_2(w_1 w_2)$, $\ldots$, $h_2(w_1 w_2 \cdots w_n)$. Thus the sum computed by the semilinear set is $h_1(w_1) + h_2(w_1) \cdot h_1(w_2) + \cdots + h_2(w_1 w_2 \cdots w_n) \cdot h_1(w_n)$, taken at the point 1. This is precisely $[h_1(w)](1)$, and thus checking whether it belongs to T is equivalent to checking whether $h(w) \in \mathcal{T}$. Hence $L = L(A, C)$. □

Now $\mathcal{L}_{\mathrm{DetCA}}$ is a variety of languages and we may naturally ask whether the smallest variety containing $\mathbf{Z}^+ \wr \mathbf{M}$, which recognizes only the languages of $\mathcal{L}_{\mathrm{DetCA}}$ by Theorem 1, is closed under iterated wreath product. We note this is not the case. Let $U_1 = (\{0, 1\}, \times)$, then:

Theorem 3. *There is a language $L \notin \mathcal{L}_{\mathrm{CA}}$ recognized by $U_1 \wr (\mathbb{Z}, \mathbb{Z}^+)$ and by $(\mathbb{Z}, \mathbb{Z}^+) \wr (\mathbb{Z}, \mathbb{Z}^+)$.*

We now turn to unambiguous CA:

Theorem 4. $\mathcal{L}(\mathbf{Z}^+ \Box \mathbf{M}) = \mathcal{L}_{\mathrm{UnCA}}$.

Proof. ($\mathcal{L}_{\mathrm{UnCA}} \subseteq \mathcal{L}(\mathbf{Z}^+ \Box \mathbf{M})$) We first note that $\mathcal{L}(\mathbf{Z}^+ \Box \mathbf{M})$ is closed under union and intersection; this is the same proof as in Theorem 2 except that f_a is now defined as:

$$f_a((m_1, m_2), (m_1', m_2')) = ([\Psi_1(h_1(a))](m_1, m_1'), [\Psi_1(h_2(a))](m_2, m_2')) .$$

Next consider an UnCA (A, C) with $A = (Q, \Sigma, \delta, q_0, F)$, and suppose (using Lemma 1) that $L(A) = \Sigma^*$ and that the constraint set is expressed by a positive Boolean combination of clauses of the form $\sum_{t \in \delta} \alpha_t x_t > 0$. Closure of $\mathcal{L}(\mathbf{Z}^+ \Box \mathbf{M})$ under $\cup$ and $\cap$ together with Lemma 2 imply that it is enough to argue the case in which C is defined by a single such clause.

Let M be the transition monoid of the deterministic version of A, obtained using the powerset construction. Let A' be defined as A with all transitions inverted (i.e., (p, a, q) is in A iff (q, a, p) is in A'). Let M' be the transition monoid of the deterministic version of A', using again the powerset construction, and let M^{c} be the monoid defined on the same elements as M' but with the operation reversed (i.e., $m_1 \circ_{M'} m_2$ in M' is $m_2 \circ_{M^{\mathrm{c}}} m_1$ in M^{c}; this is still a monoid as $\circ_{M^c}$ is still associative). We will show that $L(A, C)$ is recognized by $(S, \mathfrak{S}) = (\mathbb{Z}, \mathbb{Z}^+)\Box(M \times M^c)$.

Write η and η^c for the canonical morphisms associated with M and M^c; for $m \in M$ and $R \subseteq Q$, write $R.m$ for the action of m on R, and likewise for M^c. We first note that for $w \in \Sigma^*$, $\{q_0\}.\eta(w)$ is the set of states of A that can be reached in A reading w from q_0, and, likewise, that $F.\eta^c(w)$ is the set of states in A from which reading w leads to a final state.

Now for $m_1 \in M$, $a \in \Sigma$, and $m_2 \in M^c$, let $\tau(m_1, a, m_2)$ be the unique transition in A from a state in $\{q_0\}.m_1$ to a state in $F.m_2$ labeled a. We show that τ is well-defined. Let w_1, w_2 such that $\eta(w_1) = m_1$ and $\eta^c(w_2) = m_2$; this means that there are w_1-labeled paths in A from q_0 to *any* state in $\{q_0\}.m_1$, and, likewise, w_2-labeled paths in A from *any* state in $F.m_2$ to a final state. *(Existence):* as $w_1 a w_2$ is in $\Sigma^* = L(A)$, there is a transition in A from a state in $\{q_0\}.m_1$ to a state in $F.m_2$ labeled a. *(Uniqueness):* if two transitions (p, a, p') and (q, a, q') are such that $p, q \in \{q_0\}.m_1$ and $p', q' \in F.m_2$, this means that there are multiple accepting paths in A labeled $w_1 a w_2$, contradicting the unambiguity of A. We now define the morphism $h \colon \Sigma^* \to S$ by:

$$h(a) = (f_a, (\eta(a), \eta^c(a))), \text{ where}$$
$$f_a((m_1, m_2), (m'_1, m'_2)) = \alpha_{\tau(m_1, a, m'_2)} \ .$$

Now let $w = w_1 w_2 \cdots w_n \in \Sigma^*$, $w_i \in \Sigma$ for every $1 \le i \le n$, and π be the unique path in A from q_0 to a final state labeled w. Then:

$$\pi = \pi_1 \pi_2 \cdots \pi_n \quad \text{where}$$
$$\pi_i = \tau(\eta(w_1 w_2 \cdots w_{i-1}), w_i, \eta^c(w_{i+1} w_{i+2} \cdots w_n)) \ ,$$

and thus:

$$[\Psi_1(h(w))]((\eta(1), \eta^c(1))) = \sum_{t \in \delta} |\pi|_t \times \alpha_t \ .$$

Thus $\mathsf{Pkh}(\pi) \in C$ iff $[\Psi_1(h(w))]((\eta(1), \eta^c(1))) > 0$. Hence with the type $\mathcal{S} = \{(f, m) \in S \mid f((\eta(1), \eta^c(1))) \in \mathbb{Z}^+\}$, which is indeed a type in $\mathfrak{S}$ as $\mathbb{Z}^+$ is a type of $(\mathbb{Z}, \mathbb{Z}^+)$, we have that $h^{-1}(\mathcal{S}) = L(A, C)$.

($\mathcal{L}(\mathbf{Z}^+\Box\mathbf{M}) \subseteq \mathcal{L}_{\text{UnCA}}$) Let $L \subseteq \Sigma^*$ be recognized by $(\mathbb{Z}, \mathbb{Z}^+)^k\Box M$ using a type $\mathcal{T}$ and a morphism $h \colon \Sigma^* \to \mathbb{Z}^{M \times M} \times M$, and write for convenience $h_i(w) = \Psi_i(h(w))$, $i = 1, 2$. For any $(s_1, s_2) \in M \times M$, Let $A(s_1, s_2)$ be the automaton $(M \times M, \Sigma, \delta, (s_1, s_2), M \times \{1\})$ where:

$$\begin{aligned}\delta =&\{((m_1, m_2), a, (m'_1, m'_2)) \mid \\ &m'_1 = m_1 h_2(a) \text{ and } h_2(a) m'_2 = m_2 \in M \text{ and } a \in \Sigma\} \ .\end{aligned}$$

Note that $w \in L(A(s_1, s_2))$ implies $h_2(w) = s_2$. We argue that $A(s_1, s_2)$ is unambiguous for any $(s_1, s_2) \in M \times M$. We show that for any $w \in \Sigma^*$ and any $(s_1, s_2) \in M \times M$, w is the label of at most one accepting path in $A(s_1, s_2)$, by induction on $|w|$. If $w = 1$, then every $A(s_1, s_2)$ has at most one accepting path labeled w. Now let $w = a \cdot v$ for $v \in \Sigma^*$. Suppose $w \in L(A(s_1, s_2))$. This implies that $h_2(w) = s_2$. The states that can be reached from $(s_1, h_2(w))$ reading a are all of the form $(s_1 h_2(a), m)$, $m \in M$. Now v should be accepted by the automaton A where the initial state is set to one of these states; thus there is only one state fitting, $(s_1 h_2(a), h_2(v))$. By induction hypothesis, there is only one path in $A(s_1 h_2(a), h_2(v))$ recognizing v, thus there is only one path in $A(s_1, h_2(w))$ recognizing w. This shows that for any s_1, s_2, $A(s_1, s_2)$ is unambiguous.

Now, with $e = (1, 1)$, and as $\mathcal{L}_{\text{UnCA}}$ is closed under union and intersection, we may suppose that the type $\mathcal{T}$ is of the following form:

$$\mathcal{T} = \prod_{i=1}^{k} \{(f, m) \mid f(e, e) \in \mathcal{T}_i\} ,$$

where each $\mathcal{T}_i \in \{\emptyset, \mathbb{Z}_0^-, \mathbb{Z}^+, \mathbb{Z}\}$. Define $T = \mathcal{T}_1 \times \mathcal{T}_2 \times \cdots \times \mathcal{T}_k$, and the semilinear set C consisting of elements:

$$(x_{t_1}, x_{t_2}, \ldots, x_{t_{|\delta|}}) \text{ s.t. } \sum_{t \in \delta} x_t \times [h_1(\mathsf{Label}(t))](\Psi_1(\mathsf{From}(t)), \Psi_2(\mathsf{To}(t))) \in T .$$

We show that $\bigcup_{m \in M} L(A(1, m), C)$ is L. Let $w = w_1 w_2 \cdots w_n \in \Sigma^*$. There is a unique accepting path in $A(1, h_2(w))$ (and in no other $A(1, m)$) labeled w, and it is going successively through the states $(h_2(1), h_2(w)) = (1, h_2(w))$, $(h_2(w_1), h_2(w_2 \cdots w_n))$, …, $(h_2(w), 1) = (h_2(w), h_2(1))$. For this path, the sum computed by the semilinear set is:

$$\sum_{i=1}^{n} h_2(w_1 \cdots w_{i-1}) \cdot h_1(w_i) \cdot h_2(w_{i+1} \cdots w_n) ,$$

at the point $(1, 1)$. This is precisely $[h_1(w)](1, 1)$, and checking whether it is in T amounts to checking whether $h(w) \in \mathcal{T}$, thus $L = \bigcup_{m \in M} L(A(1, m), C)$. □

We derive an interesting property of the logical characterization and circuit complexity of UnCA. Let MSO[<] be the monadic second-order logic with $<$ as the unique numerical predicate, and FO+G[<] be the first-order logic with group quantifiers and $<$ as the unique numerical predicate. Both logics express exactly the regular languages (these are respectively the classical results of Büchi [8] and Barrington, Immerman, Straubing [4]). Now define the *extended majority* quantifier $\widehat{\text{Maj}}$, introduced in [5], as: $w \models \widehat{\text{Maj}}\ x\ \langle \varphi_i \rangle_{i=1,\ldots,m}$ iff $\sum_{j=1}^{|w|} |\{i \mid w_{x=j} \models \varphi_i\}| - |\{i \mid w_{x=j} \not\models \varphi_i\}| > 0$. Then:

Corollary 1. *A language is in $\mathcal{L}_{\text{UnCA}}$ iff it can be expressed as a Boolean combination of formulas of the form:*

$$\widehat{\text{Maj}}\ x\ \langle \varphi_i \rangle_{i=1,\ldots,m}$$

where each φ_i is an MSO[<] *formula or an* FO+G[<] *formula. Hence, $\mathcal{L}_{\text{UnCA}} \subsetneq \text{NC}^1$.*

3.3 Capturing DetAPA and UnAPA

Write $\mathfrak{Z}^+(k)$ for the type set of $(\mathbb{Z}, \mathbb{Z}^+)^k$, that is, the sets expressible as a Boolean combination of expressions of the form $x_i > 0$. Let **ZMat**$^+$ be the set of typed monoids $\{(\mathcal{M}_\mathbb{Z}(k), \mathfrak{Z}^+(k \times k)) \mid k \geq 1\}$, then:

Theorem 5. $\mathcal{L}(\mathbf{ZMat}^+) = \mathcal{L}_{\mathrm{DetAPA}}$.

Proof. $(\mathcal{L}_{\mathrm{DetAPA}} \subseteq \mathcal{L}(\mathbf{ZMat}^+))$ This is a direct consequence of Lemma 3.

$(\mathcal{L}(\mathbf{ZMat}^+) \subseteq \mathcal{L}_{\mathrm{DetAPA}})$ Given $k \geq 1$, a type $\mathcal{Z}$ of $(\mathbb{Z}, \mathbb{Z}^+)^{k \times k}$, and a morphism $h\colon \Sigma^* \to \mathcal{M}_\mathbb{Z}(k)$, we build a two-state DetAPA of dimension k^2 for $h^{-1}(\mathcal{Z})$. First, let $h'\colon \Sigma^* \to \mathcal{M}_\mathbb{Z}(k^2)$ be such that $h'(a)$ is the Kronecker product of the identity matrix of dimension k and $h(a)$. Define $\mathbf{e} = (\mathbf{e}_1, \mathbf{e}_2, \ldots, \mathbf{e}_k)$ where each $\mathbf{e}_i$ is of dimension k. Then for any word w, $h(w) \in \mathcal{Z}$ iff $h'(w)\mathbf{e} \in \mathcal{Z}$. Now let $A = (\{r, s\}, \Sigma, \delta, r, \{s\})$, with $\delta = \{r, s\} \times \Sigma \times \{s\}$. Then let $U\colon \delta^* \to \mathcal{F}_{k^2}$ for $q \in \{r, s\}$, $a \in \Sigma$, and $\mathbf{x} \in \mathbb{Z}^{k^2}$ be defined by:

$$[U((q, a, s))](\mathbf{x}) = \begin{cases} h'(a)\mathbf{e} & \text{if } q = r, \\ h'(a)\mathbf{x} & \text{otherwise.} \end{cases}$$

This implies that for $w \in \Sigma^+$ and π its unique accepting path in A, it holds that $[U(\pi)](\mathbf{0}) = h'(w)\mathbf{e}$. Thus $L(A, U, \mathcal{Z}) = h^{-1}(\mathcal{Z})$. □

Theorem 6. $\mathcal{L}(\mathbf{ZMat}^+ \Box \mathbf{M}) = \mathcal{L}_{\mathrm{UnAPA}}$.

Proof. $\mathcal{L}_{\mathrm{UnAPA}} \subseteq \mathcal{L}(\mathbf{ZMat}^+ \Box \mathbf{M})$ is the same as $\mathcal{L}_{\mathrm{UnCA}} \subseteq \mathcal{L}(\mathbf{Z}^+ \Box \mathbf{M})$ in Theorem 4, thanks to Lemma 3.

$\mathcal{L}(\mathbf{ZMat}^+ \Box \mathbf{M}) \subseteq \mathcal{L}_{\mathrm{UnAPA}}$ is the same as $\mathcal{L}(\mathbf{Z}^+ \Box \mathbf{M}) \subseteq \mathcal{L}_{\mathrm{UnCA}}$ in Theorem 4 for the automaton part, and the same as Theorem 5 for the constraint set and affine function parts. □

Now, applying the same arguments as in [9, Lemma 5], we have that DetAPA can simulate unambiguity, and thus $\mathcal{L}_{\mathrm{UnAPA}} = \mathcal{L}_{\mathrm{DetAPA}}$. This translates nicely in the algebraic framework thanks to Theorem 1:

Theorem 7. *The smallest variety containing* **ZMat**$^+\Box$**M** *is equal to that containing* **ZMat**$^+$.

4 Formal Power Series

In this section, we show that the languages of DetAPA are those expressible as a Boolean combination of positive supports of $\mathbb{Z}$-valued rational series. This helps us derive a separation over the unary languages between $\mathcal{L}_{\mathrm{CA}}$ and $\mathcal{L}_{\mathrm{DetAPA}}$ — the separation was known ([10, Proposition 28]), but not over unary languages.

Definition 4 (e.g., [7]). *Functions from Σ^* into $\mathbb{Z}$ are called* ($\mathbb{Z}$-)series*. For such a series r, it is customary to write (r, w) for $r(w)$. We write* $\mathsf{supp}_+(r)$ *for the* positive support *of r, i.e., $\{w \mid (r, w) > 0\}$.*

A linear representation *of dimension* $k \geq 1$ *is a triple* $(\mathbf{s}, h, \mathbf{g})$ *such that* $\mathbf{s} \in \mathbb{Z}^k$ *is a* row vector*,* $\mathbf{g} \in \mathbb{Z}^k$ *is a* column vector*, and* $h\colon \Sigma^* \to \mathbb{Z}^{k\times k}$ *is a monoid morphism, where the operation of the matrix monoid is the* usual *matrix multiplication. It defines the series* $r = ||(\mathbf{s}, h, \mathbf{g})||$ *with* $(r, w) = \mathbf{s}h(w)\mathbf{g}$.

A series is said to be rational *if it is defined by a linear representation. We write* $\mathbb{Z}^{\text{rat}}\langle\!\langle \Sigma^* \rangle\!\rangle$ *for the set of rational series.*

For a class $\mathcal{C}$ of languages, write $\mathsf{BC}(\mathcal{C})$ for the Boolean closure of $\mathcal{C}$. Arguments similar to those used in proving Theorem 5 allow us to show:

Theorem 8. *Over any alphabet* Σ, $\mathcal{L}_{\text{DetAPA}} = \mathsf{BC}(\mathsf{supp}_+(\mathbb{Z}^{\text{rat}}\langle\!\langle \Sigma^* \rangle\!\rangle))$.

Proof. ($\mathcal{L}_{\text{DetAPA}} \subseteq \mathsf{BC}(\mathsf{supp}_+(\mathbb{Z}^{\text{rat}}\langle\!\langle \Sigma^* \rangle\!\rangle))$) First note that there is a rational series r such that $\mathsf{supp}_+(r) = \{1\}$. Let L be in $\mathcal{L}_{\text{DetAPA}}$; we may thus suppose that $1 \notin L$. By the same token as in the proof of Theorem 5, there is a morphism $h\colon \Sigma^* \to \mathcal{M}_{\mathbb{Z}}(k)$, for some k, a vector $\mathbf{v} \in \{0,1\}^k$, and a type $\mathcal{Z}$ of $(\mathbb{Z}, \mathbb{Z}^+)^k$ such that:

$$L = \{w \mid h(w)\mathbf{v} \in \mathcal{Z}\} \ .$$

Further, similar to Lemma 2, $L(A, U, C_1 \oslash C_2) = L(A, U, C_1) \oslash L(A, U, C_2)$, for $\oslash \in \{\cup, \cap\}$ and any DetAPA $(A, U, C_1 \oslash C_2)$. Moreover, $L(A, U, \overline{C}) = \overline{L(A, U, C)} \cap L(A)$. We may thus suppose that $\mathcal{Z}$ is reduced to $\mathbb{Z}^{i-1} \times \mathbb{Z}^+ \times \mathbb{Z}^{k-i}$ for some i.

Now let h' be the morphism from Σ^* to $\mathbb{Z}^{k\times k}$ (with the *usual* matrix multiplication as operation), where $h'(a) = (h(a))^{\mathrm{T}}$, with $a \in \Sigma$ and M^{T} the transpose of M. Note that $h(a_1a_2) = h(a_2)h(a_1) = ((h(a_1))^{\mathrm{T}}(h(a_2))^{\mathrm{T}})^{\mathrm{T}}$, which is $(h'(a_1a_2))^{\mathrm{T}}$; more generally, $h(w) = (h'(w))^{\mathrm{T}}$. Thus we have that $\mathbf{v}^{\mathrm{T}}h'(w) = (h(w)\mathbf{v})^{\mathrm{T}}$. Hence with $\mathbf{s} = \mathbf{v}^{\mathrm{T}}$ and $\mathbf{g}$ the column vector $\mathbf{e}_i$, $\mathbf{s}h'(w)\mathbf{g} > 0$ iff $h(w)\mathbf{v} \in \mathcal{Z}$.

Now the triple $(\mathbf{s}, h', \mathbf{g})$ is a linear representation of a rational series which associates w to $\mathbf{s}h'(w)\mathbf{g}$, and this concludes the proof.

($\mathsf{BC}(\mathsf{supp}_+(\mathbb{Z}^{\text{rat}}\langle\!\langle \Sigma^* \rangle\!\rangle)) \subseteq \mathcal{L}_{\text{DetAPA}}$) As $\mathcal{L}_{\text{DetAPA}}$ is closed under union, complement, and intersection, we need only show that $\mathsf{supp}_+(\mathbb{Z}^{\text{rat}}\langle\!\langle \Sigma^* \rangle\!\rangle) \subseteq \mathcal{L}_{\text{DetAPA}}$.

Let $(\mathbf{s}, h, \mathbf{g})$ be a linear representation of dimension k of a rational series r over the alphabet Σ. Define $h'\colon \Sigma^* \to \mathcal{M}_{\mathbb{Z}}(k)$ by letting $h'(a) = (h(a))^{\mathrm{T}}$, for $a \in \Sigma$. Then for $w \in \Sigma^*$, $h(w) = (h'(w))^{\mathrm{T}}$. Now the rest of the proof is similar to that of Theorem 5: define $A = (\{r,t\}, \Sigma, \delta, r, \{r,t\})$, with $\delta = \{r,t\} \times \Sigma \times \{t\}$. Then let $U\colon \delta^* \to \mathcal{F}_k$ for $q \in \{r,t\}$, $a \in \Sigma$, and $\mathbf{x} \in \mathbb{Z}^k$, be defined by:

$$[U((q,a,t))](\mathbf{x}) = \begin{cases} h'(a)\mathbf{s} & \text{if } q = r, \\ h'(a)\mathbf{x} & \text{otherwise.} \end{cases}$$

This implies that for $w \in \Sigma^*$ and π its unique accepting path in A, it holds that $[U(\pi)](\mathbf{0}) = \mathbf{s}h(w)$. Thus letting $C = \{\mathbf{x} \mid \mathbf{x}\mathbf{g} > 0\}$, with $\mathbf{x}$ a row vector and $\mathbf{g}$ a column vector, we have that $L(A, U, C) = \mathsf{supp}_+(r)$. □

Remark 1. The class of positive supports of $\mathbb{Z}$-rational series is the class of $\mathbb{Q}$-*stochastic languages* (see, e.g., [23]). As we are interested in showing that $\mathcal{L}_{\mathrm{DetAPA}}$ is not closed under concatenation, it is worth noting that $\mathbb{Q}$-stochastic languages are not closed under concatenation. We mention three proofs of this fact. Two proofs [15,23] show that $\mathbb{Q}$-stochastic languages are not closed under concatenation with a *finite* language; such a concatenation is expressible as a finite union of $\mathbb{Q}$-stochastic languages, and is thus not directly applicable to our case. A third proof [25] shows that the $\mathbb{Q}$-stochastic language $L = \{a^i\#(a+\#)^*\#a^i \mid i \in \mathbb{N}\}$ is such that $L \cdot \{a, \#\}^*$ is not $\mathbb{Q}$-stochastic. We conjecture that $L \cdot \{a, \#\}^*$ is neither in $\mathcal{L}_{\mathrm{DetAPA}}$, but the proof given in [25] does not apply directly to our case. Finally, we note that the fact that unary $\mathbb{Q}$-stochastic languages are not closed under union [23] implies, as any regular language is $\mathbb{Q}$-stochastic, that there are nonregular unary languages in $\mathcal{L}_{\mathrm{DetAPA}}$.

Let $\#\mathrm{NC}^1$ be the class of functions computed by DLOGTIME-uniform arithmetic circuits of polynomial size and logarithmic depth and PNC^1 be the class of languages expressible as $\{w \mid f(w) > 0\}$ for $f \in \#\mathrm{NC}^1$ (see [12]). Note that this class is included in L. As iterated matrix multiplication can be done in $\#\mathrm{NC}^1$ and PNC^1 is closed under the Boolean operations, it is readily seen from Theorem 8 that:

Corollary 2. $\mathcal{L}_{\mathrm{DetAPA}} \subseteq \mathrm{PNC}^1$.

Conclusion

Connections between variants of the Parikh automaton and complexity classes were investigated. In particular, natural characterizations of the language classes defined by deterministic and unambiguous constrained automata, in the theory of typed monoids, were obtained. We hope that these characterizations will suggest refinements that may help to better understand classes such as PNC^1 and NC^1.

We note in conclusion that the unary languages in $\mathcal{L}_{\mathrm{DetAPA}}$, and indeed the bounded languages in $\mathcal{L}_{\mathrm{DetAPA}}$, can be shown to belong to the DLOGTIME-DCL-uniform variant of NC^1. Recall that the latter is not known to equal what is commonly referred to as DLOGTIME-uniform NC^1 (see [26, p. 162]), or ALOGTIME. Yet we were unable to show that the unary languages in $\mathcal{L}_{\mathrm{DetAPA}}$ belong to the latter. Do they?

References

1. Allender, E., Arvind, V., Mahajan, M.: Arithmetic complexity, Kleene closure, and formal power series. Theory Comput. Syst. 36(4), 303–328 (2003)
2. Barrington, D.A.M.: Bounded-width polynomial size branching programs recognize exactly those languages in NC^1. J. Computer and System Sciences 38, 150–164 (1989)

3. Barrington, D.A.M., Thérien, D.: Finite monoids and the fine structure of NC^1. J. Association of Computing Machinery 35, 941–952 (1988)
4. Barrington, D.A.M., Immerman, N., Straubing, H.: On uniformity within NC^1. J. Comput. Syst. Sci. 41(3), 274–306 (1990)
5. Behle, C., Krebs, A., Mercer, M.: Linear circuits, two-variable logic and weakly blocked monoids. In: Kučera, L., Kučera, A. (eds.) MFCS 2007. LNCS, vol. 4708, pp. 147–158. Springer, Heidelberg (2007)
6. Behle, C., Krebs, A., Reifferscheid, S.: Typed monoids - an Eilenberg-like theorem for non regular languages. In: Winkler, F. (ed.) CAI 2011. LNCS, vol. 6742, pp. 97–114. Springer, Heidelberg (2011)
7. Berstel, J., Reutenauer, C.: Noncommutative Rational Series with Applications. Encyclopedia of Mathematics and its Applications. Cambridge University Press (2010)
8. Büchi, J.R.: Weak second order arithmetic and finite automata. Z. Math. Logik Grundlagen Math. 6, 66–92 (1960)
9. Cadilhac, M., Finkel, A., McKenzie, P.: Bounded Parikh automata. In: Proc. 8th Internat. Conference on Words. EPTCS, vol. 63, pp. 93–102 (2011)
10. Cadilhac, M., Finkel, A., McKenzie, P.: Affine Parikh automata. RAIRO - Theoretical Informatics and Applications 46(4), 511–545 (2012)
11. Cadilhac, M., Finkel, A., McKenzie, P.: Unambiguous constrained automata. In: Yen, H.-C., Ibarra, O.H. (eds.) DLT 2012. LNCS, vol. 7410, pp. 239–250. Springer, Heidelberg (2012)
12. Caussinus, H., McKenzie, P., Thérien, D., Vollmer, H.: Nondeterministic NC^1 computation. J. Computer and System Sciences 57, 200–212 (1998)
13. Eilenberg, S.: Automata, Languages, and Machines. Pure and Applied Mathematics, vol. B. Academic Press (1976)
14. Enderton, H.B.: A Mathematical Introduction to Logic. Academic Press (1972)
15. Fliess, M.: Propriétés booléennes des langages stochastiques. Mathematical Systems Theory 7(4), 353–359 (1974)
16. Gill, J.: Computational complexity of probabilistic turing machines. SIAM J. Computing 6, 675–695 (1977)
17. Ginsburg, S., Spanier, E.H.: Semigroups, Presburger formulas and languages. Pacific J. Mathematics 16(2), 285–296 (1966)
18. Karianto, W.: Parikh automata with pushdown stack. PhD thesis, RWTH Aachen (2004)
19. Klaedtke, F., Rueß, H.: Monadic second-order logics with cardinalities. In: Baeten, J.C.M., Lenstra, J.K., Parrow, J., Woeginger, G.J. (eds.) ICALP 2003. LNCS, vol. 2719, pp. 681–696. Springer, Heidelberg (2003)
20. Krebs, A., Lange, K.J., Reifferscheid, S.: Characterizing TC^0 in terms of infinite groups. Theory of Computing Systems 40(4), 303–325 (2007)
21. Limaye, N., Mahajan, M., Rao, B.V.R.: Arithmetizing classes around NC^1 and L. Theory Comput. Syst. 46(3), 499–522 (2010)
22. Parikh, R.J.: On context-free languages. J. ACM 13(4), 570–581 (1966)
23. Salomaa, A., Soittola, M.: Automata-Theoretic Aspects of Formal Power Series. Springer, New York (1978)
24. Straubing, H.: Finite Automata, Formal Logic, and Circuit Complexity. Birkhäuser, Basel (1994)
25. Turakainen, P.: Some closure properties of the family of stochastic languages. Information and Control 18(3), 253–256 (1971)
26. Vollmer, H.: Introduction to Circuit Complexity – A Uniform Approach. Texts in Theoretical Computer Science. Springer (1999)

Generalized AG Codes as Evaluation Codes

Marco Calderini and Massimiliano Sala

Department of Mathematics, University of Trento, Italy
marco.calderini@unitn.it,
maxsalacodes@gmail.com

Abstract. We extend the construction of GAG codes to the case of evaluation codes. We estimate the minimum distance of these extended evaluation codes and we describe the connection to the one-point GAG codes.

Keywords: Evaluation codes, Affine-variety codes, AG codes, Generalized AG codes.

1 Introduction

In 1999, Xing, Niederreiter and Lam proposed [1,2] two constructions of linear codes based on algebraic curves using points of arbitrary degree. These generalize the construction of Algebraic Geometry (AG) codes introduced by Goppa [3,4]. Özbudak and Stichtenoth [5] showed that there is essentially only one new construction, namely that of Generalized Algebraic Geometric (GAG) codes, and introduced the notion of designed minimum distance for GAG codes.

Until now several papers have studied GAG codes in an algebraic geometry way, see e.g. [6], [7], [8], [9].

Høholdt, van Lint and Pellikaan [10] founded the theory of order domains and of the order domain codes (or evaluation codes) to simplify the description of one-point AG codes. The minimum distance of evaluation codes can be found by applying bound that relies only on some relatively simple theory [10].

Affine-variety codes, introduced by Fitzgerald and Lax in [11], are particularly interesting for their parameters and for a new efficient decoding system [12]. Geil, in [13], presents the AG codes as an example of affine-variety codes and their relation with evaluation codes.

In this paper we will extend the construction of affine-variety codes to introduce the GAG codes as a particular example of these family of codes. We extend, also, the construction of the evaluation codes and we analyze a particular case of the one-point GAG codes into the setting of these new codes. The remainder of this paper contains the following sections.

- In Section 2 we recall definitions and theorems about the minimum distance for affine-variety codes, order domain codes and generalized algebraic geometric codes.

T. Muntean, D. Poulakis, and R. Rolland (Eds.): CAI 2013, LNCS 8080, pp. 74–82, 2013.

- In Section 3 we introduce two constructions of linear codes, the extended affine-variety codes and the extended order domain codes, and we estimate a lower bound on the minimum distance for these families of codes.
- In section 4 we analyze the relation between an extended order domain code and a GAG code constructed from a rational point and we compare the relevant bounds on the minimum distance of the code.

2 Preliminaries

2.1 Affine-Variety Codes

Let $I \subseteq \mathbb{F}_q[X_1, \ldots, X_m]$ be an ideal, we define

$$I_q = I + \langle X_1^q - X_1, \ldots, X_m^q - X_m \rangle$$

$$R_q = \mathbb{F}_q[X_1, \ldots, X_m]/I_q$$

Let

$$V = \{P_1, \ldots, P_n\} = \mathcal{V}_{\mathbb{F}_q}(I) = \mathcal{V}_{\overline{\mathbb{F}}_q}(I_q)$$

be the variety of I over $\mathbb{F}_q$. Here $\overline{\mathbb{F}}$ means the algebraic closure of the field $\mathbb{F}$.

Define the evaluation map $ev : R_q \to {\mathbb{F}_q}^n$, the $\mathbb{F}_q$-linear map such that

$$ev(f + I_q) = (f(P_1), \ldots, f(P_n)). \tag{1}$$

The evaluation map is a vector space isomorphism.

Definition 1. *Let L be an $\mathbb{F}_q$- vector subspace of R_q. We define the affine variety code*

$$C(I, L) = ev(L).$$

The notation of this subsection comes from [11], where also the code $C(I, L)^\perp$ is called an affine-variety code. In this paper we will not consider this type of codes.

2.2 Order Domain Conditions

Let $J \subseteq \mathbb{F}[X_1, \ldots, X_m]$ be an ideal and let $\prec$ be a fixed monomial ordering. Denote by $\mathcal{M}(X_1, \ldots, X_m)$ the set of all monomials in the variables $X_1, \ldots, X_m$. The footprint of J (or Hilbert staircase) with respect to $\prec$ is the set

$$\Delta_\prec(J) = \{p \in \mathcal{M}(X_1, \ldots, X_m) \mid p \text{ is not the leading monomial of any polynomial in } J\}.$$

Definition 2. *Let $I \subseteq \mathbb{F}[X_1, \ldots, X_m]$ be an ideal. Let $\prec_w$ be a generalized weighted degree ordering, $w : \mathcal{M} \to \mathbb{N}_0^r$. Assume I possesses a Gröbner basis $\mathcal{G}$ such that:*

(i) any $g \in \mathcal{G}$ has exactly two monomials of highest weight in its support.
(ii) no two monomials in $\Delta_{\prec_w}(I)$ are of the same weight.

Then we say that $(I, \prec_w)$ satisfies the order domain conditions.

Let $L \subseteq R_q$ be a subspace. By using Gaussian elimination any basis of L can be transformed into a basis of the following form.

Definition 3. *Let $\prec$ be a fixed monomial ordering and $k = \dim(L)$. A basis $\{b_1 + I_q, \ldots, b_k + I_q\}$ for L such that Supp$(b_i) \subseteq \Delta_\prec(I_q)$ for $i = 1, \ldots, k$ and $\mathrm{lm}(b_1) \prec \cdots \prec \mathrm{lm}(b_k)$ is said to be well-behaving with respect to $\prec$. Here Supp(f) means the support of f, and $\mathrm{lm}(f)$ means the leading monomial of f.*

The sequence $(\mathrm{lm}(b_1), \ldots, \mathrm{lm}(b_k))$ is the same for all choices of well-behaving basis of L. So we define the set

$$\square_\prec(L) = \{\mathrm{lm}(b_1), \ldots, \mathrm{lm}(b_k)\}.$$

Definition 4. *Assume I and $\prec_w$ satisfy the order domain conditions. Let $\Gamma = w(\Delta_{\prec_w}(I)) \subseteq \mathbb{N}_0^r$ and $\Delta = \Delta_{\prec_w}(I_q)$. For any $\lambda \in w(\Delta)$ we define*

$$\sigma_\Delta(\lambda) = \sigma(\lambda) = |\{\eta \in w(\Delta) \mid \eta - \lambda \in \Gamma\}|.$$

Theorem 1 (Th. 4.27 in [13]). *Assume $(I, \prec_w)$ satisfies the order domain condition and let L subspace of R_q with $\{b_1 + I_q, \ldots, b_{\dim(L)} + I_q\}$ well-behaving basis. Then the minimum distance of $C(I, L)$ is at least*

$$\min\{\sigma(w(\alpha)) \mid \alpha \in \square_{\prec_w}(L)\}.$$

Remark 1 (Remark 4.29 in [13]). Assume that the pair $(I, \prec_w)$ satisfies the order domain conditions. Let $U \subseteq \mathcal{V}_{\mathbb{F}_q}(I)$. Every finite set of points is a variety and therefore there exists polynomials $h_1, \ldots, h_r$ such that the vanishing ideal of U equals

$$I_U = I + \langle h_1, \ldots, h_r \rangle.$$

The estimates of the minimum distances of $C(I, L)$ can be adapted if these codes are made by evaluating in U rather than in the entire variety, but we need to replace I_q with I_U.

2.3 Weight Functions and Order Domains

The concept of a weight function was introduced by Høholdt et al. in [10] to simplify the treatment of one-point geometric AG codes and to propose a generalization to objects of higher dimensions than curves.

Let (R, ρ, Γ) be an order domain, where $\Gamma \subseteq \mathbb{N}^r$ is a semigroup and $\rho : R \to \Gamma \cup \{-\infty\}$ is a weight function.

From [14][Th. 10.4] we know that every order domain with a finitely generated semigroup, Γ, can be constructed as a factor ring, $\mathbb{F}[X_1, \ldots, X_m]/I$. Therefore it can be described in the language of Gröbner basis theory.

Definition 5. *Let R be an $\mathbb{F}_q$-algebra. A surjective map $\phi : R \to {\mathbb{F}_q}^n$ is called a morphism of $\mathbb{F}_q$-algebras if ϕ is $\mathbb{F}_q$-linear and if*

$$\phi(fg) = \phi(f) * \phi(g)$$

for all $f, g \in R$. Here $*$ is the component-wise product.

Definition 6. *Let (R, ρ, Γ) be an order domain over $\mathbb{F}_q$ and $\{f_\lambda \mid \rho(f_\lambda) = \lambda, \lambda \in \Gamma\}$ be a basis. Let $\phi : R \to {\mathbb{F}_q}^n$ be a morphism as in Definition 5. Define $\alpha(1) = 0$. For $i = 2, \ldots, n$ define recursively $\alpha(i)$ to be the smallest element in Γ that is greater than $\alpha(1), \ldots, \alpha(i-1)$ and satisfies*

$$\phi(f_{\alpha(i)}) \notin Span_{\mathbb{F}_q}\{\phi(f_\lambda) \mid \lambda \prec_{\mathbb{N}^r} \alpha(i)\}.$$

Write $\Delta(R, \rho, \phi) = \{\alpha(1), \ldots, \alpha(n)\}$.

Definition 7. *Let R be an order domain over $\mathbb{F}_q$ and let ϕ be a morphism . Fix a basis $\{f_\lambda \mid \rho(f_\lambda) = \lambda, \lambda \in \Gamma\}$ and let $\Delta = \Delta(R, \rho, \phi)$. For $\lambda \in \Gamma$ and $\delta \in \mathbb{N}$ consider the codes*

$$E(\lambda) = Span_{\mathbb{F}_q}\{\phi(f_\eta) \mid \eta \preceq_{\mathbb{N}^r} \lambda\}$$
$$\tilde{E}(\delta) = Span_{\mathbb{F}_q}\{\phi(f_\eta) \mid \eta \in \Delta \text{ and } \sigma_\Delta(\eta) \geq \delta\}.$$

Theorem 2 (Th. 2 in [15]). *The minimum distance of $E(\lambda)$ is at least*

$$\min\{\sigma_\Delta(\eta) \mid \eta \preceq_{\mathbb{N}^r} \lambda\}$$

and the minimum distance of $\tilde{E}(\delta)$ is at least δ.

2.4 GAG Codes

Let $\mathcal{X}$ be a projective, geometrically irreducible, non-singular algebraic curve defined over the finite field $\mathbb{F}_q$. Let g be the genus of $\mathcal{X}$. Let Φ be the Frobenius map on $\mathcal{X}$, namely the map sending a point P with homogeneous coordinates $(a_0, \ldots, a_r)$ to the point $\Phi(P)$ with coordinates $(a_0^q, \ldots, a_r^q)$.

Let P be a point of $\mathcal{X}$. Then $\deg(P)$ denotes the degree of P, namely the least positive integer n such that P is $\mathbb{F}_{q^n}$-rational, and the closed point of P is the set $O_\Phi(P) = \{P, \Phi(P), \ldots, \Phi^{n-1}(P)\}$.

Let $\mathcal{X}$ be a curve, let $P_1, \ldots, P_s$ be points of $\mathcal{X}$ such that for every $i \neq j$ the closed points $O_\Phi(P_i)$ and $O_\Phi(P_j)$ are disjoint. Let G be an $\mathbb{F}_q$-rational divisor that has support disjoint from any closed point $O_\Phi(P_i)$. Let $k_i := \deg(P_i)$. For $i = 1, \ldots, s$ let $\pi_i : \mathbb{F}_{q^{k_i}} \to C_i$ be an $\mathbb{F}_q$-linear isomorphism from the finite field $\mathbb{F}_{q^{k_i}}$ onto a linear $[n_i, k_i, d_i]$ code $C_i \subseteq {\mathbb{F}_q}^{n_i}$.

Definition 8. *Let $n = \sum_{i=1}^{s} n_i$, and consider the $\mathbb{F}_q$-linear map*

$$\pi : \begin{cases} \mathcal{L}(G) \to & {\mathbb{F}_q}^n \\ \quad f \;\; \mapsto & (\pi_1(f(P_1)), \ldots, \pi_s(f(P_s))) \end{cases}$$

The image of π is a Generalized Algebraic Geometric code

$$C(P_1, \ldots, P_s; G; C_1, \ldots, C_s) = \pi(\mathcal{L}(G)).$$

Here $\mathcal{L}(G)$ denotes the Riemann-Roch space of G over $\mathbb{F}_q$.

The designed minimum distance $\bar{d}$ of $C(P_1,\dots,P_s;G;C_1,\dots,C_s)$ is defined as follows (see [5]): let

$$X = \left\{ S \subseteq \{1,\dots,s\} \mid \sum_{i\in S} k_i \leq \deg(G) \right\}.$$

Then

$$\bar{d} := \min \left\{ \sum_{i\notin S} d_i \mid S \in X \right\}$$

Proposition 1 (Prop. 4.1 in [5]). *If* $\sum_{i=1}^{s} k_i > \deg(G)$, *then*

$$C(P_1,\dots,P_s;G;C_1,\dots,C_s)$$

is an $[n,k,d]$ *code with parameters*

$$k = \dim(\mathcal{L}(G)) \geq \deg(G) + 1 - g \quad \text{and} \quad d \geq \bar{d}.$$

Throughout this paper, the codes C_i will be called the inner codes of the GAG code.

Remark 2. If we construct the GAG code using $P_1,\dots,P_s$ points of which h are $\mathbb{F}_q$-rational, a divisor G with $\deg(G) \leq h$ and inner codes having minimum distance all equals to 1, then the designed minimum distance is equal to $s - \deg(G)$.

3 New Construction of Codes

For any $v \in {\mathbb{F}_q}^n$, let $\mathrm{w}_H(v) = |\{i \mid v_i \neq 0\}|$.

3.1 Extended Affine-Variety Codes

Let $(I, \prec_w)$ satisfying the order domain condition and let $\mathcal{P} = \{P_1,\dots,P_h\} \subseteq \mathcal{V}_{\overline{\mathbb{F}}_q}(I)$, with $\deg(P_i) = r_i$ for $i = 1,\dots,h$. As in Remark 1 there is an ideal $J \subseteq \mathbb{F}_q[X_1,\dots,X_m]$ such that $\mathcal{P} = \mathcal{V}_{\overline{\mathbb{F}}_q}(I+J)$. Let $I + J = I_{\mathcal{P}}$.

Let $L \subseteq R_q$ be a space over $\mathbb{F}_q$ with well-behaving basis $B = \{b_1 + I_{\mathcal{P}},\dots,b_k + I_{\mathcal{P}}\}$, and for $i = 1,\dots,h$ let $\pi_i : \mathbb{F}_{q^{r_i}} \to C_i$ be an $\mathbb{F}_q$-linear isomorphism from the finite field $\mathbb{F}_{q^{r_i}}$ onto the inner code C_i over $\mathbb{F}_q$ with parameters $[n_i, r_i, d_i]$.

Definition 9. *Let* $n = \sum_{i=1}^{h} n_i$, $\mathcal{P} = \{P_1,\dots,P_h\}$ *and* $\mathcal{C} = \{C_1,\dots,C_h\}$. *Consider the* $\mathbb{F}_q$*-linear map,*

$$\overline{ev} : \begin{cases} L \to & {\mathbb{F}_q}^n \\ f \mapsto (\pi_1(f(P_1)),\dots,\pi_h(f(P_h))) \end{cases}$$

Then the extended affine-variety code is

$$\overline{ev}(L) = C(I, L, \mathcal{P}, \mathcal{C}).$$

Theorem 3. *Let $\Delta = \Delta_{\prec_w}(I_{\mathcal{P}})$, then $C(I, L, \mathcal{P}, \mathcal{C})$ has minimum distance at least*

$$\delta\hat{d},$$

where $\delta = \min\{\sigma(w(\alpha) \mid \alpha \in \square(L)\}$ and $\hat{d} = \min\{d_1, \ldots, d_h\}$.

Proof. Let $r = m.c.m.\{r_1, \ldots, r_h\}$ and B be a well-behaving basis for L. Consider

$$L' = Span_{\mathbb{F}_{q^r}} B$$

and let $ev(L') \subseteq (\mathbb{F}_{q^r})^h$ (where ev is as in (1)) be the affine variety code over $\mathbb{F}_{q^r}$ restricted at the points $P_1, \ldots, P_h$. From Theorem 1, the minimum distance of this code is at least δ.

Note that $L \subseteq L'$, then for every non zero $c \in ev(L)$ we have $\mathrm{w}_H(c) \geq \delta$.

Let $\bar{c} \in C(I, L, \mathcal{P}, \mathcal{C}) \setminus \{0\}$, then $\bar{c} = (\pi_1(f(P_1), \ldots, \pi_h(f(P_h)))$ for some f. So let $S = \{i \mid f(P_i) \neq 0\}$, we have

$$\mathrm{w}_H(c) = \sum_{i=1}^{r} \mathrm{w}_H(\pi_i(f(P_i))) = \sum_{i \in S} d_i \geq \delta\hat{d}.$$

Remark 3. We also can estimate the minimum distance of the extended code $C(I, L, \mathcal{P}, \mathcal{C})$ if the order domain conditions are not satisfy. We can look at the number of one-way well-behaving pairs (see Def. 4.8 in [13]) as in Th. 4.9 in [13]. So we are able to obtain a bound similar to Theorem 3.

3.2 Extended Order Domain Codes

Let (R, ρ, Γ) be an order domain over $\mathbb{F}_q$ and B be a well-behaving basis for R. Consider $R' = Span_{\mathbb{F}_{q^r}} B$, then (R', ρ, Γ) is an order domain over $\mathbb{F}_{q^r}$. Note that $R \subseteq R'$.

Now let $\phi : R' \to \mathbb{F}_{q^r}^h$ be a morphism $\phi = (\phi_1, \ldots, \phi_h)$. For $i = 1, \ldots, h$ define $r_i = \min\{l \mid \phi_i(R) \subseteq \mathbb{F}_{q^l}\}$.

Let $\Delta = \Delta(R', \rho, \Gamma)$ be as in Definition 6. For $i = 1, \ldots, h$ let $\pi_i : \mathbb{F}_{q^{r_i}} \to C_i$ be an $\mathbb{F}_q$-linear isomorphism from the finite field $\mathbb{F}_{q^{r_i}}$ onto the inner code C_i over $\mathbb{F}_q$ with parameters $[n_i, r_i, d_i]$.

Definition 10. *Let $\mathcal{C} = \{C_1, \ldots, C_h\}$ and $\mathcal{R} = \{r_1, \ldots, r_h\}$. For $\lambda \in \Gamma$ and $\delta \in \mathbb{N}$ consider the codes*

$$E(\lambda, \mathcal{R}, \mathcal{C}) = Span_{\mathbb{F}_q}\{(\pi_1(\phi_1(f_\eta)), \ldots, \pi_h(\phi_h(f_\eta))) \mid \eta \preceq_{\mathbb{N}^r} \lambda\}$$

$$\hat{E}(\delta, \mathcal{R}, \mathcal{C}) = Span_{\mathbb{F}_q}\{(\pi_1(\phi_1(f_\eta)), \ldots, \pi_h(\phi_h(f_\eta))) \mid \eta \in \Delta \text{ and } \sigma_\Delta(\eta) \geq \delta\}.$$

Theorem 4. *The minimum distance of $E(\lambda, \mathcal{R}, \mathcal{C})$ is at least*

$$\gamma\hat{d},$$

where $\gamma = \min\{\sigma_\Delta(\eta) \mid \eta \preceq_{\mathbb{N}^r} \lambda\}$ and $\hat{d} = \min\{d_1, \ldots, d_h\}$.

The minimum distance of $\hat{E}(\delta, \mathcal{R}, \mathcal{C})$ is at least $\delta\hat{d}$.

Proof. Obvious adaption of the proof at Theorem 3.

4 One-Point GAG Codes as Extended Order Domain Codes

Now we consider the GAG codes constructed from a rational point of the curve, using as inner code $C_i = \mathbb{F}_q^{r_i}$ for $i = 1, \ldots, h$. We refer to these as one-point GAG codes.

Let P be a rational point of a curve $\mathcal{X}$ defined over a field $\mathbb{F}_q$. Let ν_P be the valuation corresponding to P. Consider the algebraic structure

$$R = \bigcup_{m=0}^{\infty} \mathcal{L}(mP). \tag{2}$$

Defining $\rho = -\nu_P$ we have $\rho(R) = \Gamma \cup \{-\infty\}$ where $\Gamma \subseteq \mathbb{N}$ is known as the Weierstrass semigroup corresponding to P. By inspection (R, ρ, Γ) is an order domain over $\mathbb{F}_q$.

Let $P_1, \ldots, P_h$ be distinct points, with distinct closed points, and of degree $r_1, \ldots, r_h$, respectively. Let B be a well-behaving basis for R. Define $R' = Span_{\mathbb{F}_{q^r}} B$ and let $\phi : R' \to \mathbb{F}_{q^r}^h$ be a morphism with $\phi(f) = (f(P_1), \ldots, f(P_h))$. Then we have

$$C(P_1, \ldots, P_h, \lambda P, C_1, \ldots, C_h) = C(I, L, \mathcal{P}, \mathcal{C}) = E(\lambda, \mathcal{R}, \mathcal{C}),$$

where $L = \{f \,|\, \rho(f) \leq \lambda\}$, $\mathcal{P} = \{P_1, \ldots, P_h\}$, $\mathcal{R} = \{r_1, \ldots, r_h\}$ and $\mathcal{C} = \{C_1, \ldots, C_h\}$.

Lemma 1 (Lemma 2 in [15]). *Let $\Gamma = \{\lambda_1, \lambda_2, \ldots\}$ with $\lambda_1 < \lambda_2 < \ldots$ be a numerical semigroup with finitely many gaps. For any λ_i we have*

$$\#(\Gamma \setminus (\lambda_i + \Gamma)) = \lambda_i.$$

Theorem 5. *The minimum distance of $E(\lambda, \mathcal{R}, \mathcal{C})$ is at least*

$$\min\{\sigma_\Delta(\eta) \,|\, \eta \leq \lambda\} \geq h - \lambda$$

where $\Delta = \Delta(R', \rho, \phi)$.

Proof. The distances of the inner codes are all equal to 1. Consider $\lambda_i \in \Delta$, with $\lambda_i \leq \lambda$. We have $\sigma(\lambda_i) = \#(\Delta \cap (\lambda_i + \Gamma))$, the elements in Δ that are not in $\lambda_i + \Gamma$ are at most λ_i. Then $\sigma(\lambda_i) \geq h - \lambda_i \geq h - \lambda$.

Remark 4. With order domain code it is possible, sometimes, to have a bound on the minimum distance of a one-point Algebraic Geometry code better than the Goppa bound [15]. So also for GAG codes, if we are in the case as in the Remark 2, using the order domains is possible to obtain a bound always at least as good as (and sometimes better than) the bound in the Proposition 1.

Example 1. Let $\mathbb{F}_4 = \{0, 1, \alpha, \alpha^2\}$, where α is a primitive element. Consider the plane curve of affine equation $\mathcal{X} : X^6 + Y^5 + Y$. Let $\prec$ be the weighted degree lexicographic ordering given by $w(X) = 5, w(Y) = 6$. Let $I = \langle X^6 + Y^5 + Y \rangle$, then $(I, \prec)$ satisfies the order domain conditions and $w(\Delta(I))$ is the semigroup $\langle 5, 6 \rangle$.

We have 8 $\mathbb{F}_4$-rational points

$$\mathcal{V}(I_4) = \{(0,0), (0,1), (1,\alpha), (1,\alpha^2), (\alpha,\alpha), (\alpha,\alpha^2), (\alpha^2,\alpha), (\alpha^2,\alpha^2)\}$$

and $\mathcal{G} = \{Y^2 + X^3 + Y, XY^2 + XY + X, Y^4 + Y\}$ is a Gröbner basis for I_4. The monomials in the footprint of I_4 are

$$\Delta(I_4) = \{1, X, Y, X^2, XY, Y^2, X^2Y, Y^3\}$$

and its corresponding weights are

$$w(\Delta(I_4)) = \{0, 5, 6, 10, 11, 12, 16, 18\}.$$

Now we consider a point of the variety $\mathcal{V}_{\mathbb{F}_4}(I)$ of degree 3 (there are not points of degree 2). Let $\mathbb{F}_{64} = \mathbb{F}_4[Z]/\langle Z^3 + Z + 1 \rangle$ and let $\beta^3 = \beta + 1$. The point that we consider is $(1, \beta^3)$. Using Buchberger-Möller's algorithm we can compute the Gröbner basis of the vanishing ideal of the nine points, so we adjoint the monomial X^3 at the footprint and the weight 15 to $w(\Delta(I_4))$.

Consider now $L = Span_{\mathbb{F}_q}\{1, X, Y\}$, then the minimum distance of the code $C(I, L, \mathcal{P}, \mathcal{C})$, where the inner codes used are $C_1 = \cdots = C_8 = \mathbb{F}_4$ and $C_9 = \mathbb{F}_4^3$, is at least $\min\{\sigma(0), \sigma(5), \sigma(6)\} = 5$. This value improves on what obtainable from the GAG construction, as follows.

Looking at this code as a one-point GAG code we can note that the semigroup $w(\Delta(I))$ is the Weiestrass semigroup of the unique rational point at infinity, P_∞, of the curve and $L = \mathcal{L}(6P_\infty)$. Therefore the bound on minimum distance of the GAG code as in Proposition 1 is equal to 3.

In [16] was shown that an order domain with numerical weight function (i.e. the weights are in $\mathbb{N}_0$) is a sub algebra of a structure as in (2). If the semigroup related to the order domain are not numerical then they are related to structures of transcendence degree greater than one, that is, these structures are curves no longer ([14] Sec. 11). Examples of evaluation codes coming from higher dimensional objects than curves are given in [17] and these codes can be viewed as generalizations of one-point AG codes. Then our extension can be consider a generalization of the one-point GAG codes.

References

1. Niederreiter, H., Xing, C., Lam, K.Y.: New construction of algebraic-geometry codes. Appl. Algebra Engrg. Comm. Comput. 9(5), 373–381 (1999)
2. Xing, C., Niederreiter, H., Lam, K.Y.: A generalization of algebraic-geometry codes. IEEE Transactions on Information Theory 45(7), 2498–2501 (1999)

3. Goppa, V.D.: Codes on algebraic curves. Dokl. Akad. NAUK, SSSR 259, 1289–1290 (1981)
4. Goppa, V.D.: Algebraico-geometric Codes. Izv. Akad. NAUK, SSSR 46, 75–91 (1982)
5. Özbudak, F., Stichtenoth, H.: Constructing codes from algebraic curves. IEEE Transactions on Information Theory 45(7), 2502–2505 (1999)
6. Heydtmann, A.E.: Generalized geometric Goppa codes. Comm. Algebra 30(6), 2763–2789 (2002)
7. Ding, C., Niederreiter, H., Xing, C.: Some new codes from algebraic curves. IEEE Transactions on Information Theory 46(7), 2638–2642 (2000)
8. Calderini, M., Faina, G.: Generalized Algebraic Geometric Codes From Maximal Curves. IEEE Transactions on Information Theory 58(4), 2386–2396 (2012)
9. Xing, C., Yeo, S.L.: New linear codes and algebraic function fields over finite fields. IEEE Transactions on Information Theory 53(12), 4822–4825 (2007)
10. Høholdt, T., van Lint, J., Pellikaan, R.: Algebraic Geometry Codes. In: Pless, V.S., Huffman, W.C. (eds.) Handbook of Coding Theory, vol. 1, pp. 871–961. Elsevier, Amsterdam (1998)
11. Fitzgerald, J., Lax, R.F.: Decoding affine variety codes using Göbner bases. Des. Codes Cryptogr. 13(2), 147–158 (1998)
12. Orsini, E., Marcolla, C., Sala, M.: Improved decoding of affine-variety codes. Journal of Pure and Applied Algebra 216(7), 1533–1565 (2012)
13. Geil, O.: Evaluation codes from an affine variety code perspective. In: Advances in Algebraic Geometry Codes. Series on Coding Theory and Cryptology, vol. 5, pp. 153–180. World Sci. Publ., Hackensack (2008)
14. Geil, O., Pellikaan, R.: On the Structure of Order Domains. Finite Fields and their Applications 8, 369–396 (2002)
15. Geil, O.: Algebraic Geometry Codes from Order Domains. In: Sala, Mora, Perret, Sakata, Traverso (eds.) Gröbner Bases. Coding, and Cryptography, pp. 121–141. Springer (2009)
16. Matsumoto, R.: Miuras Generalization of One-Point AG codes is Equivalent to Høholdt, van Lint and Pellikaan's Generalization. IEICE Trans. Fundamentals E82-A(10), 2007–2010 (1999)
17. Andersen, H.E., Geil, O.: Evaluation codes from order domain theory. Finite Fields and Their Applications 14, 92–123 (2008)

Osculating Spaces of Varieties and Linear Network Codes

Johan P. Hansen

Department of Mathematics, Aarhus University*
matjph@imf.au.dk

Abstract. We present a general theory to obtain good linear network codes utilizing the osculating nature of algebraic varieties. In particular, we obtain from the osculating spaces of Veronese varieties explicit families of equidimensional vector spaces, in which any pair of distinct vector spaces intersect in the same dimension.

Linear network coding transmits information in terms of a basis of a vector space and the information is received as a basis of a possible altered vector space. Ralf Koetter and Frank R. Kschischang [KK08] introduced a metric on the set of vector spaces and showed that a minimal distance decoder for this metric achieves correct decoding if the dimension of the intersection of the transmitted and received vector space is sufficiently large.

The proposed osculating spaces of Veronese varieties are equidistant in the above metric. The parameters of the resulting linear network codes are determined.

Notation

- $\mathbb{F}$ is the finite field with q elements of characteristic p.
- $\overline{\mathbb{F}} = \overline{\mathbb{F}_q}$ is an algebraic closure of $\mathbb{F}$.
- $R_d = \overline{\mathbb{F}}[X_0, \dots, X_n]_d$ and $R_d(\mathbb{F}) = \mathbb{F}[X_0, \dots, X_n]_d$ the homogenous polynomials of degree d with coefficients in $\overline{\mathbb{F}}$ and $\mathbb{F}$.
- $R = \overline{\mathbb{F}}[X_0, \dots, X_n] = \oplus_d R_d$ and $R(\mathbb{F}) = \mathbb{F}[X_0, \dots, X_n] = \oplus_d R_d(\mathbb{F})$
- AffCone$(Y) \subseteq \overline{\mathbb{F}}^{M+1}$ denotes the affine cone of the subvariety $Y \subseteq \mathbb{P}^M$ and AffCone$(Y)(\mathbb{F})$ its $\mathbb{F}$-rational points.
- $O_{k,X,P} \subseteq \mathbb{P}^M$ is the embedded k-osculating space of a variety $X \subseteq \mathbb{P}^M$ at the point $P \in X$ and $O_{k,X,P}(\mathbb{F})$ its $\mathbb{F}$-rational points, see 2.
- $\mathcal{V} = \sigma_d(\mathbb{P}^n) \subseteq \mathbb{P}^M$ with $M = \binom{d+n}{n} - 1$ is the Veronese variety, see 1.1.

For generalities on algebraic geometry we refer to [Har77].

* Part of this work was done while visiting Institut de Mathématiques de Luminy, MARSEILLE, France.

T. Muntean, D. Poulakis, and R. Rolland (Eds.): CAI 2013, LNCS 8080, pp. 83–88, 2013.

1 Introduction

Algebraic varieties have in general an osculating structure. By Terracini's lemma [Ter11], their embedded tangent spaces tend to be in general position. Specifically, the tangent space at a generic point $P \in \overline{Q_1 Q_2}$ on the secant variety of points on some secant is spanned by the tangent spaces at Q_1 and Q_2. In general, the secant variety of points on some secant have the expected maximal dimension and therefore the tangent spaces generically span a space of maximal dimension, see [Zak93].

This paper suggests k-osculating spaces including tangent spaces of algebraic varieties as a source for constructing linear subspaces in general position of interest for linear network coding. The k-osculating spaces are presented in 1.1.

In particular, we will present the k-osculating subspaces of Veronese varieties and apply them to obtain linear network codes generalizing the results in [Han12]. The Veronese varieties are presented in 2.

Definition 1. *Let $X \subseteq \mathbb{P}^M$ be a smooth projective variety of dimension n defined over the finite field $\mathbb{F}$ with q elements. For each positive integer k we define the k-osculating linear network code $\mathcal{C}_{k,X}$. The elements of the code are the linear subspaces in $\mathbb{F}^{M+1}$ which are the affine cones of the k-osculating subspaces $O_{k,X,P}(\mathbb{F})$ at $\mathbb{F}$-rational points P on X, as defined in 1.1.*

Specifically

$$\mathcal{C}_{k,X} = \{\mathrm{AffCone}(O_{k,X,P})(\mathbb{F}) \mid P \in X(\mathbb{F})\}\ .$$

The number of elements in $\mathcal{C}_{k,X}$ is by construction $|X(\mathbb{F})|$, the number of $\mathbb{F}$-rational points on X.

One should remark that the elements in $\mathcal{C}_{k,X}$ are not necessarily equidimensional as linear vector spaces, however, their dimension is at most $\binom{k+n}{n}$.

Applying the construction to the Veronese variety $\mathcal{X}_{n,d}$ presented in 2, we obtain a linear network code $\mathcal{C}_{k,\mathcal{X}_{n,d}}$ and the following result, which is proved in section 2.1.

Theorem 1. *Let n, d be positive integers and consider the Veronese variety $\mathcal{X}_{n,d} \subseteq \mathbb{P}^M$, with $M = \binom{d+n}{n} - 1$, defined over the finite field $\mathbb{F}$ as in 2.*

Let $\mathcal{C}_{k,\mathcal{X}_{n,d}}$ be the associated k-osculating linear network code, as defined in Definition 1.

The packet length of the linear network code is $\binom{d+n}{n}$, the dimension of the ambient vector space. The number of vector spaces in the linear network code $\mathcal{C}_{k,\mathcal{X}_{n,d}}$ is $|\mathbb{P}^n(\mathbb{F})| = 1 + q + q^2 + \cdots + q^n$, the number of $\mathbb{F}$-rational points on $\mathbb{P}^n$.

The vector spaces $V \in \mathcal{C}_{k,\mathcal{X}_{n,d}}$ in the linear network code are equidimensional of dimension $\binom{k+n}{n}$ as linear subspaces of the ambient $\binom{d+n}{n}$-dimensional $\mathbb{F}$-vector space.

The elements in the code are equidistant in the metric $\mathrm{dist}(V_1, V_2)$ *of (5) of Section 3. Specifically, we have the following results.*

For vector spaces $V_1, V_2 \in \mathcal{C}_{k,\mathcal{X}_{n,d}}$ with $V_1 \neq V_2$

i) if $2k \geq d$, *then* $\dim_{\mathbb{F}}(V_1 \cap V_2) = \binom{2k-d+n}{n}$ *and*

$$\operatorname{dist}(V_1, V_2) = 2\left(\binom{k+n}{n} - \binom{2k-d+n}{n}\right).$$

ii) if $2k \leq d$, *then* $\dim_{\mathbb{F}}(V_1 \cap V_2) = 0$ *and*

$$\operatorname{dist}(V_1, V_2) = 2\binom{k+n}{n}.$$

1.1 Osculating Spaces

Principal Parts. Let X be a smooth variety of dimension n defined over the field K and let $\mathcal{F}$ be a locally free $\mathcal{O}_X$-module. The sheaves of k-principal parts $\mathcal{P}_X^k(\mathcal{F})$ are locally free and if $\mathcal{L}$ is of rank 1, then $\mathcal{P}_X^k(\mathcal{L})$ is a locally free sheaf of rank $\binom{k+n}{n}$.

There are the fundamental exact sequences

$$0 \to S^k \Omega_X \otimes_{\mathcal{O}_X} \mathcal{F} \to \mathcal{P}_X^k(\mathcal{F}) \to \mathcal{P}_X^{k-1}(\mathcal{F}) \to 0\,,$$

where Ω_X is the sheaf of differentials on X and $S^k\Omega_X$ its kth symmetric power. These sequences can be used to give a local description of the sheaf principal parts. Specifically, if $\mathcal{L}$ is of rank 1, then $\mathcal{P}_X^k(\mathcal{L})$ is a locally free sheaf of rank $\binom{k+n}{n}$. Assume furthermore that X is affine with coordinate ring $A = K[x_1, \ldots, x_n]$, then X and $\mathcal{L}$ can be identified with A. Also $S^k\Omega_X$ can be identified with the forms of degree k in $A[dx_1, \ldots, dx_n]$ in the indeterminates $dx_1, \ldots dx_n$ and $\mathcal{P}_X^k(\mathcal{L})$ with the polynomials of total degree $\leq k$ in the indeterminates $dx_1, \ldots dx_n$. For arbitrary X, the local picture is similar, taking local coordinates $x_1, \ldots, x_n$ at the point in question replacing A by the completion of the local ring at that point.

In general, for each k there is a canonical morphism

$$d_k : \mathcal{F} \to \mathcal{P}_X^k(\mathcal{F})\,.$$

For $\mathcal{L}$ of rank 1, using local coordinates as above, d_k maps an element in A to its truncated Taylor series

$$f = f(x_1, \ldots, x_n) \mapsto \sum_{|\alpha| \leq k} \frac{1}{|\alpha|!} \frac{\partial^{|\alpha|} f}{\partial x^\alpha}\,,$$

where $\alpha = i_1 i_2 \ldots i_n$ and $|\alpha| = i_1 + i_2 + \cdots + i_n$.

Osculating Spaces. Let X be a smooth variety of dimension n and let $f : X \to \mathbb{P}^M$ be an immersion. For $\mathcal{L} = f^*\mathcal{O}_{\mathbb{P}^n}(1)$ let $\mathcal{P}_X^k(\mathcal{L})$ denote the sheaf of principal parts of order k. Then $\mathcal{P}_X^k(\mathcal{L})$ is a locally free sheaf of rank $\binom{k+n}{n}$ and there are homomorphisms

$$a^k : \mathcal{O}_X^{M+1} \to \mathcal{P}_X^k(\mathcal{L})\,.$$

For $P \in X$ the morphism $a^k(P)$ defines the k-osculating space $O_{k,X,P}$ to X at P as

$$O_{k,X,P} := \mathbb{P}(\mathrm{Im}(a^k(P))) \subseteq \mathbb{P}^M \tag{1}$$

of projective dimension at most $\binom{k+n}{n} - 1$, see [Pie77], [BPT92] and [PT90]. For $k = 1$ the osculating space is the tangent space to X at P.

2 The Veronese Variety

Let $R_1 = \mathbb{F}[X_0, \ldots, X_n]_1$ be the $n+1$ dimensional vector space of linear forms in $X_0, \ldots, X_n$ and let $\mathbb{P}^n = \mathbb{P}(R_1)$ be the associated projective n-space over $\mathbb{F}$.

For each integer $d \geq 1$, consider R_d the vector space of forms of degree d. A basis consists of the $\binom{n+d}{d}$ monomials $X_0^{d_0} X_1^{d_1} \ldots X_n^{d_n}$ with $d_0 + d_1 + \cdots + d_n = d$. Let $\mathbb{P}^M = \mathbb{P}(R_d)$ be the associated projective space of dimension $M = \binom{n+d}{d} - 1$.

The d-uple morphism of $\mathbb{P}^n = \mathbb{P}(R_1)$ to $\mathbb{P}^M = \mathbb{P}(R_d)$ is the morphism

$$\sigma_d : \mathbb{P}^n = \mathbb{P}(R_1) \to \mathbb{P}^M = \mathbb{P}(R_d)$$
$$L \mapsto L^d$$

with image the Veronese variety

$$\mathcal{X}_{n,d} = \sigma_d(\mathbb{P}^n) = \{L^d |\ L \in \mathbb{P}(R_1)\} \subseteq \mathbb{P}^M . \tag{2}$$

2.1 Osculating Subspaces of the Veronese Variety

For the Veronese variety $\mathcal{X}_{n,d}$ of (2), the k-osculating subspaces of (1) with $1 \leq k < d$, at the point $P \in \mathcal{X}_{n,d}$ corresponding to the 1-form $L \in R_1$, can be described explicitly as

$$O_{k,\mathcal{X}_{n,d},P} = \mathbb{P}(\{L^{d-k}F |\ F \in R_k\}) = \mathbb{P}(R_k) \subseteq \mathbb{P}^M \tag{3}$$

of projective dimension exactly $\binom{k+n}{n} - 1$, see [Seg46], [CGG02], [BCGI07] and [BF03]. The osculating spaces constitute a flag of linear subspaces

$$O_{1,\mathcal{X}_{n,d},P} \subseteq O_{2,\mathcal{X}_{n,d},P} \subseteq \cdots \subseteq O_{d-1,\mathcal{X}_{n,d},P} .$$

This explicit description of the k-osculating spaces allows us to establish the claims in Theorem 1.

The associated affine cone of the k-osculating space in (3) is

$$\mathrm{AffCone}(O_{k,\mathcal{X}_{n,d},P})(\mathbb{F}) = \{L^{d-k}F |\ F \in R_k\} \tag{4}$$

of dimension $\binom{k+n}{n}$, proving the claim on the dimension of the vector spaces in the linear network code $\mathcal{C}_{k,\mathcal{X}_{n,d}}$.

As there is one element in $\mathcal{C}_{k,\mathcal{X}_{n,d}}$ for each $\mathbb{F}$-rational point on $\mathbb{P}^n$, it follows that the number of elements in $\mathcal{C}_{k,\mathcal{X}_{n,d}}$ is

$$|\mathcal{C}_{k,\mathcal{X}_{n,d}}| = |\mathbb{P}^n(\mathbb{F})| = 1 + q + q^2 + \cdots + q^n .$$

Finally, let $V_1, V_2 \in \mathcal{C}_{k,\mathcal{X}_{n,d}}$ with $V_1 \neq V_2$ and

$$V_i = \{L_i^{d-k} F_i |\ F_i \in R_k\}$$

If $2k \geq d$, we have

$$\begin{aligned} V_1 \cap V_2 =& \{L_1^{d-k} F_1 |\ F_1 \in R_k\} \cap \{L_2^{d-k} F_2 |\ F_2 \in R_k\} \\ =& \{L_1^{d-k} L_2^{d-k} G |\ G \in R_{2k-d}\} . \end{aligned}$$

Otherwise the intersection is trivial, proving the claims on the dimension of the intersections and the derived distances.

3 Linear Network Coding

In linear network, coding transmission is obtained by transmitting a number of packets into the network and each packet is regarded as a vector of length N over a finite field $\mathbb{F}$. The packets travel the network through intermediate nodes, each forwarding $\mathbb{F}$-linear combinations of the packets it has available. Eventually the receiver tries to infer the originally transmitted packages from the packets that are received, see [CWJJ03] and [HMK+06].

All packets are vectors in $\mathbb{F}^N$; however, Ralf Koetter and Frank R. Kschischang [KK08] describe a transmission model in terms of linear subspaces of $\mathbb{F}^N$ spanned by the packets and they define a fixed dimension *code* as a nonempty subset $\mathcal{C} \subseteq G(n, N)(\mathbb{F})$ of the Grassmannian of n-dimensional $\mathbb{F}$-linear subspaces of $\mathbb{F}^N$. They endowed the Grassmannian $G(n, N)(\mathbb{F})$ with the metric

$$\operatorname{dist}(V_1, V_2) := \dim_{\mathbb{F}}(V_1 + V_2) - \dim_{\mathbb{F}}(V_1 \cap V_2), \tag{5}$$

where $V_1, V_2 \in G(n, N)(\mathbb{F})$.

The size of the code $\mathcal{C} \subseteq G(n, N)(\mathbb{F})$ is denoted by $|\mathcal{C}|$, the minimal distance by

$$D(\mathcal{C}) := \min_{V_1, V_2 \in \mathcal{C}, V_1 \neq V_2} \operatorname{dist}(V_1, V_2) \tag{6}$$

and $\mathcal{C}$ is said to be of type $[N, n, \log_q |\mathcal{C}|, D(\mathcal{C})]$. Its normalized weight is $\lambda = \frac{n}{N}$, its rate is $R = \frac{\log_q(|\mathcal{C}|)}{Nn}$ and its normalized minimal distance is $\delta = \frac{D(\mathcal{C})}{2n}$.

They showed that a minimal distance decoder for this metric achieves correct decoding if the dimension of the intersection of the transmitted and received vector-space is sufficiently large. Also they obtained Hamming, Gilbert-Varshamov and Singleton coding bounds.

E. Ballico [Bal13] has recently proved that every network code can be realized by the above method.

References

[BCGI07] Bernardi, A., Catalisano, M.V., Gimigliano, A., Idà, M.: Osculating varieties of Veronese varieties and their higher secant varieties. Canad. J. Math. 59(3), 488–502 (2007)

[Bal13] Ballico, E.: Any network codes comes from an algebraic curve taking osculating spaces. Design, Codes and Cryptography, Arxiv abs/1306.0992 (to appear, 2013)

[BF03] Ballico, E., Fontanari, C.: On the secant varieties to the osculating variety of a Veronese surface. Cent. Eur. J. Math. 1(3), 315–326 (2003)

[BPT92] Ballico, E., Piene, R., Tai, H.: A characterization of balanced rational normal surface scrolls in terms of their osculating spaces. II. Math. Scand. 70(2), 204–206 (1992)

[CGG02] Catalisano, M.V., Geramita, A.V., Gimigliano, A.: On the secant varieties to the tangential varieties of a Veronesean. Proc. Amer. Math. Soc. 130(4), 975–985 (2002)

[CWJJ03] Chou, P.A., Wu, Y., Jain, K.: Practical network coding (2003)

[Han12] Hansen, J.P.: Equidistant linear network codes with maximal error-protection from veronese varieties, abs/1207.2083 (2012), `http://arxiv.org`

[Har77] Hartshorne, R.: Algebraic geometry. . Graduate Texts in Mathematics, vol. 52. Springer, New York (1977)

[HMK+06] Ho, T., Médard, M., Koetter, R., Karger, D.R., Effros, M., Shi, J., Leong, B.: A random linear network coding approach to multicast. IEEE Transactions on Information Theory 52(10), 4413–4430 (2006)

[KK08] Koetter, R., Kschischang, F.R.: Coding for errors and erasures in random network coding. IEEE Transactions on Information Theory 54(8), 3579–3591 (2008)

[Pie77] Piene, R.: Numerical characters of a curve in projective n-space. In: Real and Complex Singularities (Proc. Ninth Nordic Summer School/NAVF Sympos. Math., Oslo, 1976), pp. 475–495. Sijthoff and Noordhoff, Alphen aan den Rijn (1977)

[PT90] Piene, R., Tai, H.: A characterization of balanced rational normal scrolls in terms of their osculating spaces. In: Enumerative Geometry (Sitges, 1987). Lecture Notes in Math., vol. 1436, pp. 215–224. Springer, Berlin (1990)

[Seg46] Segre, B.: Un'estensione delle varietà di Veronese, ed un principio di dualità per forme algebriche. I. Atti Accad. Naz. Lincei. Rend. Cl. Sci. Fis. Mat. Nat. 8(1), 313–318 (1946)

[Ter11] Terracini, A.: Sulle v_k per cui la variet'a degli $s_h - h + 1$ seganti ha dimensione minore dell'ordinario. Rend. Circ. Mat. Palermo 31, 392–396 (1911)

[Zak93] Zak, F.L.: Tangents and secants of algebraic varieties. Translations of Mathematical Monographs, vol. 127. American Mathematical Society, Providence (1993); Translated from the Russian manuscript by the author

On Sets of Numbers Rationally Represented in a Rational Base Number System

Victor Marsault* and Jacques Sakarovitch

Telecom-ParisTech and CNRS,
46 rue Barrault 75013 Paris, France
victor.marsault@telecom-paristech.fr

Abstract. In this work, it is proved that a set of numbers closed under addition and whose representations in a rational base numeration system is a rational language is not a finitely generated additive monoid.

A key to the proof is the definition of a strong combinatorial property on languages : the bounded left iteration property. It is both an unnatural property in usual formal language theory (as it contradicts any kind of pumping lemma) and an ideal fit to the languages defined through rational base number systems.

1 Introduction

The numeration systems in which the base is a rational number have been introduced and studied in [1]. It appeared there that the language of representations of all integers in such a system is "complicated", by reference to the classical Chomsky hierarchy and its usual iteration properties. This work is a contribution to a better understanding of the structure of this language. It consists in a result whose statement first requires some basic facts about number systems.

Given an integer p as a base, the set of non-negative integers $\mathbb{N}$ is represented by the set of words on the alphabet $A_p = \{0, 1, \ldots, (p-1)\}$ which do not begin with a 0. This set $L_p = (A_p \setminus 0)A_p^*$ is rational, that is, accepted by a finite automaton. This representation of integers has another property related to finite automata: the addition is realised by a finite 3-tape automaton.

This addition algorithm can be broken down into two steps: first a digit-wise addition which outputs a word on the double alphabet A_{2p-1} whose value in base p is the sum of the two input words; second a transformation of a word of $(A_{2p-1})^*$ into a word of A_p^* without modifying its value. This second step can be done by a finite transducer called the *converter* (see Section 2.2.2 of [3]).

Many non-standard numeration systems that have been studied so far have the property that the set of representations of the integers is a rational language. It is even *the* property that is retained in the study of the *abstract numeration systems*, even if it is not the case that addition can be realised by a finite automaton (*cf.* [6]).

---

* Corresponding author.

T. Muntean, D. Poulakis, and R. Rolland (Eds.): CAI 2013, LNCS 8080, pp. 89–100, 2013.

In the rational base numeration systems, as defined and studied in [1], the situation is reverse: the set of integers is not represented by a rational language (not even a context-free one), but nevertheless the addition is realised by a finite automaton. More precisely, let p and q be two coprime integers, with $p > q$. In the $\frac{p}{q}$-numeration system, the digit alphabet is again A_p, and the value of a word $u = a_n \cdots a_2 a_1$ in A_p^* is $\pi(u) = \frac{1}{q}\sum_{i=0}^{n} a_i(\frac{p}{q})^i$. In this system, every integer has a unique finite representation, but the set $L_{\frac{p}{q}}$ of the $\frac{p}{q}$-representations of the integers is not a rational language. The set $V_{\frac{p}{q}}$ of all numbers that can be represented in this system, $V_{\frac{p}{q}} = \pi(A_p^*)$, is closed under addition but is not finitely generated (as an additive monoid).

In this work, we establish the contradiction between being a finitely generated additive monoid and having a rational set of representations in a rational base number system.

Theorem 1. *The set of the $\frac{p}{q}$-representations of any finitely generated additive submonoid of $V_{\frac{p}{q}}$ is not a rational language.*

The proof of this statement relies on three ingredients. The first one is the description of a weak iteration property whose negation is satisfied by the language $L_{\frac{p}{q}}$. The second one is the construction of a sequential letter-to-letter right transducer that realises, on the $\frac{p}{q}$-representations, the addition of a fixed value to the elements of $V_{\frac{p}{q}}$. Finally, the third one is a characterisation of a finitely generated additive submonoid of $V_{\frac{p}{q}}$ as a finite union of translates of the set of the integers.

The paper is organised as follows: after the preliminaries, where we essentially recall the definition of transducers, we present with more details in Section 3 the numeration system in base $\frac{p}{q}$. In Section 4, we describe the Bounded Left Iteration Property (BLIP) and in Section 5, we build a transducer called incrementer. In the last section, we give the proof of a much stronger statement than Theorem 1, expressed with the BLIP property.

2 Preliminaries

We essentially follow notations and definitions of [8] for automata and transducers. An *alphabet* is a finite set of *letters*, the *free monoid* generated by A, and denoted by A^*, is the set of finite *words* over A. The *concatenation* of two words u and v of A^* is denoted by uv, or by $u.v$ when the dot adds hopefully to readability. A *language* (over A) is any subset of A^*.

A language is said to be *rational* (resp. *context-free*) if it is accepted by a finite automaton (resp. a pushdown automaton). The precise definitions of these classes of automata are however irrelevant to the present work, and can be found in [5]. Similarly, we are only considering (and thus defining) a very restricted class of transducers, namely the *sequential letter-to-letter* transducer.

Given two alphabets A and B, a sequential letter-to-letter (left) transducer $\mathcal{T}$ from A^* to B^* is a directed graph whose edges are labelled in $A \times B$. More precisely, $\mathcal{T}$ is defined by a 6-tuple $\mathcal{T} = \langle Q, A, B, \delta, \eta, i, \omega \rangle$ where Q is the set of *states*; A is the *input alphabet*; B is the *output alphabet*; $\delta : Q \times A \to Q$ is the *transition function*; $\eta : Q \times A \to B$ is the *output function*; i is the *initial state* and $\omega : Q \to B^*$ is the *final function*.

Moreover, we call *final* any state in the definition domain of ω. As usual, the function δ (resp. η) is extended to $Q \times A^* \to Q$ (resp. $Q \times A^* \to B^*$) by $\delta(p, \epsilon) = p$ (resp. $\eta(p, \epsilon) = \epsilon$) and $\delta(p, a.u) = \delta(\delta(p, a), u)$ (resp. $\eta(p, a.u) = \eta(p, a).\eta(\delta(p, a), u)$).

Given $\mathcal{T}$, we write $p \xrightarrow[\mathcal{T}]{u \mid v} q$ if, and only if, $\delta(p, u) = q$ and $\eta(p, u) = v$. By analogy, we denote by $p \xrightarrow[\mathcal{T}]{w}$ the fact that p is a final state and that $\omega(p) = w$. The *image* by $\mathcal{T}$ of a word u, denoted by $\mathcal{T}(u)$, is the word $v.w$, if $i \xrightarrow[\mathcal{T}]{u \mid v} p \xrightarrow[\mathcal{T}]{w}$.

Finally, a transducer is said to be a *right transducer*, if it reads the words from right to left; and to be *complete* if both the transition function and the output function are total functions.

In the following, every considered transducer will be complete, letter-to-letter, right and sequential.

3 Rational Base Number System

We recall here the definitions, notations and constructions of [1]. Let p and q be two coprime integers such that $p > q > 1$. Given a positive integer N, let us define $N_0 = N$ and for all $i > 0$:

$$qN_i = pN_{i+1} + a_i \tag{1}$$

where a_i is the remainder of the Euclidean division of qN_i by p, hence in A_p. Since $p > q$, the sequence $(N_i)_i$ is strictly decreasing and eventually stops at $N_{k+1} = 0$. Moreover the equation

$$N = \sum_{i=0}^{k} \frac{a_i}{q} \left(\frac{p}{q}\right)^i \tag{2}$$

holds. The evaluation function π is derived from this formula. The value of a word $u = a_n a_{n-1} \cdots a_0$ over A_p is defined as

$$\pi(a_n a_{n-1} \cdots a_0) = \sum_{i=0}^{n} \frac{a_i}{q} \left(\frac{p}{q}\right)^i \tag{3}$$

Conversely, a word u is called a $\frac{p}{q}$-representation of a number x if $\pi(u) = x$. Since the representation is unique up to leading 0's (see [1, Theorem 1]), u is denoted by $\langle x \rangle_{\frac{p}{q}}$ (or $\langle x \rangle$ for short), and in the case of integers, can be computed with the modified Euclidean division algorithm above. By convention, the representation of 0 is the empty word ϵ.

It should be noted that a rational base number systems is **not** a β-numeration (*cf.* [7, Chapter 7]) in the special case where β is rational. In the latter, the digit set is $\{0, 1, \ldots, \lceil \frac{p}{q} \rceil\}$ and the weight of the i-th leftmost digit is $(\frac{p}{q})^i$; whereas in rational base number systems, they respectively are $\{0, 1, \ldots, (p-1)\}$ and $\frac{1}{q}(\frac{p}{q})^i$.

Definition 1. *The representations of integers in the $\frac{p}{q}$-system form a language over A_p, which is denoted by $L_{\frac{p}{q}}$.*

It is immediate that $L_{\frac{p}{q}}$ is prefix-closed (since, in the modified Euclidean division algorithm $\langle N \rangle = \langle N_1 \rangle . a_0$) and prolongable (there exists an a such that q divides $(np + a)$ and then $\langle \frac{np+a}{q} \rangle = \langle n \rangle . a$).

As a consequence, $L_{\frac{p}{q}}$ can be represented as a tree whose branches are all infinite (*cf.* Figure 1).

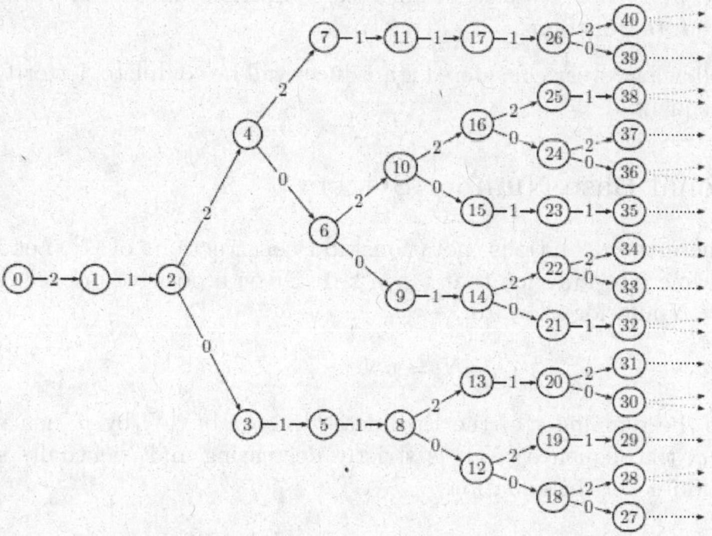

Fig. 1. The tree representation of the language $L_{\frac{3}{2}}$

On the other hand, the suffix language of $L_{\frac{p}{q}}$ is all A_p^*, and, moreover, every suffix appears periodically as established by the following:

Proposition 1 ([1, Proposition 10]). *For every word u over A_p of length k, there exists an integer $n < p^k$ such that u is a suffix of $\langle m \rangle$ if, and only if, m is congruent to n modulo p^k.*

In short, the congruence modulo p^k of n determines the suffix of length k of $\langle n \rangle$. In contrast, the congruence modulo q^k of n determines the words of length k appendable to $\langle n \rangle$ in order to stay in $L_{\frac{p}{q}}$, as is stated in the next lemma.

Lemma 1 ([1, Lemma 6]). *Given two integers n, m and a word u over A_p:*

(i) if both $\langle n \rangle.u$ and $\langle m \rangle.u$ are in $L_{\frac{p}{q}}$, then $n \equiv m\ [q^{|u|}]$
(ii) if $n \equiv m\ [q^{|u|}]$, $\langle n \rangle.u$ is in $L_{\frac{p}{q}}$ implies $\langle m \rangle.u$ is in $L_{\frac{p}{q}}$.

Proof. (i). The word $\langle n \rangle.u$ is in $L_{\frac{p}{q}}$ if, and only, if $(n(\frac{p}{q})^{|u|} + \pi(u))$ is an integer, and similarly for m. It follows that $(n - m)(\frac{p}{q})^{|u|}$ is equal to some integer z, and then $(p^{|u|})(n - m) = zq^{|u|}$, hence $n \equiv m\ [q^{|u|}]$.
(ii). Analogous to (i).

A direct consequence of this lemma is that given any two distinct words u and v of $L_{\frac{p}{q}}$, there exists a word w such that uw is in $L_{\frac{p}{q}}$ but vw is not. Hence, the set $\{u^{-1}L_{\frac{p}{q}} \mid u \in A_p^*\}$ of left quotients of $L_{\frac{p}{q}}$ is infinite, or equivalently:

Corollary 1. *The language $L_{\frac{p}{q}}$ is not rational.*

Definition 2 (The value set). *We denote by $V_{\frac{p}{q}}$ the set of numbers representable in base $\frac{p}{q}$, namely:*

$$V_{\frac{p}{q}} = \{x \mid \exists u \in A_p^*, \pi(u) = x\} \tag{4}$$

or equivalently $V_{\frac{p}{q}} = \pi(A_p^*)$

The most notable property of $V_{\frac{p}{q}}$ is that it is closed under addition, or more precisely that the addition is realised by a transducer, described in 5 (a full proof can be found in [1, Section 3.3]).

Secondly, from the definition of π, one derives easily that $V_{\frac{p}{q}} \subseteq \mathbb{Q}$. More precisely $V_{\frac{p}{q}}$ contains only numbers of the form $\frac{x}{y}$ where y divides a power of q, and conversely, for all k, $V_{\frac{p}{q}}$ contains almost every number $\frac{x}{q^k}$.

Lemma 2. *For every integer k, there exits an integer m_k such that, for every integer n greater than m_k, $\frac{n}{q^k}$ belongs to $V_{\frac{p}{q}}$.*

Proof. If $k = 0$, then one can take $m = 0$ since $\mathbb{N}$ is contained in $V_{\frac{p}{q}}$.

For $k \geq 1$, the words 1 and $1.0^{(k-1)}$ have for respective value $\frac{1}{q}$ and $\frac{p^{k-1}}{q^k}$. For every integer i and j, the number $(\frac{i \times p^{(k-1)} + j \times q^{(k-1)}}{q^k})$ is in $V_{\frac{p}{q}}$, since $V_{\frac{p}{q}}$ is closed under addition, and this can be rewritten as $(p^{(k-1)}\mathbb{N} + q^{(k-1)}\mathbb{N})\frac{1}{q^k} \subseteq V_{\frac{p}{q}}$. Since $p^{(k-1)}$ and $q^{(k-1)}$ are coprime, $(p^{(k-1)}\mathbb{N} + q^{(k-1)}\mathbb{N})$ ultimately covers $\mathbb{N}$.

Experimentally, the bound m_k is increasing with k but the expression resulting from this Lemma is far from being tight. As a consequence, it proves to be difficult to define $V_{\frac{p}{q}}$ without using the $\frac{p}{q}$-rational base number system.

4 BLIP Languages

In the previous section, an insight is given about why $L_{\frac{p}{q}}$ is not rational. It is additionally proven in [1] that $L_{\frac{p}{q}}$ is not context-free either. However, being context sensitive doesn't seem to accurately describe $L_{\frac{p}{q}}$. This section depicts a very strong language property, taylored to capture the structural complexity of $L_{\frac{p}{q}}$.

Let us first define a (very) weak iteration property for languages:

Definition 3. *A language L of A^* is said to be* left-iterable *if there exist two words u and v in A^* such that $u\,v^i$ is a prefix of words in L for an infinite number of exponents i.*

Of course, every rational or context-free language is left-iterable. The definition is indeed designed above all for stating its negation.

Definition 4. *A language L which is not left-iterable is said to have* the Bounded Left-Iteration Property, *or, for short, to be* BLIP.

Example 1. A very simple way of building BLIP languages is to consider infinitely many prefixes of an infinite and aperiodic word. For instance the language $\{u_i\}$, where $u_0 = \epsilon$ and $u_{i+1} = u_i.1.0^i$; or the language of the finite powers of the Fibonacci morphism $\{\sigma^i(0)\}$ where $\sigma(0) = 01$ and $\sigma(1) = 0$.

In order to build a less trivial example let us define the following family of functions f_i:

$$f_i \;:\; \begin{array}{l} n \mapsto n \text{ if } n \neq i \\ n \mapsto 0 \text{ if } n = i. \end{array}$$

The language $\{u_{i,j}\}$, where $u_{i,0} = 1$ and $u_{i,j+1} = u_{i,j}.1.0^{f_i(j)}$, is BLIP as can be easily checked.

Since Definition 4 was taylored for the study of $L_{\frac{p}{q}}$, the following holds, as essentially established in [1, Lemma 8].

Proposition 2. *The language $L_{\frac{p}{q}}$ is BLIP.*

Proof. If $L_{\frac{p}{q}}$ were left iterable, there would exist two nonempty words u and v such that $u\,v^i$ is prefix of a word of $L_{\frac{p}{q}}$ for infinitely many i. Since $L_{\frac{p}{q}}$ is prefix-closed, the word $u\,v^i$ would be itself in $L_{\frac{p}{q}}$, for all i. From Lemma 1, it follows that the integers $\pi(u)$ and $\pi(uv)$ are congruent modulo q^k, for all k, a contradiction.

Being BLIP is a very stable property for languages, as expressed by the following properties.

Lemma 3. *(i) Every finite language is BLIP.*
(ii) Any finite union of BLIP languages is BLIP.
(iii) Any intersection of BLIP languages is BLIP.
(iv) Any sublanguage of a BLIP language is BLIP.

Of course, BLIP languages are not closed under complementation, star or transposition.

The bounded left iteration property can be expressed with the more classical notion of IRS language (for Infinite Regular Subset) that has been introduced by Sheila Greibach in her study of the family of context-free languages ([4], *cf.* also [2]). A language is IRS if it does not contain any infinite rational sublanguage. For instance, the language $\{a^n \mid n \text{ is a prime number}\}$ is IRS (but not BLIP).

It is immediate that a BLIP language is IRS; even that a BLIP language contains no infinite context-free sublanguage. However the converse is not true as seen with the above example. More precisely, the following statement holds:

Proposition 3. *A language L is BLIP if, and only if, $Pref(L)$ is IRS.*

Proof.

$$\begin{aligned}
\mathrm{Pref}(L) \text{ is not IRS} &\Longleftrightarrow \mathrm{Pref}(L) \text{ contains a sublanguage of the form } u\,v^*w \\
&\Longleftrightarrow uv^* \text{ is a sublanguage of } \mathrm{Pref}(L) \\
&\Longleftrightarrow \text{for infinitely many } i,\ u\,v^i \text{ is prefix of a word of } L \\
&\Longleftrightarrow L \text{ is not BLIP}
\end{aligned}$$

Proposition 3 shows that BLIP and IRS are equivalent properties on prefix-closed languages, which means that IRS is indeed a very strong property for prefix-closed languages.

Even though the purpose of this work is to prove Theorem 1, we actually prove a stronger version of it:

Theorem 2. *The set of the $\frac{p}{q}$-representations of any finitely generated additive submonoid of $V_{\frac{p}{q}}$ is a BLIP language.*

This is not a minor improvement, as it shows that every language representing a finitely generated monoid is basically as complex as $L_{\frac{p}{q}}$.

5 The Incrementer

The purpose of this section is to build a letter-to-letter sequential right transducer $A_p \to A_p$ realising a constant addition: given as *parameter* a word w of A_p^* it would perform the application $u \mapsto v$, such that $\pi(v) = \pi(u)+\pi(w)$. This transducer is based on the converter defined in [3] that we recall in Definition 5, below.

Theorem 3 ([1],[3]). *Given any digit alphabet A_n, there exists a finite letter-to-letter right sequential transducer $\mathcal{C}_{\frac{p}{q},n}$ from A_n to A_p such that for every w in ${A_n}^*$, $\pi\left(\mathcal{C}_{\frac{p}{q},n}(w)\right) = \pi(w)$.*

Definition 5. *For every integer n, the* converter $\mathcal{C}_{\frac{p}{q},n} = \langle \mathbb{N}, A_n, A_p, 0, \delta, \eta, \omega \rangle$, *is the right transducer with input alphabet* A_n, *output alphabet* A_p, *and whose transition and output functions are defined by:*

$$\forall s \in \mathbb{N}, \forall a \in A_n \quad s \xrightarrow{a \mid c} s' \iff q\,s + a = p\,s' + c,$$

and final function by: $\omega(s) = \langle s \rangle_{\frac{p}{q}}$, *for every state s in* $\mathbb{N}$.

Definition 5 describes a transducer with an infinite number of states, but its reachable part is finite (*cf.* [1, Proposition 13] or [3, Section 2.2.2]). In particular, if $n = 2p - 1$, the converter is in fact an additioner: given two words $u = a_n \cdots a_2\, a_1$ and $v = b_n \cdots b_2\, b_1$ over A_p, the digit-wise addition yields the word $(a_n + b_n) \cdots (a_1 + b_1)$ over A_{2p-1} which is transformed by $\mathcal{C}_{\frac{p}{q},2p-1}$ into $\langle \pi(u) + \pi(v) \rangle_{\frac{p}{q}}$. The converter from A_5 to A_3 in base $\frac{3}{2}$ is shown at 2.

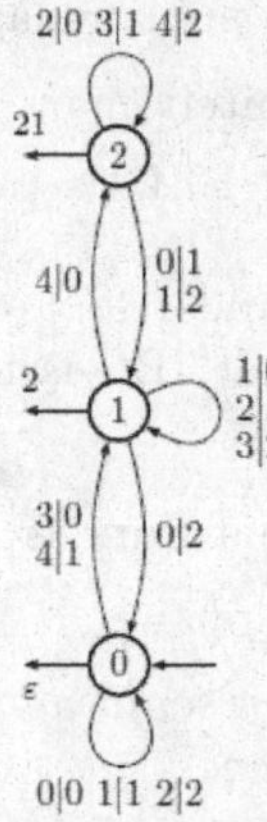

Fig. 2. The converter $\mathcal{C}_{\frac{3}{2},5}$

For every word w of A_p^*, we define a letter-to-letter sequential right transducer $\mathcal{R}_{\frac{p}{q},w}$ which increments the input by w, that is, given a word u as input, it outputs the $\frac{p}{q}$-representation $\langle \pi(u) + \pi(w) \rangle_{\frac{p}{q}}$. It is obtained as a specialisation of $\mathcal{C}_{\frac{p}{q},2p-1}$.

Definition 6. *For every* $w = b_{n-1} \cdots b_1\, b_0$ *in* A_p^*, *the* incrementer

$$\mathcal{R}_{\frac{p}{q},w} = \langle \mathbb{N} \times \{0, 1, \ldots, n\}, A_p, A_p, (0, 0), \delta', \eta', \psi \rangle$$

is the (right) transducer with input and output alphabet A_p, *and whose transition and output functions are defined by:*

$$\forall s \in \mathbb{N}, \forall a \in A_p,$$

$$\forall i < n \quad (s,i) \xrightarrow{a \mid c} (s', i+1) \Longleftrightarrow q\,s + (a + b_i) = p\,s' + c$$

$$(s,n) \xrightarrow{a \mid c} (s', n) \quad \Longleftrightarrow \quad q\,s + a = p\,s' + c$$

and whose final function is defined by:

$$\forall s \in \mathbb{N} \quad \psi((s,n)) = \langle s \rangle_{\frac{p}{q}}$$

$$\psi((s,i)) = \psi((s', i+1)).c \quad \textit{if} \quad i < n \quad \textit{and} \quad (s,i) \xrightarrow{0 \mid c} (s', i+1)$$

This last line means that if the input word is shorter than w, then the final function behaves as if the input word ended with enough 0's (on the left, since we read from right to left). Definition 6 describes a transducer with an infinite number of states but, as in the case of the converter, it is easy to verify that its reachable part is finite. The incrementer $\mathcal{R}_{\frac{3}{2},121}$ is shown at Figure 3.

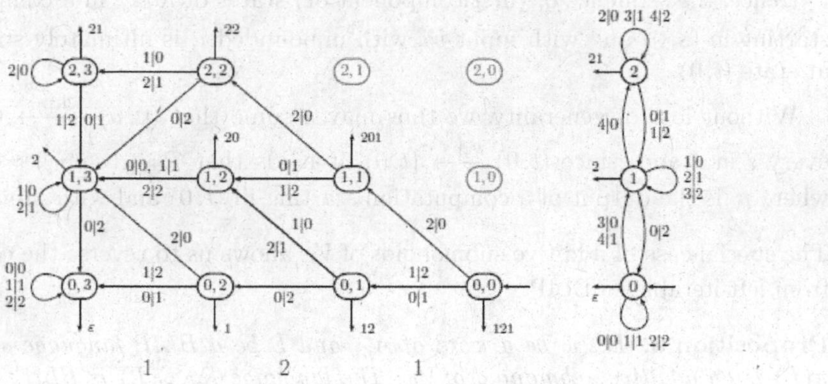

Fig. 3. The incrementer $\mathcal{R}_{\frac{3}{2},121}$

It is a simple verification that the incrementer has the expected behaviour.

Proposition 4. *For every u and w in A_p^*, $v = \mathcal{R}_{\frac{p}{q},w}(u)$ is a word in A_p^* such that $\pi(v) = \pi(u) + \pi(w)$ holds.*

6 Proof of Theorem 2

The core of the proof lies in the next statement.

Proposition 5. *For every w in A_p^*, the image of a left-iterable language by $\mathcal{R}_{\frac{p}{q},w}$ is left-iterable.*

Proof. Let u and v be in A_p^*, $I \subseteq \mathbb{N}$ an infinite set of indexes and $\{y_i\}_{i\in I}$ an infinite family of words in A_p^*. The proof consists in showing that

$$\left\{\mathcal{R}_{\frac{p}{q},w}(u\,v^i\,y_i) \mid i \in I\right\}$$

is left-iterable. Since I is infinite, we may assume, without loss of generality, that the length of the y_i's is strictly increasing hence, that all y_i's have a length greater than $n = |w|$ but also that the reading of every y_i leads $\mathcal{R}_{\frac{p}{q},w}$ to a *same* state $(s, 0)$:

$$\forall s \in \mathbb{N},\ \forall i \in I \quad (0, n) \xrightarrow[\mathcal{R}_{\frac{p}{q},w}]{y_i|y_i'} (s, 0)$$

From the definition of the transitions of $\mathcal{R}_{\frac{p}{q},w}$:

$$(s, 0) \xrightarrow{a\ |\ c} (s', 0) \quad \Longleftrightarrow \quad q\,s + a = p\,s' + c$$

follows, since $a < p$ and $q < p$, that $s \geq s'$.

Hence, the sequence of (first component of) states of $\mathcal{R}_{\frac{p}{q},w}$ in a computation starting in $(s, 0)$ and with input v^i, with unbounded i, is ultimately stationary at state $(t, 0)$.

Without loss of generality, we thus may assume that $(0, n) \xrightarrow{y_i|y_i'} (t, 0)$ for every i in I and, since $(t, 0) \xrightarrow{v\ |\ v'} (t, 0)$, it holds that $\mathcal{R}_{\frac{p}{q},w}(u\,v^i y_i) = u'\,v'^i y_i'$, where u' is the output of a computation starting in $(t, 0)$ and with input u.

The special case of additive submonoids of $V_{\frac{p}{q}}$ allows us to reverse the condition from left-iterable to BLIP:

Proposition 6. *Let w be a word of A_p^*, and L be a BLIP language such that $\pi(L)$ is an additive submonoid of $V_{\frac{p}{q}}$. The language $\mathcal{R}_{\frac{p}{q},w}(L)$ is BLIP.*

Proof. Since $\pi(L)$ is an additive submonoid of $V_{\frac{p}{q}}$, it contains $m\,\mathbb{N}$ for some m (as it must contains some number $\frac{m}{q^l}$ for some m and l).

Let n and k be the integers such that $\pi(w) = \frac{n}{q^k} = x$. From Lemma 2, it follows that there exists m_k such that for every $j > m_k$, $\frac{j}{q_k}$ is in $V_{\frac{p}{q}}$. In particular, there exists j such that $n + j \equiv 0 \mod (m\,q^k)$ and $\frac{j}{q^k}$ is in $V_{\frac{p}{q}}$. If we denote by $y = \frac{j}{q^k}$, it means that $(x + y)$ is in $m\,\mathbb{N}$. Hence, $\pi(L) + x + y$ is contained in $\pi(L)$.

Let us denote by $u = \langle y \rangle_{\frac{p}{q}}$, and $L' = \mathcal{R}_{\frac{p}{q},w}(L)$.

It follows that $\pi\left(\mathcal{R}_{\frac{p}{q},u}(L')\right) = (\pi(L) + x + y) \subseteq \pi(L)$, hence that $\mathcal{R}_{\frac{p}{q},u}(L')$ is an infinite subset of L, and as such BLIP (from Lemma 3). If L' were left-iterable, so would be $\mathcal{R}_{\frac{p}{q},u}(L')$ by Proposition 5, a contradiction.

Finally we prove a property of finitely generated submonoids of $V_{\frac{p}{q}}$.

Proposition 7. *Let M be a finitely generated additive submonoid of $V_{\frac{p}{q}}$. There exists a finite family $\{g_i\}_{i\in I}$ of elements of $V_{\frac{p}{q}}$ such that M is contained in*

$$\bigcup_{i\in I}(g_i+\mathbb{N}).$$

Proof. Let $\{y_1, y_2, \ldots, y_h\}$ be a generating family of M. Every y_j is in $V_{\frac{p}{q}}$ and it is then a rational number $\frac{n_j}{q^{k_j}}$ for some integers n_j and k_j. Let k be the largest of the k_j. Hence, every element in M is a rational number whose denominator is a divisor of q^k, and thus $M \subseteq V_{\frac{p}{q}} \cap \left(\frac{1}{q^k}\mathbb{N}\right)$.

Since every number in $\frac{1}{q^k}\mathbb{N}$ can be written as $n + \frac{i}{q^k}$ for some n in $\mathbb{N}$ and some i in $\{0, 1, \ldots, q^k - 1\}$, it follows that $\frac{1}{q^k}\mathbb{N} = \bigcup_{0\leqslant i<q^k}(\mathbb{N} + \frac{i}{q^k})$, hence

$$M \subseteq \bigcup_{0\leqslant i<q^k} (V_{\frac{p}{q}} \cap (\mathbb{N} + \frac{i}{q^k})).$$

Besides, for every i in $\{0, 1, \ldots, q^k - 1\}$, we denote by g_i the smallest number in $V_{\frac{p}{q}} \cap (\mathbb{N} + \frac{i}{q^k})$. Then, and since $V_{\frac{p}{q}} + \mathbb{N} = V_{\frac{p}{q}}$, for every i, $V_{\frac{p}{q}} \cap (\mathbb{N} + \frac{i}{q^k}) = m_i + \mathbb{N}$. Hence $M \subseteq \bigcup_{0\leqslant i<q^k}(\mathbb{N} + m_i)$.

Even though this proposition seems rather weak (it is a poor approximation from above), it is enough: it indeed reduces Theorem 2 to proving that $\langle n + \mathbb{N}\rangle$ (or equivalently $\mathcal{R}_{\frac{p}{q},w}(L_{\frac{p}{q}})$) is BLIP for any n, which was proven in Proposition 6.

Proof (of Theorem 2). Let M be a finitely generated additive submonoid of $V_{\frac{p}{q}}$. By Proposition 7, there exists a finite family $\{m_i\}_{i\in I}$ of elements of $V_{\frac{p}{q}}$ such that $M \subseteq \bigcup_{i\in I}(m_i + \mathbb{N})$.

Let $L = \langle M\rangle_{\frac{p}{q}}$ the language of the $\frac{p}{q}$-representations of the elements of M and write $w_i = \langle m_i\rangle_{\frac{p}{q}}$. Hence, L is contained in $(\bigcup_i \mathcal{R}_{\frac{p}{q},w_i}(L_{\frac{p}{q}}))$, and thus BLIP by Lemma 3.

7 Conclusion and Future Work

In this work, we have defined a new property, in an effort to capture the structural complexity of $L_{\frac{p}{q}}$. This property contradicts any form of pumping lemma, placing $L_{\frac{p}{q}}$ outside the scope of classical language theory. Even more so that every other example of BLIP languages we describe seem to be purely artificial (*cf.* Example 3)

Paradoxically, Theorem 2 shows that such examples are very common within a rational base number system. It seems that every reasonable number set is represented by a BLIP language and that every simple language represents a complicated set of numbers.

This work led us to a conjecture about rational approximations of $L_{\frac{p}{q}}$:

Conjecture 1. Let L be a rational language closed by addition and containing $L_{\frac{p}{q}}$. Then L contains $X.A_p^*$ where $X = L_{\frac{p}{q}} \cap A_p^{\leqslant k}$, for some k.

Any approximation of $L_{\frac{p}{q}}$ by a rational language L, would only keep a finite part of the structure: the automaton accepting L would be the subtree of depth k of $L_{\frac{p}{q}}$ whose leaves are all-accepting states. Figure 4 gives two examples of rational approximation of $L_{\frac{3}{2}}$, respectively when the $L_{\frac{p}{q}}$ is cut at depth $k = 2$ and $k = 5$.

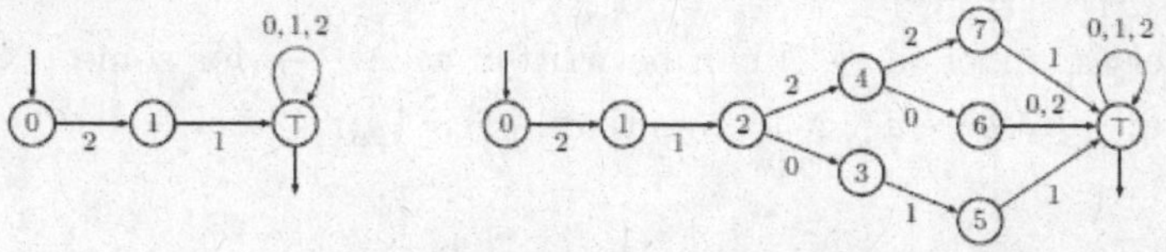

Fig. 4. Two rational approximations of $L_{\frac{3}{2}}$

References

1. Akiyama, S., Frougny, C., Sakarovitch, J.: Powers of rationals modulo 1 and rational base number systems. Israel J. Math. 168, 53–91 (2008)
2. Autebert, J.M., Beauquier, J., Boasson, L., Latteux, M.: Indécidabilité de la condition IRS. ITA 16(2), 129–138 (1982)
3. Frougny, C., Sakarovitch, J.: Number representation and finite automata. In: Berthé, V., Rigo, M. (eds.) Combinatorics, Automata and Number Theory. Encyclopedia of Mathematics and its Applications, vol. 135, pp. 34–107. Cambridge Univ. Press (2010)
4. Greibach, S.A.: One counter languages and the IRS condition. J. Comput. Syst. Sci. 10(2), 237–247 (1975)
5. Hopcroft, J.E., Motwani, R., Ullman, J.D.: Introduction to Automata Theory, Languages and Computation. Addison-Wesley (2000)
6. Lecomte, P., Rigo, M.: Abstract numeration systems. In: Berthé, V., Rigo, M. (eds.) Combinatorics, Automata and Number Theory. Encyclopedia of Mathematics and its Applications, vol. 135, pp. 108–162. Cambridge Univ. Press (2010)
7. Lothaire, M.: Algebraic Combinatorics on Words. Cambridge University Press (2002)
8. Sakarovitch, J.: Elements of Automata Theory. Cambridge University Press (2009); Corrected English translation of Éléments de théorie des automates, Vuibert (2003)

A New Bound for Cyclic Codes Beating the Roos Bound

Matteo Piva and Massimiliano Sala

Department of Mathematics, University of Trento, Italy
piva@science.unitn.it, maxsalacodes@gmail.com

Abstract. We present a lower bound for the distance of a cyclic code, which is computed in polynomial time from the defining set of the code. Our bound beats other similar bounds, including the Roos bound, in the majority of computed cases.

Keywords: Cyclic code, BCH bound, Hartmann-Tzeng bound, Roos bound.

1 Introduction

Many lower bounds exist for the distance of a cyclic code, that elaborate in polynomial time some information from the defining set of the code, e. g. the BCH bound [1], the HT bound [2], the Roos bound [4] and BS bound [5]. We present a new bound which also has polynomial-time cost, beating all other similar bounds in the majority of computed cases. We call this bound " bound C "(Theorem 2). It comes from two preliminary results: bound A (Proposition 1) and bound B (Proposition 2).

2 Preliminaries

In this section we fix some notation and we recall the method we use to prove our result.
Let $(k)_n$ be the remainder of division k by n. Let $\mathbb{F}_q$ be a finite field with q elements, C indicates an arbitrary cyclic code $[n, k, d]$ over $\mathbb{F}_q$, and we denote with g the generator polynomial of C. From now on, we always assume that $\gcd(n, q) = 1$. Let $\mathbb{F}$ be the splitting field of $x^n - 1$ and let α be a primitive $n-$th root of unity in $\mathbb{F}$ then we indicate with S_C the defining set of C:

$$S_C = \left\{ 1 \leq i \leq n-1 \mid g(\alpha^i) = 0 \right\}.$$

We collect together some definitions from [5] and [8]:

- Let $\mathcal{U}$ be a set of three symbols $\left\{ 0, \Delta, \Delta^+ \right\}$ then, with a little abuse of notation, $\mathcal{U} = (\mathcal{U}, +, \cdot)$ represents a field where we have partial information on the element value. More precisely: Δ^+ represents an element for which we

T. Muntean, D. Poulakis, and R. Rolland (Eds.): CAI 2013, LNCS 8080, pp. 101–112, 2013.

are sure it is different from zero, 0 represents an element for which we are sure it is zero, Δ represents an element for which we do not claim if it is zero or not. (The sum and the product on $\mathcal{U}$ are straightforward, but you can see [5], [8] or [6] for a complete description).
- $R(n, S_C)$ is the n–tuple $(u_0, \ldots, u_{n-1}) \in \mathcal{U}^n$ such that

$$u_i = \begin{cases} 0, & \text{if } i \in S_C \\ \Delta, & \text{otherwise.} \end{cases}$$

- $M(\mathbf{v}) \in \mathcal{U}^{n \times n}$ is the circulant matrix obtained from a $\mathbf{v} \in \mathcal{U}^n$.
- Given a $\mathbf{v} \in \mathcal{U}^n$ we denote by $\mathcal{A}(\mathbf{v})$ the set of all $\mathbf{u} \in \mathcal{U} \setminus \mathbf{0}$ s.t.

$$\begin{aligned} &\mathbf{u}[i] = 0, \text{ if } \mathbf{v}[i] = 0, \\ &\mathbf{u}[i] = \Delta^+, \text{ if } \mathbf{v}[i] = \Delta^+, \\ &\mathbf{u}[i] = \Delta^+ \text{ or } \mathbf{u}[i] = 0, \text{ if } \mathbf{v}[i] = \Delta. \end{aligned}$$

We recall the singleton procedure (see [5], [8], [9]) to verify the linear independence of a set of rows on $\mathcal{U}$. For any matrix M, $M[i,j]$ is the (i,j) entry, $M[i]$ is the i–th row and $M(j)$ is the j–th column.

Definition 1. *Let M be a matrix over $\mathcal{U}$. We say that a column $M(j)$ is a* **singleton** *if it contains only one non-zero component $M[i,j]$, i.e. $M[i,j] = \Delta^+$ and $M[t,j] = 0$ for $t \neq i$. When this happens we say that $M[i]$ is the row corresponding to the singleton.*

Any set of t rows of length n with $t \leq n$ forms a matrix $M_t \in \mathcal{U}^{t \times n}$. If a column $M(j)$ is a singleton, then the row corresponding to the singleton is clearly linear independent from the others. Then we delete the $j - th$ column and the corresponding row (we call this operation **s-deletion**), obtaining a new matrix, M_{t-1}, and we search for a new singleton in M_{t-1}. If this procedure can continue until we find a matrix M_1 with at least one Δ^+, we say that the singleton procedure is successful for the set of t rows considered.

Definition 2. *Let M be a matrix over $\mathcal{U}$, we denote by* $\text{prk}(M)$ *the* pseudo rank *of M, i.e., the largest t such that there exists a set of t rows in M for which the singleton procedure is successful.*

Our interest for the rank of a matrix on $\mathcal{U}$ is due to the following result.

Theorem 1. *Let C be a cyclic code with defining set S_C and length n. If d is the distance of the code, then*

$$d \geq \min \{ \, \text{prk}(M(\mathbf{u})) \mid \mathbf{u} \in \mathcal{A}(R(n, S_c)) \, \}$$

Proof. See [6] or [9].

3 Statement of Bound A and Bound B

Proposition 1 (bound A). *Let C be an $\mathbb{F}_q[n,k,d]$ cyclic code with defining set S_C and $\gcd(q,n)=1$. Suppose that there are $\ell,\ m,\ r,\ s \in \mathbb{N}$, $1 \le m \le \ell$ and $i_0 \in \{0,\dots,n-1\}$ such that $\gcd(n,m+r) < m$ or $\gcd(n,m+r)=1$. If:*

a) $(i_0+j)_n \in S_C$, $\forall j = 0,\dots,\ell-1$,
b) $(i_0+j)_n \in S_C$,

$$\forall j = i_0+\ell+r+h(m+r)+1,\dots,\ i_0+\ell+r+m+h(m+r)$$
$$\forall 0 \le h \le s-1$$

then

$$d \ge \ell+1+s-r\left\lfloor \frac{\ell}{m+r} \right\rfloor - \max\{\,(\ell)_{m+r}-m,0\,\}. \tag{1}$$

In other words, the assumptions of Proposition 1 are equivalent to saying that $R(n,S_C)$ contains a block of the form $(0^\ell \Delta^r)(0^m \Delta^r)^s$, i.e. :

$$\underbrace{0\dots0}_{\ell}\underbrace{\Delta\dots\Delta}_{r}(\underbrace{0\dots0}_{m}\underbrace{\Delta\dots\Delta}_{r})^s \subset R(n,S_C).$$

Remark 1. We can see Proposition 1 as generalization of the HT bound. In fact with $\ell = m$ our statement becomes the same of the general Hartmann-Tzeng bound (see [8] and [3]).

We are able to prove another bound, similar to the previous:

Proposition 2 (bound B). *Let C be an $[n,k,d]$ cyclic code over $\mathbb{F}_q$ with defining set S_C. Suppose that there are $m,\ell,s \in \mathbb{N}$, $m,\ell \ge 1$, $s \ge m+1$, $\gcd(n,\ell) < \ell-1$ or $\gcd(n,\ell)=1$. If there is $i_0 \in \{0,\dots,n-1\}$ such that:*

a) $(i_0+j)_n \in S_C$, $j = 0,\dots,m\ell-1$,
b) $(i_0+j)_n \in S_C$, $j = (m+h)\ell+1,\dots,(m+h)\ell+\ell-1$, $0 \le h \le s-1$,

Then:

$$d \ge m\ell+\ell+s-m-1.$$

In other words, the assumptions of Proposition 2 are equivalent to saying that $R(n,S_C)$ contains a block of the form $(0^{\ell m}\Delta)(0^{\ell-1}\Delta^r)^s$, i.e. :

$$\underbrace{0\dots0}_{\ell m}\Delta(\underbrace{0\dots0}_{\ell-1}\Delta)^s \subset R(n,S_C).$$

Remark 2. Proposition 2 is a generalization of the BS bound ([5]), except for the uncommon cases in which $\ell | n$, since $\gcd(n,\ell) \le \ell$ and $\gcd(n,\ell)=\ell \iff \ell | n$.

4 Proofs of Bound A and Bound B

In this section we provide the proof of Proposition 1, and we sketch the proof of Proposition 2.

Remark 3. The main tool we use to prove Proposition 1 and Proposition 2 is Theorem 1 which, in principle, allows us to work only with matrices that have as entries just 0 or Δ^+. Nevertheless during the proof we use matrices that have also Δ as entry. This fact must not worry the reader, since when a Δ appears we mean it can be indifferently 0 or Δ^+, and the correctness of the proof is not affected by such decision.

Proof (of Proposition 1). The general plan of the proof is as follow. Thanks to Theorem 1 we aim at proving that

$$\min\{\ \mathrm{prk}(M(\mathbf{v}))|\mathbf{v}\in\mathcal{A}(R(n,S_c))\ \}\geq\ell+1+s-r\left\lfloor\frac{\ell}{m+r}\right\rfloor-\max\{\ (\ell)_{m+r}-m,0\ \}.$$

In order to do that, for any $\mathbf{v}\in\mathcal{A}(n,S_C)$, we need to choose $\ell+s+1$ rows in $M(\mathbf{v})$ and we must prove that, discarding at most $r\left\lfloor\frac{\ell}{m+r}\right\rfloor+\max\{\ (\ell)_{m+r}-m,0\ \}$ rows, we actually obtain a set of rows for which the singleton procedure is successful.

We can suppose w.l.o.g. that $i_0=n-\ell$ (see Lemma 3.1 in [5]), so that:

$$\mathbf{v}=\underbrace{\Delta\ldots\Delta}_{r}(\underbrace{0\ldots0}_{m}\underbrace{\Delta\ldots\Delta}_{r})^s\ldots\underbrace{0\ldots0\ldots0}_{\ell}.$$

We introduce two notions releated to $\mathbf{v}$ (see [8]). From now on, the meaning of $\mathbf{v}$ is fixed.

Definition 3. *Let $1\leq i'\leq n$. We say that i' is the* **primary pivot** *of* $\mathbf{v}$ *if* $\mathbf{v}[i']$ *is the first Δ^+ that occurs in* $\mathbf{v}$, *i.e.*

$$i'=\min\{h\mid\mathbf{v}[h]=\Delta^+\}\,.$$

We can suppose that $1\leq i'\leq r$, otherwise $\mathbf{v}=0^r(0^m\Delta^r)^s\ldots0^\ell$ and so $(0^{\ell+r+m}\Delta^r)(0^m\Delta^r)^{s-1}\subset\mathbf{v}$ and the bound would be trivially satisfied, since it would give:

$$\begin{aligned}d&\geq\ell+r+m+1+s-1-\left\lfloor\frac{\ell+r+m}{m+r}\right\rfloor r-\max\{\ (\ell+m+r)_{m+r}-m,0\ \}\\&=\ell+r+m+s-\left\lfloor\frac{\ell}{m+r}\right\rfloor r-\max\{\ (\ell)_{m+r}-m,0\ \}\\&\geq\ell+r+1+s-\left\lfloor\frac{\ell}{m+r}\right\rfloor r-\max\{\ (\ell)_{m+r}-m,0\ \}.\end{aligned}$$

Definition 4. *Let $n,m,r,s\in\mathbb{N}$ s. t. m, $s\geq1$, $n\geq m+r$ and $(n,m+r)\leq m$. $((0)^m(\Delta)^r)^s\subset\mathbf{v}$. Then there are i'' in $\{1,\ldots,n\}$, $k\in\mathbb{N}$ and $t\in\{1,\ldots,m\}$, with the following properties:*

1. $\mathbf{v}[i''] = \Delta^+$,
2. $i'' \equiv (s+k)(m+r)+t \mod (n)$,
3. $\mathbf{v}[i] = 0$, *for any* i *s.t.*

$$i \equiv (s+k')(m+r)+j \mod (n)\,,$$

where $k' \in \{0,\dots,k-1\}$ *and* $j \in \{1,\dots,m\}$.

We call such i'' *the* **secondary pivot** *of* $\mathbf{v}$ *with respect to block* $((0)^m(\Delta)^r)^s$.

It is possible to show that if $\gcd(m+r,n) \leq m$ (which includes the classical case $\gcd(m+r,n)=1$), then the secondary pivot exists.

We can suppose $s(m+r)+r+1 \leq i'' \leq s(m+r)+r+m$, otherwise we have $(0^\ell \Delta^r)(0^m \Delta^r)^{s+1} \subset \mathbf{v}$ and the bound is trivially satisfied:

$$\begin{aligned} d &\geq \ell+1+s+1-\left\lfloor \frac{\ell}{m+r} \right\rfloor r - \max\{\,(\ell+m+r)_{m+r}-m, 0\,\} \\ &\geq \ell+1+s-\left\lfloor \frac{\ell}{m+r} \right\rfloor r - \max\{\,(\ell)_{m+r}-m, 0\,\}\,. \end{aligned}$$

We note that $\mathbf{v}[i''-z\cdot(m+r)] = 0$ for any $z=1,\dots,s$. Moreover, i' and i'' may coincide, but this is not a problem.

Now, we are going to choose $(\ell+1+s)$ rows of $M(\mathbf{v})$. We start from the $((n-i'+k)_n+1)-$ th rows with $k=1,\dots,m$, that is, we take the rows with the primary pivot in the first position and its shifts up to the $(m-1)-$th shift included. We collect these rows in submatrix T_1.

$$T_1 = \begin{pmatrix} \Delta^+ \dots 0 \dots 0\ \Delta \dots \Delta \dots 0 \dots 0\ \Delta \dots \Delta^+ \dots \dots \dots 0 \dots \dots 0 \\ 0\ \Delta^+ \dots 0 \dots 0\ \Delta \dots \Delta \dots 0 \dots 0\ \Delta \dots \Delta^+ \dots \dots \dots 0 \dots 0 \\ 0\ 0\ \Delta^+ \dots 0 \dots 0\ \Delta \dots \Delta \dots 0 \dots 0\ \Delta \dots \Delta^+ \dots \dots \dots 0 \dots \\ \vdots \\ 0 \dots 0\ \underset{\substack{\downarrow \\ m}}{\Delta^+} \dots 0 \dots 0\ \Delta \dots \Delta \dots 0 \dots 0\ \Delta \dots \Delta^+ \dots \dots 0 \dots \end{pmatrix}$$

We now consider the $(k+1)$-th rows for $k=m,\dots,\ell$, collected in submatrix T_2.

$$T_2 = \begin{pmatrix} 0 \dots 0\ \Delta \dots \Delta \dots 0 \dots 0\ \Delta \dots \Delta \dots \Delta^+ \dots \dots \dots \dots 0 \dots \dots \\ 0 \dots \dots 0\ \Delta \dots \Delta \dots 0 \dots 0\ \Delta \dots \Delta \dots \Delta^+ \dots \dots \dots \dots 0 \dots \\ 0 \dots 0 \dots 0\ \Delta \dots \Delta \dots 0 \dots 0\ \Delta \dots \Delta \dots \Delta^+ \dots \dots \dots \dots \\ \vdots \\ 0 \dots \dots 0 \dots 0 \dots 0\ \Delta \dots \Delta \dots 0 \dots 0\ \Delta \dots \Delta \dots \Delta^+ \dots \dots \\ 0 \dots \underset{\substack{\downarrow \\ m}}{0} \dots 0 \dots 0 \dots \underset{\substack{\downarrow \\ \ell}}{0}\ \Delta \dots \Delta \dots 0 \dots 0\ \Delta \dots \Delta \dots \Delta^+ \dots \end{pmatrix}$$

Note that T_1 and T_2 have no common rows. Note also that in T_2 for any row $h=1,\dots,\ell+1-m$ and any column $1 \leq j \leq (s-1)(m+r)+m$ we have:

$$T_2[h,j] = \Delta \implies T_2[h, j+(m+r)] = \Delta \tag{2}$$

Our third and last submatrix, T_3, is formed by the $((n-r-k\cdot(m+r))_n+1)-$th rows, for $k=0,\dots,(s-1)$:

$$T_3=\begin{pmatrix}
0 & \dots & 0 & \Delta & \dots & \Delta & \dots & 0 & \dots & 0 & \Delta & \dots & \Delta & 0 & \dots & 0 & \Delta & \dots & \Delta & \dots & \Delta^+ & \dots \\
0 & \dots & 0 & \Delta & \dots & \Delta & \dots & 0 & \dots & 0 & \Delta & \dots & \Delta & \dots & \Delta^+ & \dots & \dots & \dots & \dots & \dots & \dots & \dots \\
\vdots & \vdots & \vdots & \vdots & \vdots & \vdots & \vdots & \vdots & \vdots & \vdots & \vdots & \vdots & \vdots & \vdots & \vdots & \vdots & \vdots & \vdots & \vdots & \vdots & \vdots & \vdots \\
0 & \dots & 0 & \Delta & \dots & \Delta & \dots & \Delta^+ & \dots & \dots & \dots & \dots & \dots & \dots & \dots & \dots & \dots & \dots & \dots & \dots & \dots & \dots \\
 & & \downarrow & & & \downarrow & & \downarrow & & & & & & & & & & & & & \downarrow & \\
 & & m & & & m+r & & \begin{matrix} i''-r- \\ (s-1)(m+r) \end{matrix} & & & & & & & & & & & & & i''-r &
\end{pmatrix}$$

Lemma 1. *The singleton procedure is successful for T_3 and thus $\mathrm{prk}(T_3)=s$.*

Proof. *We note that the rows of T_3, by construction, have the property that $T_3[a+1,h]=T_3[a,h+(m+r)]$ because each row is a $(m+r)$ left shift of the previous one. This is sufficient to prove that $T_3(i''-r-(s-1)(m+r))$ is a singleton. We claim that the $s-$th row of T_3 corresponds to a singleton. Indeed*

$$T_3[s,i''-r-(s-1)(m+r)]=T_3[1,i''-r-(s-1)(m+r)+(s-1)(m+r)]=$$

$$T_3[1,i''-r]=\Delta^+$$

and for $k=1,\dots,s-1$:

$$T_3[k,i''-r-(s-1)(m+r)]=T_3[1,i''-r-(s-1)(m+r)+(k-1)(m+r)]=$$

$$T_3[i''-r-(s-k)(m+r)]=0$$

so we can s-delete it. Once this is done, we might also s-delete the $(s-1)-$th row, since

$$T_3[s-1,i''-r-(s-2)(m+r)]=T_3[1,i''-r-(s-2)(m+r)+(s-2)(m+r)]=$$

$$T_3[1,i''-r]=\Delta^+$$

and for $k=1,\dots,s-2$:

$$T_3[k,i''-r-(s-2)(m+r)]=T_3[i''-r-(s-2)(m+r)+(k-1)(m+r)]=$$

$$T_3[1,i''-r-(s-1-k)(m+r)]=0.$$

In this way for any row of T_3 we obtain a singleton in $T_3\,(i''-r-k(m+r))$ for $k=0,\dots,s-1$, by recursively s-deleting from the last row to the first.

Collecting all these submatrices T_1, T_2, T_3, we obtain an $(\ell+1+s)\times n$ matrix T, as follows:

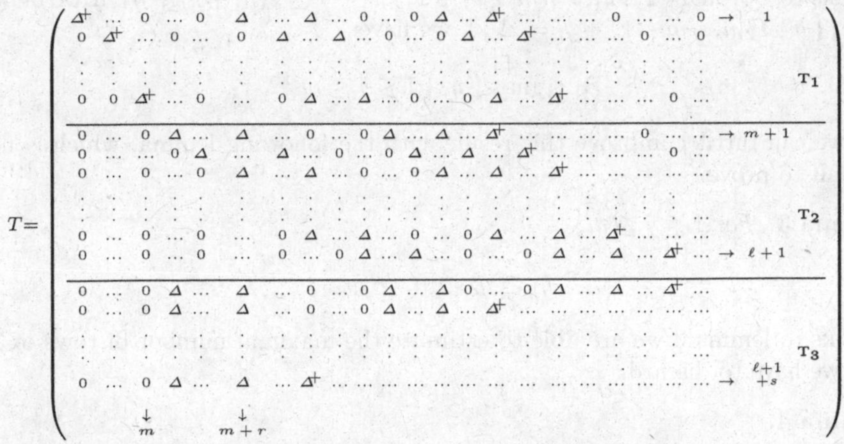

Observe that the rows from $(m+1)$ to $(\ell+s+1)$ have a block of zero in the first m positions and then we can obviously s-delete the first m rows (i.e the rows of T_1). After these first m s-deletions we obtain a matrix T' composed of the last $(\ell+1+s-m)$ rows of T, as the following:

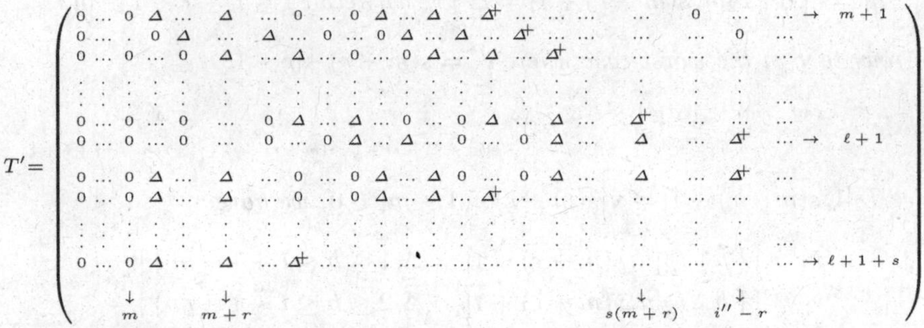

where $1+s(m+r)\le i''-r\le m+s(m+r)$ by hypothesis. We note that T' is composed by the rows of T_2 and T_3.

We use the singletons of T_3 to proceed with the singleton procedure, but in order to do that we have to discard some rows in T_2. More precisely, let us define:

$$B_k=\{\,h\mid T_2[h,i''-r-k(m+r)]=\Delta\,\}\qquad\text{for } k=0,\ldots,s-1$$

then the rows to discard in T_2 in order that $T(i''-r-k(m+r))$ becomes a singleton for $k=0,\ldots,s-1$ are:

$$\mathbf{B}=\cup_{k=0}^{s-1}B_k. \tag{3}$$

Lemma 2. *Let* $0\le k<k'\le s-1$, *then* $B_{k'}\subseteq B_k$.

Proof. Obvious from (2).

Corollary 1. $\mathbf{B} = B_0 = \{\, h \mid T_2[h, i'' - r] = \Delta \,\}$.

Thanks to Corollary 1, since $s(m+r)+1 \le i'' - r \le s(m+r)+m$, if we define $\eta_j = |\{\, h \mid T_2[h, s(m+r)+j] = \Delta \,\}|$, we have:

$$|\mathbf{B}| \le \max\{\, \eta_j \mid 1 \le j \le m \,\}.$$

and we can further improve this result with the following lemma, which is not difficult to prove.

Lemma 3. *For* $1 \le j \le m$*:*

$$\eta_1 \ge \eta_2 \ge \cdots \ge \eta_m.$$

Thanks to lemma 3 we are able to estimate the maximal number of rows of T_2 that we have to discard.

Lemma 4.

$$|\mathbf{B}| \le \eta_1 \le \left\lfloor \frac{\ell}{m+r} \right\rfloor r + \max\{\, (\ell)_{m+r} - m, 0 \,\}$$

Proof. For Corollary 1 and Lemma 3 we have $|\mathbf{B}| \le \eta_1$. *Now:*

$$\eta_1 = |\{\, h \mid T_2[h, s(m+r)+1] = \Delta \,\}|, \text{ but recall } 1 \le h \le \ell + 1 - m.$$

We rewrite $\mathbf{v}$ *in the worst case where* $i'' = s(m+r)+r+1$*:*

$$\mathbf{v} = \underset{1}{\underset{\downarrow}{\Delta}} \ldots \underset{r}{\underset{\downarrow}{\Delta}}\, 0 \ldots \underset{m+r}{\underset{\downarrow}{0}} \quad (\Delta^r 0^m)^{s-2}\ \Delta \ldots \Delta \underset{s(m+r)-m+1}{\underset{\downarrow}{0}} \quad \ldots \quad \underset{s(m+r)}{\underset{\downarrow}{0}} \quad \Delta \ldots \Delta \underset{s(m+r)+r+1}{\underset{\downarrow}{\Delta^{+}}} \ldots\ldots$$

Since $T_2[1, s(m+r)+1] = \mathbf{v}[s(m+r)+1-m] = 0$, *we have*

$$\begin{aligned} \eta_1 &= |\{\, h \mid T_2[h, s(m+r)+1] = \Delta, 1 \le h \le \ell+1-m \,\}| \\ &= |\{\, h \mid T_2[h, s(m+r)+1] = \Delta, 2 \le h \le \ell+1-m \,\}|. \end{aligned}$$

Now $T_2[h+1, j] = T_2[h, j-1]$ *(for* $h \ge 1$*) and* $T_2[1, j] = \mathbf{v}[j-m]$*, by construction of* T_2*. So:*

$$\begin{aligned} \eta_1 &= |\{\, h \mid T_2[h, s(m+r)+1] = \Delta, 2 \le h \le \ell+1-m \,\}| \\ &= |\{\, h \mid T_2[1, s(m+r)+1-(h-1)] = \Delta, 2 \le h \le \ell+1-m \,\}| \\ &= |\{\, h \mid \mathbf{v}[s(m+r)-m+2-h] = \Delta, 2 \le h \le \ell+1-m \,\}| \\ &= |\{\, h \mid \mathbf{v}[s(m+r)+2-h] = \Delta, 2 \le h \le \ell+1 \,\}| \end{aligned}$$

Thus, to compute η_1 *we have to count the number of* Δ*'s we meet,* $\mathbf{v}[s(m+r)]$ *to* $\mathbf{v}[s(m+r)-\ell+1]$ *(i.e. from* $\mathbf{v}[s(m+r)]$ *and going back of* ℓ *positions). Let us consider the worst case, which is when* $\ell \le s(m+r)$*. Passing through* *the block* $(0^m \Delta^r)$ *from right to left of* ℓ *positions, every* $m+r$ *steps we meet a block formed by* r Δ*'s and* m 0*'s, thus the contibute to* η_1 *per block is by* r*. Since we move only by* ℓ *positions, we can meet no more than* $\left\lfloor \frac{\ell}{m+r} \right\rfloor$ *blocks and so we have* $\eta_1 \leq \left\lfloor \frac{\ell}{m+r} \right\rfloor r + \eta_1'$*, where* η_1' *are the* Δ*'s coming from the last* $(\ell)_{m+r}$ *steps left. The first* m*-positions we meet doing the last* $(\ell)_{m+r}$ *steps are zero, since they correspond to the last block* $(\Delta^r 0^m)$*, thus* η_1' *can be at most* $(\ell)_{m+r} - m$ *and it is non-negative only if* $(\ell)_{m+r} \geq m$*. In conclusion:* $\eta_1 \leq \left\lfloor \frac{\ell}{m+r} \right\rfloor r + \max\{\, (\ell)_{m+r} - m, 0 \,\}$.

Thanks to Lemma 4, discarding at most $\left\lfloor \frac{\ell}{m+r} \right\rfloor r + \max\{\, (\ell)_{m+r} - m, 0 \,\}$ rows of T_2, we can remove by s-deletions T_3 from T'. The matrix that remains, $\widetilde{T}$, is a submatrix of T_2 not having row indeces in $\mathbf{B}$ which has full rank, since T_2 has full rank, adopting the singleton procedure as can be seen by Lemma 3.2 in [5].

Example 1. Let us suppose C be a cyclic code of length n, with defining set S_C satisfying the assumptions of Proposition 1 with parameters $\ell = 7$, $m = 2$, $r = 1$, $s = 5$. We want to prove that for Proposition 1 the distance of the code C is at least $d \geq 7 + 1 + 5 - \left\lfloor \frac{7}{2+1} \right\rfloor 1 - \max\{\, (7)_{3+2} - 2, 0 \,\} = 11$. Let $\mathbf{v} \in \mathcal{A}(R(n, S_C))$ with $\mathbf{v}[1] = \Delta^+$. The matrix T is:

$$
\begin{array}{ccccccccccccccccccccccccc l}
\Delta^+ & 0 & 0 & \Delta & 0 & 0 & \Delta & 0 & 0 & \Delta & 0 & 0 & \Delta & 0 & 0 & \Delta & \Delta^+ & \Delta & \Delta & \Delta & \Delta & \Delta & \Delta & \Delta & \Delta & \ldots\ \ldots \\
0 & \Delta^+ & 0 & 0 & \Delta & 0 & 0 & \Delta & 0 & 0 & \Delta & 0 & 0 & \Delta & 0 & 0 & \Delta & \Delta^+ & \Delta & \Delta & \Delta & \Delta & \Delta & \Delta & \Delta & \ldots\ \ldots \\
\hline
0 & 0 & \Delta^+ & 0 & 0 & \Delta & 0 & 0 & \Delta & 0 & 0 & \Delta & 0 & 0 & \Delta & 0 & 0 & \Delta & \Delta^+ & \Delta & \Delta & \Delta & \Delta & \Delta & \Delta & \ldots\ \ldots \\
0 & 0 & 0 & \Delta^+ & 0 & 0 & \Delta & 0 & 0 & \Delta & 0 & 0 & \Delta & 0 & 0 & \Delta & 0 & 0 & \Delta & \Delta^+ & \Delta & \Delta & \Delta & \Delta & \Delta & \ldots\ \ldots \\
0 & 0 & 0 & 0 & \Delta^+ & 0 & 0 & \Delta & 0 & 0 & \Delta & 0 & 0 & \Delta & 0 & 0 & \Delta & 0 & 0 & \Delta & \Delta^+ & \Delta & \Delta & \Delta & \Delta & \ldots\ \ldots \\
0 & 0 & 0 & 0 & 0 & \Delta^+ & 0 & 0 & \Delta & 0 & 0 & \Delta & 0 & 0 & \Delta & 0 & 0 & \Delta & 0 & 0 & \Delta & \Delta^+ & \Delta & \Delta & \Delta & \ldots\ \ldots \\
0 & 0 & 0 & 0 & 0 & 0 & \Delta^+ & 0 & 0 & \Delta & 0 & 0 & \Delta & 0 & 0 & \Delta & 0 & 0 & \Delta & 0 & 0 & \Delta & \Delta^+ & \Delta & \Delta & \ldots\ \ldots \\
0 & 0 & 0 & 0 & 0 & 0 & 0 & \Delta^+ & 0 & 0 & \Delta & 0 & 0 & \Delta & 0 & 0 & \Delta & 0 & 0 & \Delta & 0 & 0 & \Delta & \Delta^+ & \Delta & \ldots\ \ldots \\
\hline
0 & 0 & \Delta & 0 & 0 & \Delta & 0 & 0 & \Delta & 0 & 0 & \Delta & 0 & 0 & \Delta & \Delta^+ & \Delta & \Delta & \Delta & \Delta & \Delta & \Delta & \Delta & \Delta & \Delta & \ldots\ \ldots \\
0 & 0 & \Delta & 0 & 0 & \Delta & 0 & 0 & \Delta & 0 & 0 & \Delta & \Delta^+ & \Delta & \Delta & \Delta & \Delta & \Delta & \Delta & \Delta & \Delta & \Delta & \Delta & \Delta & \Delta & \ldots\ \ldots \\
0 & 0 & \Delta & 0 & 0 & \Delta & 0 & 0 & \Delta & \Delta^+ & \Delta & \Delta & \Delta & \Delta & \Delta & \Delta & \Delta & \Delta & \Delta & \Delta & \Delta & \Delta & \Delta & \Delta & \Delta & \ldots\ \ldots \\
0 & 0 & \Delta & 0 & 0 & \Delta & \Delta^+ & \Delta & \Delta & \Delta & \Delta & \Delta & \Delta & \Delta & \Delta & \Delta & \Delta & \Delta & \Delta & \Delta & \Delta & \Delta & \Delta & \Delta & \Delta & \ldots\ \ldots \\
0 & 0 & \Delta & \Delta^+ & \Delta & \ldots\ \ldots
\end{array}
$$

For the secondary pivot we have two possibilities: $i'' = 11$ or $i'' = 12$. We show that in both cases it is possible to obtain 11 s-deletions, removing at most $\left\lfloor \frac{7}{2+1} \right\rfloor 1 + \max\{\, (7)_{3+2} - 2, 0 \,\} = 2$ rows from the matrix T.

Case 1: $i'' = 11$.

$$
\begin{array}{ccccccccccccccccccccccccc l l}
\Delta^+ & 0 & 0 & \Delta & 0 & 0 & \Delta & 0 & 0 & \Delta & 0 & 0 & \Delta & 0 & 0 & \Delta & \Delta^+ & \Delta & \Delta & \Delta & \Delta & \Delta & \Delta & \Delta & \Delta & \ldots & \to \text{1st s-deletion} \\
0 & \Delta^+ & 0 & 0 & \Delta & 0 & 0 & \Delta & 0 & 0 & \Delta & 0 & 0 & \Delta & 0 & 0 & \Delta & \Delta^+ & \Delta & \Delta & \Delta & \Delta & \Delta & \Delta & \Delta & \ldots & \to \text{2nd s-deletion} \\
0 & 0 & \Delta^+ & 0 & 0 & \Delta & 0 & 0 & \Delta & 0 & 0 & \Delta & 0 & 0 & \Delta & 0 & 0 & \Delta & \Delta^+ & \Delta & \Delta & \Delta & \Delta & \Delta & \Delta & \ldots & \to \text{8th s-deletion} \\
0 & 0 & 0 & \Delta^+ & 0 & 0 & \Delta & 0 & 0 & \Delta & 0 & 0 & \Delta & 0 & 0 & \Delta & 0 & 0 & \Delta & \Delta^+ & \Delta & \Delta & \Delta & \Delta & \Delta & \ldots & \to \textbf{REMOVED} \\
0 & 0 & 0 & 0 & \Delta^+ & 0 & 0 & \Delta & 0 & 0 & \Delta & 0 & 0 & \Delta & 0 & 0 & \Delta & 0 & 0 & \Delta & \Delta^+ & \Delta & \Delta & \Delta & \Delta & \ldots & \to \text{9th s-deletion} \\
0 & 0 & 0 & 0 & 0 & \Delta^+ & 0 & 0 & \Delta & 0 & 0 & \Delta & 0 & 0 & \Delta & 0 & 0 & \Delta & 0 & 0 & \Delta & \Delta^+ & \Delta & \Delta & \Delta & \ldots & \to \text{10th s-deletion} \\
0 & 0 & 0 & 0 & 0 & 0 & \Delta^+ & 0 & 0 & \Delta & 0 & 0 & \Delta & 0 & 0 & \Delta & 0 & 0 & \Delta & 0 & 0 & \Delta & \Delta^+ & \Delta & \Delta & \ldots & \to \textbf{REMOVED} \\
0 & 0 & 0 & 0 & 0 & 0 & 0 & \Delta^+ & 0 & 0 & \Delta & 0 & 0 & \Delta & 0 & 0 & \Delta & 0 & 0 & \Delta & 0 & 0 & \Delta & \Delta^+ & \Delta & \ldots & \to \text{11th s-deletion} \\
0 & 0 & \Delta & 0 & 0 & \Delta & 0 & 0 & \Delta & 0 & 0 & \Delta & 0 & 0 & \Delta & \Delta^+ & \Delta & \Delta & \Delta & \Delta & \Delta & \Delta & \Delta & \Delta & \Delta & \ldots & \to \text{7th s-deletion} \\
0 & 0 & \Delta & 0 & 0 & \Delta & 0 & 0 & \Delta & 0 & 0 & \Delta & \Delta^+ & \Delta & \Delta & \Delta & \Delta & \Delta & \Delta & \Delta & \Delta & \Delta & \Delta & \Delta & \Delta & \ldots & \to \text{6th s-deletion} \\
0 & 0 & \Delta & 0 & 0 & \Delta & 0 & 0 & \Delta & \Delta^+ & \Delta & \Delta & \Delta & \Delta & \Delta & \Delta & \Delta & \Delta & \Delta & \Delta & \Delta & \Delta & \Delta & \Delta & \Delta & \ldots & \to \text{5th s-deletion} \\
0 & 0 & \Delta & 0 & 0 & \Delta & \Delta^+ & \Delta & \Delta & \Delta & \Delta & \Delta & \Delta & \Delta & \Delta & \Delta & \Delta & \Delta & \Delta & \Delta & \Delta & \Delta & \Delta & \Delta & \Delta & \ldots & \to \text{4th s-deletion} \\
0 & 0 & \Delta & \Delta^+ & \Delta & \ldots & \to \text{3rd s-deletion}
\end{array}
$$

Case 2: $i'' = 12$.

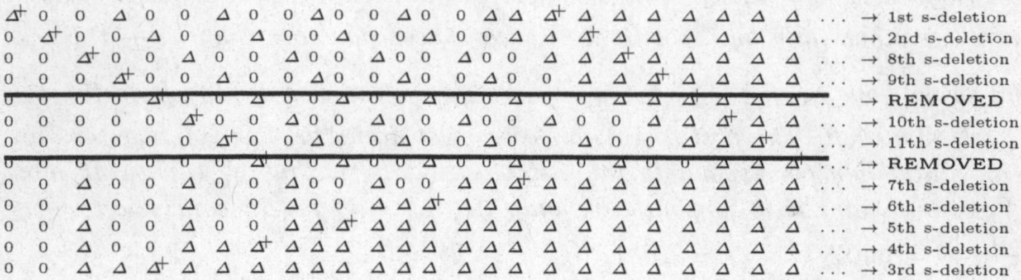

For Proposition 2 the proof proceeds similarly.

Proof (of Proposition 2). We can suppose:

(i) $\mathbf{v} = \underbrace{0\ldots0\Delta}_{\ell m+1}\overbrace{\underbrace{0\ldots0\Delta}_{\ell}\ldots\ldots\underbrace{0\ldots0\Delta}_{\ell}}^{s-\text{times}}\ldots$;

(ii) $i' = \ell m + 1$;

(iii) $m\ell + s(\ell) + 2 \le i'' \le m\ell + s(\ell) + 1 + m$.

We take the rows $((n - (\ell m + 1) + k)_n + 1)-$ th rows, with $k = 1, \ldots, m\ell + \ell$: we take the rows with the primary pivot in first position and its shifts until the $(m\ell + \ell - 1)-$th shift:

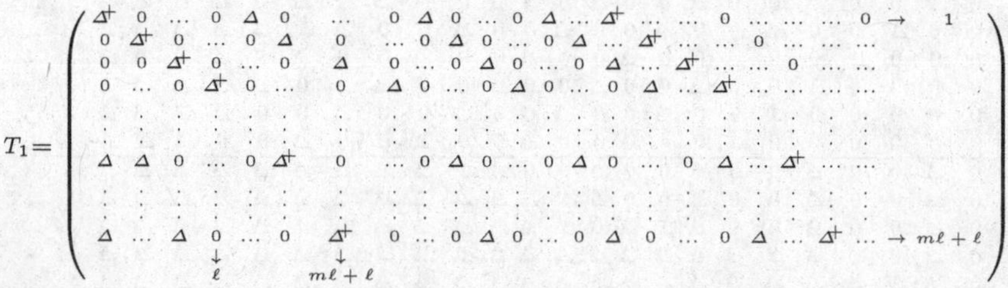

and then we add $s - 1$ rows: the $((n - i'' - (s\ell + 1))_n + (k + 1)\ell)$ $-$th rows with $k = 1, \ldots, s - 1$, which are the rows with the secondary pivot in position $(k+1)\ell$ with $k = 1, \ldots, s - 1$.

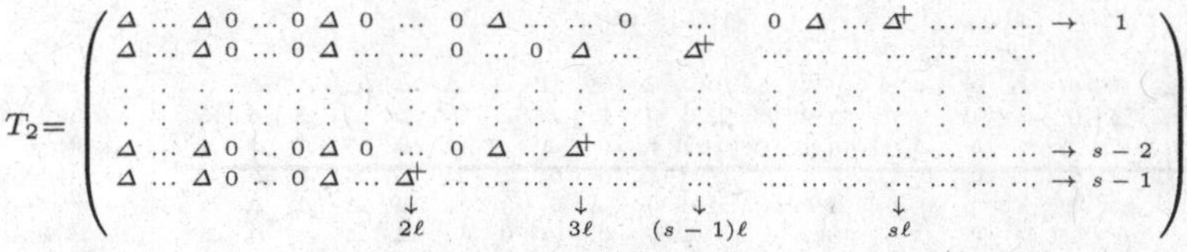

And collecting together the rows of T_1 and T_2, the proof concludes as in case of Proposition 1.

We summarize the results of Proposition 1 and Proposition 2 in a unique form that constitutes the statement of bound C.

Theorem 2 (bound C). *Let C be an $\mathbb{F}_q[n,k,d]$ cyclic code with defining set S_C and $\gcd(q,n)=1$. Suppose that there are ℓ, m, r, $s \in \mathbb{N}$, $1 \leq m \leq \ell$ and $i_0 \in \{0,\ldots,n-1\}$ such that $\gcd(n,m+r) < m$ or $\gcd(n,m+r)=1$. If*

a) $(i_0+j)_n \in S_C$, $\forall j = 0,\ldots,\ell-1$,
b) $(i_0+j)_n \in S_C$,

$$\forall j = i_0+\ell+r+h(m+r),\ldots,\ i_0+\ell+r+m-1+h(m+r)$$
$$\forall 0 \leq h \leq s-1.$$

Then

$$d \geq \ell+1+s-r\left\lfloor\frac{\ell}{m+r}\right\rfloor - \max\{(\ell)_{m+r}-m,0\}. \tag{4}$$

In the particular case that for some ℓ' and m', $\ell = m'\ell'$, $m = \ell'-1$, $s \geq m'+1$ and $r=1$ we also have:

$$d \geq \ell' m' + \ell' + s - m' - 1. \tag{5}$$

5 Computational Results and Costs

As explained in Remark 1 and in Remark 2 bound C is both a generalization of HT bound and BS bound (except when $\ell|n$) and so it is sharper and tighter. The relation between our bound and the Roos bound is not clear: sometimes our bound is sharper and tighter than Roos or but for other codes it is the opposite. However, from the computed codes it appears that bound C works better than the Roos bound in general. Although the BS bound sometimes beats the Roos bound, in the majority of computed cases the Roos bound is better, as reported in [5] and checked by us. Bound C is the first polynomial-time bound outperforming the Roos bound on a significant sample of codes.
As regards computational costs, bound C requires:

- n operations for i_0
- n operations for ℓ
- n operations for m
- n operations for r
- n operations for s

and so it costs $O(n^5)$ which is slightly more than the Roos bound which needs $O(n^4)$, in fact the latter requires at most:

- n operations for i_0,
- n operations for m,
- n operations for r,
- n operations for s

while the other bounds cost less: BCH-$O(n^2)$, HT-$O(n^3)$, bound BS-$O(n^{2.5})$. We tested all cyclic codes in the following range: on $\mathbb{F}_2$ with $15 \leq n \leq 125$, on $\mathbb{F}_3$ with $8 \leq n \leq 79$ and $82 \leq n \leq 89$, on $\mathbb{F}_5$ with $8 \leq n \leq 61$, on $\mathbb{F}_7$ with $8 \leq n \leq 47$. We have chosen the largest ranges that we could compute in a reasonable time.

In the following table we report the number of codes on which each bound considered is not tight.

Table 1. Bound tightness

	$\mathbb{F}_2$	$\mathbb{F}_3$	$\mathbb{F}_5$	$\mathbb{F}_7$	total
number of codes	70488	93960	1163176	106804	1434428
BCH	11192	16376	151219	13696	182483
HT	10531	15334	139161	11093	176119
BS	10959	15545	139783	11283	177570
ROOS	10014	14583	133546	10709	168852
bound C	10306	14565	131072	9541	165484

Acknowledgments. These bounds appear in the 2010 Master's thesis of the first author [10], who thanks his supervisor (the second author).

References

1. Bose, R.C., Ray Chaudhuri, D.K.: On a class of error correcting binary group codes. Information and Control 3, 68–79 (1960)
2. Hartmann, C.R.P., Tzeng, K.K.: Generalizations of the BCH bound. Information and Control 20, 489–498 (1972)
3. Roos, C.: A generalization of the BCH bound for cyclic codes, including the Hartmann-Tzeng bound. Journal of Combinatorial Theory Ser. A 33, 229–232 (1982)
4. Roos, C.: A new lower bound for the minimum distance of a cyclic code. IEEE Trans. on Inf. Th. 29(3), 330–332 (1983)
5. Betti, E., Sala, M.: A New Bound for the Minimum Distance of a Cyclic Code From Its Defining Set. IEEE Trans. on Inf. Th. 52(8), 3700–3706 (2006)
6. Schaub, T.: A Linear Complexity Approach to Cyclic Codes. PhD thesis, Swiss Federal Inst. of Tech. Zurich (1988)
7. van Lint, J.H., Wilson, R.M.: On the minimum distance of cyclic codes. IEEE Trans. on Inf. Th. 32(1), 23–40 (1986)
8. Betti, E., Sala, M.: A theory for distance bounding cyclic codes. BCRI preprint 63 (2007), `www.bcri.ucc.ie`
9. Ponchio, F., Sala, M.: A lower bound on the distance of cyclic codes. BCRI preprint 7 (2003), `http://www.bcri.ucc.ie`
10. Piva, M.: Probabilità d'errore in decodifica con un nuovo bound. Master's thesis (Laurea), University of Trento, Department of Mathematics (2010)

On a Conjecture of Helleseth

Yves Aubry[1,2] and Philippe Langevin[1,*]

[1] Institut de Mathématiques de Toulon, Université du Sud Toulon-Var, France
[2] Insitut de Mathématiques de Luminy, Université d'Aix-Marseille, France

Abstract. We are concerned about a conjecture proposed in the middle of the seventies by Hellesseth in the framework of maximal sequences and theirs cross-correlations. The conjecture claims the existence of a zero outphase Fourier coefficient. We give some divisibility properties in this direction.

1 Two Conjectures of Helleseth

Let L be a finite field of order $q > 2$ and characteristic p. Let μ be the canonical additive character of L i.e.

$$\mu(x) = \exp(2i\pi \mathrm{Tr}\,(x)/p)$$

where Tr is the trace function with respect to the finite field extension $L/\mathbb{F}_p$. The *Fourier coefficient* of a mapping $f : L \to L$ is defined at $a \in L$ by

$$\widehat{f}(a) = \sum_{x \in L} \mu(ax + f(x)). \tag{1}$$

The distribution of these values is called the *Fourier spectrum* of f. Note that when f is a permutation the *phase* Fourier coefficient $\widehat{f}(0)$ is equal to 0.

The mapping $f(x) = x^s$ is called the power function of exponent s, and it is a permutation if and only if $(s, q-1) = 1$. Moreover, if $s \equiv 1 \mod (p-1)$ the Fourier coefficients of f are rational integers. Helleseth made in [3] the following conjecture on the quantity (related to Dedekind determinant, see [9])

$$\mathfrak{D}(f) = \prod_{a \in L^\times} \widehat{f}(a). \tag{2}$$

Conjecture 1 (Helleseth). Let L be a field of cardinal $q > 2$. If f is a power permutation of L of exponent $s \equiv 1 \mod (p-1)$ then $\mathfrak{D}(f) = 0$.

For $p = 2$, it generalizes Dillon's conjecture (see [2]) which corresponds to the case $s = q - 2 \equiv -1 \pmod{q-1}$, and known to be true because it is related to the vanishing of Kloosterman sums and the class number h_q of the imaginary quadratic number field $\mathbb{Q}(\sqrt{1-4q})$ (see [5,8]). Note also that in odd characteristic the Kloosterman sums do not vanish (see [7]) except if $p = 3$ (see [5]).

In the same paper [3], Helleseth proposed a second conjecture:

* The authors would like to thank the anonymous reviewers for their valuable comments and suggestions to improve this manuscript.

T. Muntean, D. Poulakis, and R. Rolland (Eds.): CAI 2013, LNCS 8080, pp. 113–118, 2013.

Conjecture 2. If $[L : \mathbb{F}_p]$ is a power of 2 then the spectrum of a power permutation of exponent not a power of p modulo $q-1$ takes at least four values.

In this note, we prove some results concerning the divisibility properties of the Fourier coefficients of a power permutation in connection with Conjecture 1. Our results can be seen as a proof "modulo ℓ" of Conjecture 1 for certain primes ℓ.

2 Boolean Function Case

In this section, we assume $p = 2$. In [10], the second author has computed the Fourier spectra of power permutations for all the fields of characteristic 2 with degree less or equal to 25 without finding any counter-example to the above conjectures. More curiously, if we denote by nbz (s) the number of vanishing Fourier coefficients of the power function of exponent s then the numerical experience suggests that:

$$\text{nbz}\,(s) \geq \text{nbz}\,(-1) = h_q.$$

At this point, it is interesting to notice that Helleseth's conjecture can not be extended to the set of all permutations. Indeed, let m be a positive integer and let $g\colon \mathbb{F}_2^m \to \mathbb{F}_2$ be a Boolean function in m variables. One defines the Walsh coefficient of g at $a \in \mathbb{F}_2^m$ by :

$$g^{\mathcal{W}}(a) = \sum_{x \in \mathbb{F}_2^m} (-1)^{a.x + g(x)}.$$

Identifying L with the $\mathbb{F}_2$-vector space $\mathbb{F}_2^m$, the Boolean function g has a trace representation i.e. there exists a mapping $f\colon L \to L$ such that $g(x) = \text{Tr}_L(f(x))$ for all x in L. Of course, the trace representation is not unique. Moreover, if g is balanced then g can be represented by a permutation of L. In all the cases, the Walsh spectrum of g and the Fourier spectrum of f are identical.

In [6], an example of a ten-variables Boolean function with a very atypical Walsh spectrum (see Tab. 1) is given. This Boolean function is balanced and its Walsh coefficients vanish only once. This numerical example, say g, implies the existence of a permutation f of $\mathbb{F}_{1024}$ (not a power permutation) such that

$$g(x) = \text{Tr}_{\mathbb{F}_{1024}} f(x),$$

whence the Fourier spectrum of f is equal to the Walsh spectrum of g, and thus $\sum_{x \in \mathbb{F}_{1024}} \mu(ax + f(x)) \neq 0$ for all $a \in \mathbb{F}_{1024}^{\times}$.

A possible generalization of the conjecture of Helleseth could be the following one:

Conjecture 3. If f is a permutation of L then $\prod_{\lambda \in L^{times}} \mathfrak{D}(\lambda f) = 0$.

Note that Conjecture 2 is know to be true in characteristic 2 since recent works of Daniel Katz in [4] and Tao Feng in [12]. The next conjecture that appeared in the paper by Pursley and Sarwate (see [11]) is still open

Table 1. An example of Walsh spectrum having only one Walsh coefficient equal to zero (see [6])

Walsh	-48	-44	-40	-36	-32	-28	-24	- 20	-16	-12
mult.	5	30	85	70	115	100	31	62	20	10
Walsh	0	8	16	20	24	28	32	36	40	44
mult.	1	5	25	20	85	90	90	80	50	50

Conjecture 4. If f is a power permutation of L where $[L : \mathbb{F}_2]$ is even then $\sup_{a\in L} \widehat{f}(a) \geq 2\sqrt{q}$.

In the sequel, if $\lambda \in L$ then we denote by $\widehat{f}(a)$ the Fourier coefficient of $x \mapsto \lambda f(x)$. If f is a power permutation of exponent s, denoting by t the inverse of s modulo $q-1$, for all $y \in L^\times$, we have :

$$\widehat{f_\lambda}(a) = \sum_{x\in L} \mu(\lambda x^s + ax) = \sum_{x\in L} \mu(\lambda y^s x^s + axy) = \widehat{f}(a\lambda^{-t}). \quad (3)$$

Hence, one of the specifities of power permutations among the permutations of L is that the spectrum of λf does not depend on $\lambda \in L^\times$.

We conclude this section by giving a divisibility result. Recall that a function f defined over a field L of characteristic 2 is said to be almost perfect nonlinear (APN) if for all $u \in L^\times$ the derivative $x \mapsto f(x+u) + f(x)$ is two-to-one. It is for example the case of $f(x) = x^3$ over any field L and of $f(x) = x^{-1}$ when $[L : \mathbb{F}_2]$ is odd.

Theorem 1. *Let f be a power permutation over a field L of even characteristic of cardinal $q \not\equiv 2, 4 \mod 5$. If f is almost perfect nonlinear then there exists $a \in L^\times$ such that $\widehat{f}(a) \equiv 0 \mod 5$ i.e.*

$$\mathfrak{D}(f) \equiv 0 \mod 5.$$

Proof. It is well-known (see [1]) that an APN function f satisfies

$$\sum_{\lambda\in L^\times} \sum_{a\in L} \widehat{f_\lambda}(a)^4 = 2q^3(q-1). \quad (4)$$

Since the spectrum of f does not depend on λ, it implies that:

$$\sum_{a\in L} \widehat{f_\lambda}(a)^4 = 2q^3. \quad (5)$$

Assuming $\mathfrak{D}(f) \not\equiv 0 \mod 5$, we get the congruence

$$q - 1 = 2q^3 \pmod 5$$

implying $q \equiv 2, 4 \mod 5$.

3 Hyperplane Section

The key point of view of this note is to consider the number, say $N_n(u,v)$, of solutions in L^n of the system

$$\begin{cases} x_1 + x_2 + \ldots + x_n = u \\ f(x_1) + f(x_2) + \ldots + f(x_n) = v. \end{cases} \tag{6}$$

By a counting principle using characters, we can state:

Lemma 1. *Let f be a permutation of L. The number $N_n(u,v)$ of solutions in L^n of the system (6) verifies*

$$q^2 N_n(u,v) = q^n + \sum_{\alpha \in L^\times} \sum_{\beta \in L^\times} \widehat{f_\beta}(\alpha)^n \bar{\mu}(\alpha u + \beta v).$$

Proof. For any function $f\colon X \longrightarrow G$ where X is a set and G is a finite abelian group, the number N of solutions in X of $f(x) = y$ for $y \in G$ is

$$N = \frac{1}{|G|} \sum_{x \in X} \sum_{\chi \in \widehat{G}} \chi(f(x) - y)$$

where $\widehat{G}$ denotes the group of characters of G.

For any $\alpha \in L$, we denote by μ_α the additive character of L defined by $\mu_\alpha(x) = \mu(\alpha x)$, then we have:

$$\begin{aligned} q^2 N_n(u,v) &= \sum_{x_1, x_2, \ldots, x_n} \sum_{\beta \in L} \sum_{\alpha \in L} \bar{\mu}_\beta(v - \sum_{i=1}^n f(x_i)) \bar{\mu}_\alpha(u - \sum_{i=1}^n x_i) \\ &= \sum_\beta \sum_\alpha \Big(\sum_{y \in L} \mu(\beta f(y) + \alpha y) \Big)^n \bar{\mu}(\alpha u + \beta v) \\ &= \sum_\beta \sum_\alpha \widehat{f_\beta}(\alpha)^n \bar{\mu}(\alpha u + \beta v) \\ &= \sum_\alpha \widehat{f_0}(\alpha)^n \bar{\mu}(\alpha u) + \sum_{\beta \neq 0} \sum_\alpha \widehat{f_\beta}(\alpha)^n \bar{\mu}(\alpha u + \beta v) \\ &= q^n + \sum_{\alpha \neq 0} \sum_{\beta \neq 0} \widehat{f_\beta}(\alpha)^n \bar{\mu}(\alpha u + \beta v). \end{aligned}$$

Proposition 1. *Assuming the Fourier coefficients of λf, $\lambda \in L$, are integers. Let $\ell \neq p$ be a prime such that $\prod_{\lambda \in L^\times} \mathfrak{D}(\lambda f) \not\equiv 0 \mod \ell$. Then*

$$q^2 N_{\ell-1}(u,v) \equiv 1 + (q\delta_0(u) - 1)(q\delta_0(v) - 1) \mod \ell$$

where $\delta_a(b)$ is equal to 1 if $b = a$ and 0 otherwise.

Proof. By the Fermat's little Theorem, we have the congruence

$$\widehat{f_\lambda}(a)^{\ell-1} \equiv 1 - \delta_0(a) \mod \ell.$$

Hence, by Lemma (1), we have:

$$\begin{aligned} q^2 N_{\ell-1}(u,v) &= q^{\ell-1} + \sum_{\alpha\neq 0}\sum_{\beta\neq 0} \widehat{f_\beta}(\alpha)^{\ell-1}\bar{\mu}(\alpha u + \beta v) \\ &\equiv 1 + \sum_{\alpha\neq 0}\sum_{\beta\neq 0} \bar{\mu}(\alpha u + \beta v) \mod \ell \end{aligned}$$

and we conclude remarking that $\sum_{\alpha\in L^\times}\bar{\mu}(\alpha u) = q\delta_0(u) - 1$.

4 Divisibility of Fourier Coefficients

In [3], it is proved that for the exponents $s \equiv 1 \pmod{p-1}$, the Fourier coefficients are multiple of p. In this section, we are interested in divisibility properties modulo a prime $\ell \neq p$.

Assuming that the Fourier coefficients of any permutation f are rational integers, we can see that if 3 does not divide $\mathfrak{D}(f)$ then we have necessarily $q \equiv 2 \mod 3$. Indeed, using Parseval relation, we can write

$$1 \equiv q^2 = \sum_{a\in L} |\widehat{f}(a)|^2 \equiv q - 1 \mod 3.$$

Theorem 2. *Let f be a power permutation of $\mathbb{F}_{p^n}$ (with $p^n > 2$) of exponent $s = 1 \mod (p-1)$. Then*

$$\mathfrak{D}(f) \equiv 0 \mod 3.$$

Moreover, if n is a power of a prime ℓ and $p \not\equiv 2 \mod \ell$ then

$$\mathfrak{D}(f) \equiv 0 \mod \ell.$$

Proof. First point. Since p divides $\mathfrak{D}(f)$, we may assume that $p \neq 3$. Suppose that $\mathfrak{D}(f) \not\equiv 0 \mod 3$. Applying Proposition 1 with $\ell = 3$, we get

$$\forall u \in L^\times, \quad \forall v \in L^\times, \qquad N_2(u,v) \not\equiv 0 \pmod{\ell}. \tag{7}$$

In order to obtain a contradiction, we prove the existence of $v \in L^\times$ such that $N_2(1,v) = 0$. The mapping $x \mapsto (1-x)^s + x^s$ sends x and $1-x$ to the same point. An element v in the image has at least 2 preimages except when $x = 1-x$, which can only happen when p is odd and $x = 1/2$. So this means that if $p = 2$, the cardinality of the image is less or equal to $q/2$ elements, while if p is odd, the image of the map has at most $(q+1)/2$ elements. If $q > 3$ the complementary of the image contains at least two elements whence a nonzero v such that $N(1,v) = 0$.

Second point. Suppose now that n is a power of a prime ℓ and $p \not\equiv 2 \mod \ell$. The Frobenius automorphism acts on the solutions of the system (6) with $u = 0$, $v = 1$. Since $s \equiv 1 \mod (p-1)$, the system has no $\mathbb{F}_p$-solutions, thus $N_{\ell-1}(0,1) \equiv 0 \mod \ell$. On the other hand, by Proposition 1, if $\mathfrak{D}(f) \not\equiv 0 \mod \ell$ then

$$q^2 N_{\ell-1}(0,1) \equiv 2 - q \equiv 2 - p \mod \ell.$$

References

1. Chabaud, F., Vaudenay, S.: Links between differential and linear cryptanalysis. In: De Santis, A. (ed.) EUROCRYPT 1994. LNCS, vol. 950, pp. 356–365. Springer, Heidelberg (1995)
2. Dillon, J.F.: Elementary Hadamard Difference Sets. PhD thesis, Univ. of Maryland (1974)
3. Helleseth, T.: Some results about the cross-correlation function between two maximal linear sequences. Discrete Math. 16(3), 209–232 (1976)
4. Katz, D.J.: Weil sums of binomials, three-level cross-correlation, and a conjecture of Helleseth. J. Comb. Theory, Ser. A 119(8), 1644–1659 (2012)
5. Katz, N., Livné, R.: Sommes de Kloosterman et courbes elliptiques universelles en caractéristiques 2 et 3. C. R. Acad. Sci. Paris Sér. I Math. 309(11), 723–726 (1989)
6. Kavut, S., Maitra, S., Yücel, M.D.: Search for boolean functions with excellent profiles in the rotation symmetric class. IEEE Transactions on Information Theory 53(5), 1743–1751 (2007)
7. Keijo, K., Marko, R.A., Keijoe, V.: On integer value of Kloosterman sums. IEEE Trans. Info. Theory (2010)
8. Lachaud, G., Wolfmann, J.: Sommes de Kloosterman, courbes elliptiques et codes cycliques en caractéristique 2. C. R. Acad. Sci. Paris Sér. I Math. 305, 881–883 (1987)
9. Lang, S.: Cyclotomic fields I and II, 2nd edn. Graduate Texts in Mathematics, vol. 121. Springer, New York (1990), With an appendix by Karl Rubin
10. Langevin, P.: Numerical projects page (2007), `http://langevin.univ-tln.fr/project/spectrum`
11. Pursley, M.B., Sarwate, D.V.: Cross correlation properties of pseudo-random and related sequences. Proc. IEEE 68, 593–619 (1980)
12. Tao, F.: On cyclic codes of length $2^{2^r} - 1$ with two zeros whose dual codes have three weights. Designs, Codes and Cryptography 62(3) (2012)

Lattice Attacks on DSA Schemes Based on Lagrange's Algorithm

Konstantinos Draziotis[1] and Dimitrios Poulakis[2]

[1] Department of Informatics, Aristotle University of Thessaloniki, P.O. Box 114, Thessaloniki 54124, Greece
drazioti@csd.auth.gr

[2] Department of Mathematics, Aristotle University of Thessaloniki, Thessaloniki 54124, Greece
poulakis@math.auth.gr

Abstract. Using Lagrange's algorithm for the computation of a basis of a 2-dimensional lattice formed by two successive minima, we present some attacks on DSA and ECDSA which permit us, under some assumptions, to compute the secret key of the scheme provided that one or two signed messages are given.

MSC 2010: 94A60, 11T71, 11Y16.

Keywords: Public Key Cryptography, Digital Signature Algorithm, Elliptic Curve Digital Signature Algorithm, Lagrange's Algorithm, Lattice Reduction, Lattice attack.

1 Introduction

The signature schemes DSA and ECDSA. In 1985, T. ElGamal published the first digital signature scheme based on the discrete logarithm problem in the finite prime fields $\mathbb{F}_p$ [5]. Since then several variants of this scheme have been proposed. In 1991, the U.S. government's National Institute of Standards and Technology (NIST) proposed an efficient variant of the ElGamal signature scheme known as DSA, for Digital Signature Algorithm [6,12,14]. In 1998, an elliptic curve analogue called Elliptic Curve Digital Signature Algorithm (ECDSA) was proposed and standardized [8,11,12].

First, let us summarize DSA. The signer chooses a prime p of size between 1024 and 3072 bits with increments of 1024, as recommended in FIPS 186-3 [6, page 15]. Also he chooses a prime q of size 160, 224 or 256 bits, with $q|p-1$ and a generator g of the unique order q subgroup G of $\mathbb{F}_p^*$. Furthermore, he selects a random integer $a \in \{1,\ldots,q-1\}$ and computes $R = g^a \bmod p$. The public key of the signer is (p,q,g,R) and his private key a. He also publishes a hash function $h : \{0,1\}^* \rightarrow \{0,\ldots,q-1\}$. To sign a message $m \in \{0,1\}^*$, he selects a random number $k \in \{1,\ldots,q-1\}$ which is the ephemeral key, and computes

$$r = (g^k \bmod p) \bmod q, \quad s = k^{-1}(h(m) + ar) \bmod q.$$

T. Muntean, D. Poulakis, and R. Rolland (Eds.): CAI 2013, LNCS 8080, pp. 119–131, 2013.

The signature of m is the pair (r, s). The verification of the signature is performed by checking

$$r = ((g^{s^{-1}h(m) \bmod q} R^{s^{-1} r \bmod q}) \bmod p) \bmod q.$$

For the ECDSA the signer selects an elliptic curve E over the finite prime field $\mathbb{F}_p$, a point $P \in E(\mathbb{F}_p)$ with order a prime q of size at least 160 bits. According to FIPS 186-3, the prime p must be in the set $\{160, 224, 256, 512\}$. Further, he chooses a random integer $a \in \{1, \ldots, q-1\}$ and computes $Q = aP$. The public key of the signer is (E, p, q, P, Q) and his private key a. He also publishes a hash function $h : \{0,1\}^* \to \{0, \ldots, q-1\}$. To sign a message m, he selects a random number $k \in \{1, \ldots, q-1\}$ which is the ephemeral key and computes $kP = (x, y)$ (where x and y are regarded as integers between 0 and $p-1$). Next, he computes

$$r = x \bmod q \quad \text{and} \quad s = k^{-1}(h(m) + ar) \bmod q.$$

The signature of m is (r, s). For its verification one computes

$$u_1 = s^{-1}h(m) \bmod q, \;\; u_2 = s^{-1}r \bmod q, \;\; u_1P + u_2Q = (x_0, y_0).$$

He accepts the signature if and only if $r = x_0 \bmod q$.
The security of the two systems is relied on the assumption that the only way to forge the signature is to recover either the secret key a, or the ephemeral key k (in this case is very easy to compute a). Thus, the parameters of these systems were chosen in such a way that the computation of discrete logarithms is computationally infeasible.

Related Work. A very important tool for attacking public-key cryptosystems are the lattices and the so-called LLL reduction method [13]. Attacks to DSA and to ECDSA based on these techniques and using he equality $s = k^{-1}(h(m) + ar) \bmod q$ are described in [1,3,10,16,17,18,19]. In [1], it was shown that the DSA secret key a can be computed provided the ephemeral key k is produced by Knuth's linear congruential generator with known parameters, or variants. In [10] the authors pointed out that Babai's method can be used heuristically in order to recover the secret key a, if enough signatures and some bits of the ephemeral keys are known. A polynomial time attack which computes the DSA secret key a is proposed in [15], in case the following holds: The size of q is not too small compared with p, the probability of collisions for the hash function is not too large compared to $1/q$ and for a polynomially bounded number of messages, about $\lfloor \sqrt{\log_2 q} \rfloor$ of the least significant bits of the ephemeral keys are known. This attack is adapted to the case of ECDSA [18]. In [3], the authors compute, using the LLL reduction method, two short vectors of a three-dimensional lattice and in case where the second shortest vector is sufficiently short, they deduce two lines which intersect in (a, k), provided that a and k are sufficiently small. Finally, in [19], an attack is described in case where the secret and the ephemeral key of a signed message or their modular inverses are sufficiently small.

Our Contribution. In this paper we present some attacks on DSA and ECDSA taking advantage of the equation $s = k^{-1}(h(m) + ar) \bmod q$. We use a two-dimensional lattice $\mathcal{L}$ defined by a signed message. Lagrange Lattice Reduction algorithm, provide us with a basis of $\mathcal{L}$ formed by two successive minima $\mathbf{b}_1, \mathbf{b}_2$. Using this basis we construct two linear polynomials $f_i(x,y)$ such that (a,k) is the intersection point of two straight lines of the form $f_i(x,y) = c_i q$, where $c_i \in \mathbb{Z}$ $(i = 1, 2)$. If a and k are sufficiently small, then c_i belong to a small set and so we can compute the secret key a in polynomial time. Similar attacks hold for the pairs $(k^{-1} \bmod q, k^{-1}a \bmod q)$ and $(a^{-1} \bmod q, a^{-1}k \bmod q)$.

If two signed messages with ephemeral keys k_1 and k_2 are available, then we obtain an equation of the same form relating k_1 and k_2 and so, earlier statements about secret and ephemeral keys also apply to k_1 and k_2. In case we know some bits of the secret and ephemeral keys and their sum is almost the half of sum of the bits of secret and ephemeral keys, then we can compute efficiently the secret key. Similar results also hold for all the aforementioned cases.

The paper is organized as follows. In Section 2, we recall some basic results about Lagrange lattice reduction algorithm. Our attacks are presented in Sections 3 - 10. An example is given in Section 11 and Section 12 concludes the paper.

2 Lagrange Lattice Reduction

Let $\mathbb{R}^n$ denote the n-dimensional real Euclidean space. The *inner product* of two elements $\mathbf{u} = (u_1, \ldots, u_n)$ and $\mathbf{v} = (v_1, \ldots, v_n)$ of $\mathbb{R}^n$ is defined to be the quantity $\mathbf{u} \cdot \mathbf{v} = u_1 v_1 + \cdots + u_n v_n$ and the *Euclidean norm* of a vector $\mathbf{v} = (v_1, \ldots, v_n)$ the quantity $\|\mathbf{v}\| = (v_1^2 + \cdots + v_n^2)^{1/2}$.

Let $B = \{\mathbf{b}_1, \ldots, \mathbf{b}_n\}$ be a basis of $\mathbb{R}^n$. A *n-dimensional lattice* spanned by B is the set

$$\mathcal{L} = \{z_1 \mathbf{b}_1 + \cdots + z_n \mathbf{b}_n / \ z_1, \ldots, z_n \in \mathbb{Z}\}.$$

If $\mathbf{b}_i = (b_{i1}, \ldots, b_{in})$ $(i = 1, \ldots, n)$, then the *determinant* $\det\mathcal{L}$ of $\mathcal{L}$ is the absolute value of the determinant whose (i,j) element is $b_{i,j}$.

Let $\mathcal{L} \subset \mathbb{R}^2$ be a 2-dimensional lattice. We say that a basis $\{\mathbf{v}_1, \mathbf{v}_2\}$ of $\mathcal{L}$ is formed by two *successive minima* of $\mathcal{L}$, if $\mathbf{v}_1$ is the shortest nonzero vector of $\mathcal{L}$ and $\mathbf{v}_2$ is the shortest vector of $\mathcal{L}$ among all vectors of $\mathcal{L}$ linearly independent from $\mathbf{v}_1$. In this section we recall an algorithm due to Lagrange (1773) which computes a basis of $\mathcal{L}$ formed by two successive minima of $\mathcal{L}$. This algorithm was also described later by Gauss (1801), and is often erroneously called Gauss' algorithm. If $x \in \mathbb{R}$, then we set $\lfloor x \rceil = \lfloor x + 0.5 \rfloor$.

LAGRANGE'S ALGORITHM
Input: A basis $\{\mathbf{v}_1, \mathbf{v}_2\}$ of $\mathcal{L}$.
Output: A basis $\{\mathbf{b}_1, \mathbf{b}_2\}$ formed by two successive minima of $\mathcal{L}$.
Loop
    If $\|\mathbf{v}_2\| < \|\mathbf{v}_1\|$, then swap $\mathbf{v}_1$ and $\mathbf{v}_2$.
    Compute $m = \lfloor \mathbf{v}_1 \cdot \mathbf{v}_2 / \|\mathbf{v}_1\|^2 \rceil$.
    If $m = 0$, return the basis vectors $\mathbf{v}_1$ and $\mathbf{v}_2$.
    Replace $\mathbf{v}_2$ with $\mathbf{v}_2 - m\mathbf{v}_1$.
Continue Loop

Theorem 1. *The above algorithm computes a basis* $\{\mathbf{b}_1, \mathbf{b}_2\}$ *formed by two successive minima of* $\mathcal{L}$ *in time* $O((\log M)^2)$ *bit operations, where* M *is defined by* $M = \max\{\|\mathbf{v}_1\|, \|\mathbf{v}_2\|\}$. *If* θ *is the angle between* $\mathbf{b}_1$ *and* $\mathbf{b}_2$, *then* $\pi/3 \leq \theta \leq 2\pi/3$ *and* $|cos\theta| < \|\mathbf{b}_1\|/2\|\mathbf{b}_2\|$. *Further,* $\|\mathbf{b}_1\| \leq \sqrt{4/3}\,(\det \mathcal{L})^{1/2}$.

Proof. See [15], [9, Section 6.12.1] and [9, Theorem 6.25].

3 The DSA-ATTACK-1 Algorithm

Let m be a message and (r, s) its signature with DSA (resp. ECDSA). Then there is $k \in \{1, \ldots, q-1\}$ such that $r = (g^k \bmod p) \bmod q$ (resp. $kP = (x, y)$) and $s = k^{-1}(h(m) + ar) \bmod q$. We conserve these notations for the rest of the paper. In this section we present our first attack using the signed message.

Let $\mathbb{Z}_{(q)}$ be the subring of $\mathbb{Q}$ formed by the fractions u/v, where u and v are integers with $\gcd(u, v) = 1$ and $q \nmid v$. If $x \in \mathbb{Z}$, then we denote by $\bar{x}$ the class of x in $\mathbb{F}_q$. We have a ring homomorphism $\phi : \mathbb{Z}_{(q)} \to \mathbb{F}_q$ defined by $\phi(u/v) = \bar{u}(\bar{v})^{-1}$. We denote by $[u/v]_q$ the representative of the class $\phi(u/v)$ having the smallest absolute value. We have chosen this presentation for the elements of $\mathbb{F}_q$ since our method involves the absolute values of the unknown quantities which give the keys and our attacks are practical when these quantities are below a certain bound.

The following algorithm computes efficiently a in case where a and k are sufficiently small.

DSA-ATTACK-1
Input: A signed message (m, r, s) and integers $0 < X, Y < q$ with $X > \sqrt{q}$.

1. Compute $A = -rs^{-1} \bmod q$ and $B = -s^{-1}h(m) \bmod q$.
2. Compute, using Lagrange's algorithm, a basis formed by two successive minima $\mathbf{b}_1 = (b_{11}X, b_{12}Y)$ and $\mathbf{b}_2 = (b_{21}X, b_{22}Y)$ of the lattice $\mathcal{L}$ spanned by the vectors (AX, Y) and $(qX, 0)$.
3. Compute $b_{i3} = b_{i2}B \bmod q$.
4. Compute the quantities

$$\beta_i = \left\lfloor \frac{b_{i3} + \sqrt{2}\,\|\mathbf{b}_i\|}{q} \right\rfloor \quad (i = 1, 2).$$

5. For every $c_1 \in \{0, \pm 1, \ldots, \pm\beta_1\}$ and $c_2 \in \{0, \pm 1, \ldots, \pm\beta_2\}$,
   (a) compute
   $$x_0 = c_1 b_{22} - c_2 b_{12} + \frac{b_{23}b_{12} - b_{13}b_{22}}{q},$$
   (b) compute $g^{x_0} \bmod p$ (resp. x_0P).
   (c) If $g^{x_0} = R \bmod p$ (resp. $x_0P = Q$), then output $a = x_0 \bmod q$ and stop. Otherwise, output "Failure".

Proof of Correctness. We shall prove that this algorithm computes correctly a in case $|[a]_q| < X$ and $|[k]_q| < Y$. For this purpose, we shall construct, using Lagrange's algorithm, a family of pairs of independent bivariate linear equations such that $([a]_q, [k]_q)$ is the common solution of one such pair, and hence we shall be able to determine it.

The determinant of the lattice $\mathcal{L}$ spanned by (AX, Y) and $(qX, 0)$ is $\det \mathcal{L} = qXY$. By Theorem 1, Lagrange's algorithm gives a basis $\{\mathbf{b}_1, \mathbf{b}_2\}$ of $\mathcal{L}$ formed by two successive minima of $\mathcal{L}$ satisfying

$$2\ |cos\theta|\ \|\mathbf{b}_2\| < \|\mathbf{b}_1\| \leq \sqrt{4/3}\ (qXY)^{1/2}, \tag{1}$$

where θ is the angle between $\mathbf{b}_1$ and $\mathbf{b}_2$.

Put $\mathbf{b}_i = (b_{i1}X, b_{i2}Y)$ $(i = 1, 2)$. Then there are λ_{i1}, $\lambda_{i2} \in \mathbb{Z}$ such that $b_{i1} = \lambda_{i1}q + \lambda_{i2}A$ and $b_{i2} = \lambda_{i2}$ $(i = 1, 2)$. Suppose that $q|\lambda_{i2}$. Then $\lambda_{i2} = \lambda'_{i2}q$, where $\lambda'_{i2} \in \mathbb{Z}$ $(i = 1, 2)$. We have

$$\|\mathbf{b}_i\|^2 = q^2\Big((\lambda_{i1} + \lambda'_{i2}A)^2X^2 + \lambda'^2_{i2}Y^2\Big) > \|(qX, 0)\|.$$

Since $X > \sqrt{q}$ and $0 < X, Y < q$, we have $Y/X < \sqrt{q}$ and so

$$\|(AX, Y)\|^2 \leq ((q-1)^2 + q)X^2 < \|(qX, 0)\|^2.$$

Hence, we get

$$\|(AX, Y)\| < \|(qX, 0)\| < \|\mathbf{b}_i\| \quad (i = 1, 2)$$

which is a contradiction, since $\mathbf{b}_1$, $\mathbf{b}_2$ are two successive minima of $\mathcal{L}$. Hence, we have $\lambda_{i2} \not\equiv 0 \pmod{q}$. Further, by the Euclidean division, there are $\pi_i, b_{i3} \in \mathbb{Z}$ such that $\lambda_{i2}B = \pi_i q + b_{i3}$ and $0 < b_{i3} < q$ $(i = 1, 2)$.

Put $h_i(x, y) = b_{i1}x + b_{i2}y + b_{i3}$ $(i = 1, 2)$. We have

$$h_i(x, y) \equiv \lambda_{i2}(y + Ax + B) \pmod{q} \quad (i = 1, 2).$$

Since $\lambda_{i2} \not\equiv 0 \pmod{q}$, we deduce that $h_i(x, y)$ $(i = 1, 2)$ are not identical to zero over $\mathbb{F}_q$. Moreover, the congruence $[k]_q + A[a]_q + B \equiv 0 \pmod{q}$ implies $h_i([a]_q, [k]_q) \equiv 0 \pmod{q}$ $(i = 1, 2)$. On the other hand, we deduce

$$|h_i([a]_q, [k]_q)| \leq b_{i3} + |\mathbf{b}_i \cdot ([a]_q/X, [k]_q/Y)| \leq b_{i3} + \sqrt{2}\ \|\mathbf{b}_i\| \quad (i = 1, 2).$$

Setting

$$\beta_i = \left\lfloor \frac{b_{i3} + \sqrt{2}\ \|\mathbf{b}_i\|}{q} \right\rfloor \quad (i = 1, 2),$$

we have $h_i([a]_q, [k]_q) = c_i q$ where $|c_i| \in \{0, 1, \ldots, \beta_i\}$ $(i = 1, 2)$.

Since the vectors $\mathbf{b}_1$ and $\mathbf{b}_2$ are linearly independent, the lines defined by the equations $h_1(x, y) = c_1 q$ and $h_2(x, y) = c_2 q$ are not parallels. Thus there are c_1, c_2 as previously such that $([a]_q, [k]_q)$ is the intersection point of the lines $h_1(x, y) = c_1 q$ and $h_2(x, y) = c_2 q$. Hence we have

$$[a]_q = c_1 b_{22} - c_2 b_{12} + \frac{b_{23}b_{12} - b_{13}b_{22}}{q}, \quad [k]_q = c_2 b_{11} - c_1 b_{21} + \frac{b_{13}b_{21} - b_{11}b_{23}}{q},$$

where $|c_i| \in \{0, 1, \ldots, \beta_i\}$ $(i = 1, 2)$.

Time Complexity. In Step 1, the application of the Extended Euclidean algorithm for the computation of s^{-1} mod q and next the modular multiplications for the computations of A and B require $O((\log q)^2)$ bit operations. By Theorem 1, the complexity of Step 2 is $O((\log qXY)^2)$ bit operations. Step 3 and 4 need $O((\log q)^2)$ bit operations. Put $\beta_i^* = \max\{1, \beta_i\}$ $(i = 1, 2)$. Step 5 needs

$$O(\beta_1^* \beta_2^* (\log p)^2 \log(q \beta_1^* \beta_2^*))$$

bit operations, in case of DSA and

$$O(\beta_1^* \beta_2^* \log q \log(q \beta_1^* \beta_2^*))$$

bit operations and

$$O(\beta_1^* \beta_2^* \log q)$$

elliptic curve group operations, in case of ECDSA. Further, note that the bit complexity analysis can be improved by using the Shoenhage-Strassen multiplication algorithm.

From the above discussion it is clear that the time complexity of our algorithm it heavily depends on the size of the quantities β_1 and β_2. As large these quantities are as less practical our attack is. We shall compute bounds for β_1 and β_2 in order to evaluate more closely the time complexity of our attack. Using the bounds for b_{13} and $\|\mathbf{b}_1\|$ we obtain $\beta_1 < 3(XY/q)^{1/2}$. In case where θ is not very close to 90°, using the inequalities (1) and $0 < b_{23} < q$, we get $\beta_2 < 1 + C(\theta)(XY/q)^{1/2}$, where $C(\theta) = \sqrt{2}/(\sqrt{3}|cos\theta|)$. We have $\pi/3 \leq \theta \leq 2\pi/3$. Furthermore, suppose that θ is not between 89.999^o and 90.001^o. It follows that $C(\theta) < 46785$. Hence $\beta_2 < 46785(XY/q)^{1/2}$. Therefore, in case where the keys a and k are quite small, say $[a]_q[k]_q < q(\log q)^4$ and θ is not very close to 90°, we can efficiently compute a. Note that the assumption about the angle θ is not really essential (see Appendix).

It is worthwhile to remark that the step 5 of our algorithm can be parallelized and so its running time can be reduced.

4 The DSA ATTACK-2 Algorithm

Put $A = -h(m)s^{-1}$ mod q and $B = -rs^{-1}$ mod q. Then $([a^{-1}]_q, [a^{-1}k]_q)$ is a solution to the congruence $y + Ax + B \equiv 0 \pmod q$. So we can develop an algorithm similar to the previous one, which computes the secret key a in case where $|[a^{-1}]_q| < X$, $|[a^{-1}k]_q| < Y$. The only differences are in Step 5. More precisely, in Steps 5(b) and 5(c) instead of x_0 we set x_0^{-1} mod q.

5 The DSA-ATTACK-3 Algorithm

We set $A = h(m)r^{-1}$ mod q and $B = -sr^{-1}$ mod q. Then, $([k^{-1}]_q, [k^{-1}a]_q)$ is a solution to the congruence $y + Ax + B \equiv 0 \pmod q$. Thus, we can develop an algorithm, similar to the first one, which computes the secret key a in case where $|[k^{-1}]_q| < X$ and $|[k^{-1}a]_q| < Y$. The only differences are in Step 5. More precisely, in Step 5(a) we have not only to compute x_0 but also $y_0 = -Ax_0 - B \bmod q$, and in Steps 5(b) and 5(c) instead of x_0 (of the first attack) we consider the value $y_0 x_0^{-1} \bmod q$.

6 The DSA-ATTACK-4 Algorithm

In this section we give a version of DSA-Attack 1 which the computation of its complexity in terms of q does not involve the angle between the vectors of the basis that we use.

DSA-ATTACK-4
Input: One signed message (m, r, s) and integers X, Y with $1 < X, Y < q$.

1. Compute $A = \lceil (\sqrt{12qXY - 9Y^2})/3X \rceil$. Put $\mathbf{a} = (AX, Y)$.
2. Compute $\eta \in \{1, \dots, q-1\}$ such that $-\eta s^{-1} r = A \bmod q$ and next compute $B = -\eta s^{-1} h(m) \bmod q$.
3. Compute, using Lagrange's algorithm, a vector $\mathbf{b} = (b_1 X, b_2 Y)$ having the smallest size among the nonzero vectors of the lattice spanned by $(qX, 0)$ and $\mathbf{a}$.
4. Compute $b_3 = b_2 B \bmod q$.
5. Compute the quantities

$$\alpha = \left\lfloor \frac{B + \sqrt{2}\,\|\mathbf{a}\|}{q} \right\rfloor, \ \beta = \left\lfloor \frac{b_3 + \sqrt{2}\,\|\mathbf{b}\|}{q} \right\rfloor.$$

6. For every $c \in \{0, \pm 1, \dots, \pm\alpha\}$ and $d \in \{0, \pm 1, \dots, \pm\beta\}$,
   (a) compute $x_0 = cb_2 - d + (b_3 - Bb_2)/q$,
   (b) compute $g^{x_0} \bmod p$ (resp. $x_0 P$).
   (c) If $g^{x_0} = R \bmod p$ (resp. $x_0 P = Q$), then output $a = x_0 \bmod q$ and stop.
   Otherwise output "Failure".

Proof of Correctness. Since $k - s^{-1}ra - \eta s^{-1} h(m) \equiv 0 \pmod q$, setting $l = \eta k \bmod q$ and $f(x, y) = y + Ax + B$, we have that $([a]_q, [l]_q)$ is a solution of $f(x, y) \equiv 0 \pmod q$. We shall prove that this algorithm computes correctly a in case where $|[a]_q| \le X$ and $|[l]_q| \le Y$.

Denote by $\mathcal{L}$ the lattice spanned by the linearly independent vectors $\mathbf{a}$ and $(qX, 0)$. By Theorem 1, Lagrange's algorithm gives a vector $\mathbf{b} = (b_1 X, b_2 Y)$ of $\mathcal{L}$ with size

$$\|\mathbf{b}\| \le \sqrt{4/3}\ \det \mathcal{L}^{1/2} \le \sqrt{4/3}\ (qXY)^{1/2}.$$

Further, we have $\|\mathbf{a}\| > \sqrt{4/3}(qXY)^{1/2}$ and so, we get $\|\mathbf{b}\| < \|\mathbf{a}\|$. If $\mathbf{a}$ and $\mathbf{b}$ are linear dependent, then $b_1 = b_2 A$, whence $\mathbf{b} = b_2 \mathbf{a}$. Hence $\|\mathbf{b}\| \ge \|\mathbf{a}\|$ which is a contradiction. Thus, $\mathbf{a}$ and $\mathbf{b}$ are linearly independent.

Let $\mathbf{b} = \lambda\mathbf{a} + \mu(qX, 0)$, where $\lambda, \mu \in \mathbb{Z}$. Then $\lambda = b_2$. By the Euclidean division of λB by q, we have $\lambda B = \pi q + b_3$, where $\pi, b_3 \in \mathbb{Z}$ and $0 \leq b_3 < q$. Put $h(x, y) = b_1 x + b_2 y + b_3$. Therefore

$$h(x, y) \equiv \lambda f(x, y) \pmod{q}.$$

If $q|\lambda$, then we deduce $\|\mathbf{b}\| \geq qY$ and so, $qY < 2(qXY)^{1/2}$, whence $qY < 2X$. Since $Y > 1$ and $X < q$ we have a contradiction. Therefore $q \not|\lambda$ and so, $h(x, y)$ is not identical to zero over $\mathbb{F}_q$. Furthermore, we have $h([a]_q, [l]_q) \equiv 0 \pmod{q}$. On the other hand, we deduce

$$|f([a]_q, [l]_q)| \leq B + \sqrt{2}\,\|\mathbf{a}\|, \quad |h([a]_q, [l]_q)| \leq b_3 + \sqrt{2}\,\|\mathbf{b}\|.$$

Since $h([a]_q, [l]_q) \equiv f([a]_q, [l]_q) \equiv 0 \pmod{q}$, setting

$$\alpha = \left\lfloor \frac{B + \sqrt{2}\,\|\mathbf{a}\|}{q} \right\rfloor, \quad \beta = \left\lfloor \frac{|b_3| + \sqrt{2}\,\|\mathbf{b}\|}{q} \right\rfloor,$$

we get $f([a]_q, [l]_q) = cq$ and $h([a]_q, [l]_q) = dq$, where $c \in \{0, \pm 1, \ldots, \pm\alpha\}$ and $d \in \{0, \pm 1, \ldots, \pm\beta\}$. Since $\mathbf{a}$ and $\mathbf{b}$ are linear independent, we have that the polynomials $f(x, y) - cq$ and $h(x, y) - dq$ are independent and so we obtain $[a]_q = cb_2 - d + (b_3 - Bb_2)/q$.

Time Complexity. The Step 1 requires $O((\log qXY)^2)$ and Step 2, $O((\log q)^2)$ bit operations. By Theorem 1, the time complexity of Step 3 is $O((\log qXY)^2)$ bit operations. The use of Euclidean division in Step 4 needs $O((\log q)^2)$ bit operations. Since $\|\mathbf{a}\| \leq 2(qXY)^{1/2}$ and $\|\mathbf{b}\| < \sqrt{4/3}\,(qXY)^{1/2}$, Step 5 needs $O((\log qXY)^2)$ bit operations. Furthermore, we deduce $\alpha, \beta < 3(XY/q)^{1/2}$. So, Step 6, in case of DSA, requires $O((XY/q)(\log p)^2 \log(pXY/q))$ bit operations and, in case of ECDSA, $O((XY/q) \log q \log(XY))$ bit operations and $O(XY(\log q)/q)$ elliptic curve group operations. Therefore, for DSA, the time complexity of our algorithm is $O((XY/q) \log p \log(pXY/q))$ bit operations, and for ECDSA, $O((XY/q) \log q \log(XY))$ bit operations and $O(XY(\log q)/q)$ elliptic curve group operations. Note that when $XY < q(\log q)^4$ our attack is practical. Moreover, the computation of the complexity in terms of q does not use the angle between the vectors $\mathbf{a}$ and $\mathbf{b}$, as in the previous section.

7 The DSA-ATTACK-5 Algorithm

Set $B = \lceil (\sqrt{12qXY - 9Y^2})/3X \rceil$. We determine $\eta \in \{1, \ldots, q-1\}$ such that $-\eta s^{-1} h(m_1) \equiv B \pmod{q}$ and next we compute $A = -\eta s^{-1} r \bmod q$. Then taking $l = [\eta k]_q$, we have $l + B[a^{-1}]_q + A \equiv 0 \pmod{q}$. Thus, we can develop an algorithm similar of the above for the computation of $[a^{-1}]_q$ and hence a.

8 Heuristic Attacks

Every loop of Lagrange's algorithm provide us with linear polynomials $h_1(x, y)$ and $h_2(x, y)$. Taking advantage of this fact we develop an heuristic attack. Put

$A = -rs^{-1} \bmod q$ and $B = -s^{-1}h(m) \bmod q$ and fix bounds M_1 and M_2. We apply Lagrange's algorithm on the lattice $\mathcal{L}$ spanned by $\mathbf{v}_1 = (A, 1)$ and $\mathbf{v}_2 = (q, 0)$. In each loop the algorithm provide us with a basis $\mathbf{b}_1 = (b_{11}, b_{12})$ and $\mathbf{b}_2 = (b_{21}, b_{22})$ of $\mathcal{L}$. Further, we compute $b_{i3} = b_{i2}B \bmod q$. For every $c_i \in \{0, \pm 1, \ldots, \pm M_i\}$ $(i = 1, 2)$ we compute

$$x_0 = c_1 b_{22} - c_2 b_{12} + \frac{b_{23}b_{12} - b_{13}b_{22}}{q},$$

and next $g^{x_0} \bmod p$ (resp. x_0P). If $g^{x_0} = R \bmod p$ (resp. $x_0P = Q$), then output $a = x_0 \bmod q$. Similar attacks hold for the other two cases.

9 Attacks Using Some Known Bits

Suppose now, as in [10,17,18,3], that some side channel attack or another attack has provided us with some bits of the two keys which consist of some blocks of non-contiguous bits. Then the unknown bits are in, say d blocks, of contiguous bits. So $a = u + a'$ where $a' = \sum_{j=1}^{d} 2^{\lambda_j} a_j$, with u, λ_j are known and a_j $(j > 0)$ are unknown. Similarly, $k = v + k'$, where v and k' are positive integers. Then (a', k') is a solution of $y + Ax + B' \equiv 0 \pmod q$, where $A = -rs^{-1} \bmod q$ and $B' = -h(m)s^{-1} + Au + v \bmod q$. If $[a]_q[k]_q < q(\log q)^4$, then we can use the algorithm DSA-ATTACK-1 in order to determine efficiently a' and so a. Similarly, we proceed for the other cases.

10 Attacks Using Two Signed Messages

Suppose now that we possess two signed messages (m_1, r_1, s_1), (m_2, r_2, s_2) signed with ephemeral keys k_1 and k_2 respectively. Eliminating a from the resulting congruences, we obtain a new congruence in k_1 and k_2, and so we can use the above attacks in order to compute k_1 or k_2 and hence a.

11 An Example

We consider the elliptic curve E given in [2, Example 3, p. 182] defined over the finite field $\mathbb{F}_p$, where $p = 2^{160} + 7$ is a prime, by the equation

$$y^2 = x^3 + 10x + 1343632762150092499701637438970764818528075565078.$$

The number of points of $E(\mathbb{F}_p)$ is the 160-bit prime

$$q = 1461501637330902918203683518218126812711137002561.$$

Consider the point $P = (x(P), y(P))$ of $E(\mathbb{F}_p)$ of order q, where

$$x(P) = 858713481053070278779168032920613680360047535271,$$

$$y(P) = 364938321350392265038182051503279726748224184066.$$

We take as private key the 159– bit integer

$$a = 632790624926327566095930109784509564088868719925$$

and so, the public key is $Q = aP = (x(Q), y(Q))$ where

$$x(Q) = 323143541338258768513217516737351163416580216802,$$

$$y(Q) = 1082565742094335698214311210481468633571227672 05.$$

Let m be a message with hash value

$$h(m) = 142942603167529132165819177831617629501580942.$$

We shall produce two different signatures to m and we shall apply our attacks on the signed messages.

(1) We choose the 160–bit integer

$$k = 1379446954338586918993483976734898999722877873502$$

as an ephemeral key and we produce the signature (r, s) for m, where

$$r = 1068783781268713267197646474159527053924987016485,$$

$$s = 1378560981845469016588440780739126281039851915903.$$

We have

$$a^{-1} \bmod q = 18014398509481986 < \sqrt{q} \log q,$$

$$a^{-1}k \bmod q = 4345987167328739085676500 8 < \sqrt{q} \log q.$$

We apply DSA-ATTACK-2 with $X = Y = \lfloor \sqrt{q} \log q \rfloor$ and, since $\beta_1 = 141$, $\beta_2 = 176$, we have to check 99899 values.

(2) Next, we choose the 147–bit integer

$$k = 166512230815695319456800029769107765911320439$$

as an ephemeral key and we produce the signature (r, s) for m, where

$$s = 1306196044283986966897196176500887228348994582394,$$

$$r = 724139290201395716568648544869144698056699972161.$$

We have $|[a]_q|, |[k]_q| > q^{0.9}$. Also, $a^{-1} \bmod q < \sqrt{q} \log q$ and $|[a^{-1}k]_q| > q^{0.9}$. Further, $|[k^{-1}]_q|$ and $|[k^{-1}a]_q|$ are $> q^{0.9}$ and so, the attacks of Section 3, 4 and 5 do not work. We try the heuristic attacks of Section 3.4. We set $A = -h(m)s^{-1} \bmod q$, $B = -rs^{-1} \bmod q$ and we take $M = 292820000 \simeq 2(\log q)^4$. In 12th loop we obtain $\mathbf{b}_1 = (b_{11}, b_{12})$ and $\mathbf{b}_2 = (b_{21}, b_{22})$, where

$$b_{11} = -100818561570428157294605379284231893689 02, \quad b_{12} = -84226299,$$

$$b_{21} = 1012670394183657723999456780974912306541, \quad b_{22} = -136503451,$$

which form a basis of the lattice spanned by $(q,0)$, $(A,1)$. We compute

$$Bb_{12} \bmod q = -781982649082028777596295134710513450423832598231$$

$$Bb_{22} \bmod q = -206354035197723594446285891701235586069110289872.$$

which are b_{13} and b_{23}, respectively. For $c_1 = -197283790$ and $c_2 = -105851969$ we obtain $a^{-1} \bmod \ q$ and hence a.

The above attack can be easily parallelized. We divide the interval $[-M, M]$ to $\kappa = 1000 \approx \lceil(\log q)^{2.54}\rceil$ subintervals of length about $\mu = \lceil 2M/\kappa \rceil = 312745$. For $\varrho = 0, 1, 2, ..., \kappa - 1$, we consider the intervals $I_\varrho = [-M + \varrho\mu, -M + (\varrho+1)\mu]$ and we compute x_0 for each integer pair $(c_1, c_2) \in I_i \times I_j$. Thus, this step can be executed in $O(\mu^2)$ steps. For this parallel computation we need $\kappa^2 = 10^6$ processors. Note that the bitcoin grid uses at least $10^{5.2}$ processors and a modern supercomputer uses over 10^5 processors.

All the computations are executed in a computer with Pentium 2GHz and memory 3Gb. Further for the elliptic curve computations we have used SAGE [20], for the arithmetic computations MAPLE 12 and for the choice of $h(m)$ and k the B.B.S. pseudo-random generator [4]. The secret key a is not chosen at random but in order to support a case where our attack works.

12 Conclusion

In this paper we have presented rigorous attacks on the DSA and ECDSA based on Lagrange's algorithm, which recover the secret key in polynomial time, provided that the sizes of the numbers in one of the pairs $([a]_q, [k]_q)$, $([k^{-1}]_q, [k^{-1}a]_q)$ and $([a^{-1}]_q, [a^{-1}k]_q)$ are sufficiently small. Thus, a signer has to take care about the size of all the above quantities and not only of a and k. We have also presented an heuristic variant of this attack using all the loops of Lagrange's algorithm and attacks in case where two signed messages are known. An interesting feature of all these attacks are that can be parallelized. If we know a sufficient number of bits of a and k, then our attacks can compute a. Note that the needed number of known bits is fewer than in [3]. Our attacks can also be applied on other signature schemes where the secret and the ephemeral keys are solutions of a modular bivariate linear equation. Such schemes are Schnorr' signature, Heyst-Pedersen signature, GPS, etc [7,14,21].

References

1. Bellare, M., Goldwasser, S., Micciancio, D.: "Pseudo-random" number generation within cryptographic algorithms: The DSS case. In: Kaliski Jr., B.S. (ed.) CRYPTO 1997. LNCS, vol. 1294, pp. 277–291. Springer, Heidelberg (1997)
2. Blake, I.F., Seroussi, G., Smart, N.: Elliptic Curves in Cryptography. Cambridge University Press (2000)

3. Blake, I.F., Garefalakis, T.: On the security of the digital signature algorithm. Des. Codes Cryptogr. 26(1-3), 87–96 (2002)
4. Blum, L., Blum, M., Shub, M.: A Simple Unpredictable Pseudo-Random Number Generator. SIAM Journal on Computing 15, 364–383 (1986)
5. ElGamal, T.: A public key cryptosystem and a signature scheme based on discrete logarithm. IEEE Transactions on Information Theory 31, 469–472 (1985)
6. FIPS PUB 186-3: Federal Information Processing Standards Publication, Digital Signature Standard (DSS)
7. Girault, M., Poupard, G., Stern, J.: Global Payment System (GPS): un protocole de signature à la volée. In: Proceedings of Trusting Electronic Trade (1999)
8. Johnson, D., Menezes, A.J., Vanstone, S.A.: The elliptic curve digital signature algorithm (ECDSA). Intern. J. of Information Security 1, 36–63 (2001)
9. Hoffstein, J., Pipher, J., Silverman, J.: An Introduction to Mathematical Cryptography. Springer (2008)
10. Howgrave-Graham, N.A., Smart, N.P.: Lattice Attacks on Digital Signature Schemes. Des. Codes Cryptogr. 23, 283–290 (2001)
11. Koblitz, N., Menezes, A.J., Vanstone, S.A.: The state of elliptic curve cryptography. Des. Codes Cryptogr. 19, 173–193 (2000)
12. Koblitz, N., Menezes, A.J.: A survey of Public-Key Cryptosystems. SIAM Review 46(4), 599–634 (2004)
13. Lenstra, A.K., Lenstra Jr., H.W., Lovász, L.: Factoring polynomials with rational coefficients. Math. Ann. 261, 513–534 (1982)
14. Menezes, A.J., van Oorschot, P.C., Vanstone, S.A.: Handbook of Applied Cryptography. CRC Press, Boca Raton (1997)
15. Nguyen, P.Q., Stehlé, D.: Low-Dimensional Lattice Basis Reduction Revisited. ACM Transactions on Algorithms 5(4), Article 46 (2009)
16. Nguyên, P.Q., Stern, J.: The Two Faces of Lattices in Cryptology. In: Silverman, J.H. (ed.) CaLC 2001. LNCS, vol. 2146, pp. 146–180. Springer, Heidelberg (2001)
17. Nguyen, P.Q., Shparlinski, I.E.: The Insecurity of the Digital Signature Algorithm with Partially Known Nonces. J. Cryptology 15, 151–176 (2002)
18. Nguyen, P.Q., Shparlinski, I.E.: The Insecurity of the Elliptic Curve Digital Signature Algorithm with Partially Known Nonces. Des. Codes Cryptogr. 30, 201–217 (2003)
19. Poulakis, D.: Some Lattice Attacks on DSA and ECDSA Applicable Algebra in Engineering. Communication and Computing 22, 347–358 (2011)
20. Stein, W.A., et al.: Sage Mathematics Software (Version 4.6), The Sage Development Team, `http://www.sagemath.org`
21. Stinson, D.R.: Cryptography, Theory and Practice, 2nd edn. Chapman & Hall/CRC (2002)

Appendix

Here we give some experiential results for the angle θ between the two reduced vectors of a lattice of the form $L_q = \{(qX, 0), (AX, Y)\}$ after we applied Lagrange algorithm. We considered five random choices for the prime q of 160-bits. For each prime q we have chosen randomly 200 triples of X, Y, A in the interval $(\sqrt{q}, q)$, and then we applied Lagrange algorithm to the lattice L_q. In all cases, the angle between the two reduced vectors was not in the interval $I = [89.999^o, 90.001^o]$.

Now we fix (q, A, B) and we let X, Y to vary under the constraints $X > \sqrt{q}$ and $XY < q(\log q)^4$. In each row of the matrix below, we chose a random triple (q, A, B) with q having 160 bits and A, B random integers in the interval $(\sqrt{q}, q)$. We chose 200 random values of (X, Y) under the previous constraints and for each instance (X, Y) we executed Lagrange algorithm. Then we compute the pair (X, Y) which give us the closest angle to 90° (of the output vectors). Finally, for the specific value of (X, Y) we compare the real bound β_2 with the theoretical bound $C(\theta) \cdot (\log q)^2 \simeq C(\theta) \cdot 160^2$. We summarize the results in the matrix below. We observe that the bound β_2 is far smaller than the theoretical bound $C(\theta) \cdot (\log q)^2$. This suggests that the algorithm has complexity far smaller than the theoretical given under the assumptions that the angle of the output vectors must be in the interval I. Also, if for some choice of X, Y we got an angle in the interval I, then a new choice is very likely to give a new anlge not in I.

(q, A, B)	β_2	angle θ	$C(\theta) \cdot (\log q)^2$
1	14375	90.04^o	$1170 \cdot 160^2 = 259531 \cdot 10^2$
2	15460	89.99^o	$4678 \cdot 160^2 = 1197743 \cdot 10^2$
3	64888	90.001^o	$46782 \cdot 160^2 = 11976330 \cdot 10^2$
4	52999	89.984^o	$2924 \cdot 160^2 = 748636 \cdot 10^2$

Side Channel Attacks against Pairing over Theta Functions

Nadia El Mrabet*

LIASD - University Paris 8
elmrabet@ai.univ-paris8.fr

Abstract. In [17], Lubicz and Robert generalized the Tate pairing over any abelian variety and more precisely over Theta functions. The security of the new algorithms is an important issue for the use of practical cryptography. Side channel attacks are powerful attacks, using the leakage of information to reveal sensitive data. The pairings over elliptic curves were sensitive to side channel attacks. In this article, we study the weaknesses of the Tate pairing over Theta functions when submitted to side channel attacks.

Keywords: pairing based cryptography, Theta function, side channel attacks, differential power analysis, fault attacks.

1 Introduction

Since they appeared in cryptography, the efficient computation of pairings is a very active area of research. Originally defined over elliptic curves in Weierstrass model [19], pairings have been computed in other models of elliptic curves (for example Edwards [13], Huff [14], Jacobi [5]). They have also been studied in different systems of coordinates such as affine [16], Jacobian, projective, Chudnovsky [3] or in original representation of finite fields RNS [2]. The main algorithm to compute pairings is the Miller algorithm [19]. It is based on a double and add scheme. Several works aimed to reduce the number of iterations of Miller's algorithm and to develop the notion of optimal pairings [12]. In both Optimal Pairings [21] and Pairings Lattices [11] the authors present methods to find the Miller algorithm with the smallest number of iterations. All these works deal with a computation of pairing over elliptic (or hyper elliptic) curves.

The latest improvement in the computation of pairing was the description of efficient pairing computation in a more general case for any algebraic variety; and in particular pairings over Theta functions. In [17], Lubicz and Robert generalize the notion of the Weil and the Tate pairings to any abelian variety. To do so, they made an explicit link between the Weil and the Tate pairings and the intersection pairing on the degree 1 homology of an abelian variety. The result is a general definition of pairings and they explicit the formulas for the case of level 2 and

* The author wishes to acknowledge support from French project ANR INS 2012 SIMPATIC.

T. Muntean, D. Poulakis, and R. Rolland (Eds.): CAI 2013, LNCS 8080, pp. 132–146, 2013.

4 Theta functions in order to obtain the most efficient algorithm, considering time and memory consumption. Their algorithm to compute pairing is based on a Montgomery Ladder's approach.

Each time new formulas for pairing are proposed, the security and implementation of the new algorithms are an important issue for the use in practical cryptography. As for every cryptographic protocol constructed nowadays, the size of groups involved in pairing computation are chosen to be large enough to avoid the discrete logarithm attack. Consequently, pairing implementations are secured against mathematical attacks. Nevertheless, considering side channel attacks, we cannot predict if an algorithm is more or less secure than another given the representation of the groups. Weaknesses to side channel attacks of pairing based cryptography over elliptic curve have been highlighted [20,23,22,7,8]. Then, wondering if a pairing implemented in Theta function would be vulnerable to side channel attacks is an important issue for pairing based cryptography and this is the main objective of this contribution. The remaining of the article is organized as follows. The Section 2 is devoted to the definition of pairings over Theta functions. In Section 3 we describe the application of side channel attacks to pairing over Theta functions, we highlight the weaknesses of the pairing and provide countermeasures to secure the computation. We conclude in Section 4.

2 Pairings over Theta Function

This Section is a brief review of the results in [17]. We present the notations and background of Theta functions in Section 2.1. We give the definition of the Weil and the Tate pairings and of the algorithm to compute the Tate pairing in Section 2.2.

2.1 Background on Theta Function

Let $\mathbb{H}_g$ be the g dimensional Siegel upper-half space which is the set of $g \times g$ symmetric matrices Ω whose imaginary part is positive definite. For $\Omega \in \mathbb{H}_g$, let $\Lambda_\Omega = \Omega\mathbb{Z}^g \times \mathbb{Z}^g$ the lattice of $\mathbb{C}^g$ defined by Ω. If A is an abelian variety of dimension g over the number field K with a principal polarization then A is analytically isomorphic to $\mathbb{C}^g/\Lambda_\Omega$. Let $\Pi : \mathbb{C} \to \mathbb{C}^g/\Lambda_\Omega = A$ be the canonical projection. The classical theory of Theta functions gives a lot of functions on $\mathbb{C}^g$ that are pseudo-periodic with respect to Λ_Ω and can be used as a projective coordinate system for A. For $a, b \in \mathbb{Q}^g$, the Theta function with rational characteristics (a, b) is an analytic function on $\mathbb{C}^g \times \mathbb{H}_g$ given by

$$\theta\begin{bmatrix} a \\ b \end{bmatrix}(z, \Omega) = \sum_{n\in\mathbb{Z}^g} \exp\left[\Pi i^t(n+a).\Omega.(n+a) + 2\Pi i^t(n+a).(z+b)\right],$$

where ^{t} represents the transpose of a vector.

In order to describe the pseudo-periodicity relations verified by the Theta function, we introduce a certain pairing on $\mathbb{C}^g$. We have that $\mathbb{C}^g$ is isomorphic to $\mathbb{R}^{2g}$ via the map

$$\begin{cases} \mathbb{R}^{2g} \longrightarrow \mathbb{C}^g \\ (x_1, x_2) \longrightarrow \Omega x_1 + x_2. \end{cases}$$

For $\alpha, \beta \in \mathbb{R}^{2g}$, let $\alpha = (\alpha_1, \alpha_2)$ and $\beta = (\beta_1, \beta_2)$, we define $e_\Omega : \mathbb{R}^{2g} \to \mathbb{C}$ by $e_\Omega(\alpha, \beta) = \exp(2i\Pi(\alpha_1\beta_2 - \alpha_2\beta_1))$.

The pseudo periodicity of θ is given by

$$\theta \begin{bmatrix} a \\ b \end{bmatrix} (z + \Omega.m + n, \Omega) =$$

$$e_\Omega(\Omega a + b, \Omega m + n) \times e^{(-\Pi i^t m \Omega m - 2\Pi i^t m z)} \times \theta \begin{bmatrix} a \\ b \end{bmatrix} (z, \Omega).$$

A function f on $\mathbb{C}^g$ is Λ_Ω-Theta-periodic of level $l \in \mathbb{N}$ if for all $z \in \mathbb{C}^g$ and $m \in \mathbb{Z}^g$, we have

$$f(z+m) = f(z),\ f(z + \Omega.m) = \exp(-\Pi i l^t m.\Omega.m - 2\Pi i l^t z.m) f(z).$$

For any $l \in \mathbb{N}^\star$, the set $H_{\Omega,l}$ of Λ_Ω-quasi-periodic functions of level l is a finite dimensional $\mathbb{C}$-vector space whose basis can be given by the Theta functions with characteristics $\left(\theta \begin{bmatrix} 0 \\ b/l \end{bmatrix} (z, l^{-1}.\Omega)\right)_{b \in [0,\ldots,l-1]^g}$. If $l = k^2$, then an alternative basis of $H_{\Omega,l}$ is $\left(\theta \begin{bmatrix} a/k \\ b/k \end{bmatrix} (kz, \Omega)\right)_{a,b \in [0,\ldots,k-1]^g}$.

Once the level $l \in \mathbb{N}$ is fixed, the following conventions are adopted $\mathbb{Z}(\overline{l}) = (\mathbb{Z}/l\mathbb{Z})^g$ and for a point $z_P \in \mathbb{C}^g$ and $i \in \mathbb{Z}(\overline{l})$ let

$$\theta_i(z_P) = \theta \begin{bmatrix} 0 \\ i/l \end{bmatrix} (z_P, \Omega/l).$$

If $l = k^2$, for $i, j \in \mathbb{Z}(\overline{k})$, let $\theta_{i,j}(z_P) = \theta \begin{bmatrix} i/k \\ j/k \end{bmatrix} (k.z_P, \Omega)$.

Let $\widetilde{P}$ denote the element of $\mathbb{A}^{l^g}(\mathbb{C})$ with coordinates $\widetilde{P}_i = \theta_i(z_P)$. Let P be the associated point of A that will be be considered depending on the situation as embedded in $\mathbb{P}^{l^g-1}$ or as a point on the analytic variety $\mathbb{C}^g/\Lambda_\Omega$. For $n, l \in \mathbb{N}$, if n divides l then $\mathbb{Z}(\overline{n})$ will be considered as a subgroup of $\mathbb{Z}(\overline{l})$ via the morphism $x \to (l/n).x$. Let Ξ be the Theta divisor of level l on A, i.e. Ξ is the divisor of zero of $\left(\theta \begin{bmatrix} 0 \\ 0 \end{bmatrix} (z, l^{-1}\Omega)\right)$. There is an isogeny $\phi_l : A \to \widehat{A} = \mathrm{Pic}^0_A$, defined by $x \to \tau_x^* \Xi_l - \Xi_l$ where τ_x is the translation by x morphism on A. Let $A[l]$ be the kernel of ϕ_l. Let $K(A)$ be the function field of A and (f) be the divisor of a function $f \in K(A)$. We then present the definition of the Weil and the Tate pairing.

2.2 Definition and Computation of Pairings over Theta Function

The Weil Pairing. For $\Omega \in \mathbb{H}_g$, let $A = \mathbb{C}^g/\Lambda_\Omega$ be the complex abelian variety and denote by $\pi : \mathbb{C}^g \to A$ the natural projection. Let l be a positive integer and μ_l be the subgroup of $\mathbb{C}^\star$ of lth roots of unity. For $z_P, z_Q \in \mathbb{C}^g$, let P, Q be the associated points of A. The Weil pairing is the map $e_W : A[l] \times A[l] \to \mu_l$, $(P,Q) \to e_\Omega(z_P, z_Q)^l$. The value $e_W(P,Q)$ does not depend on the choice of z_P and z_Q representing P and Q and e_W is a non-degenerate skew linear form. This pairing can be expressed using certain Theta functions.

Definition 1. *Let $\Omega \in \mathbb{H}_g$, $a, b \in \mathbb{Q}^g$, l be a positive integer and let z_P, $z_Q \in \mathbb{C}^g$ be such that $l.z_P = l.z_Q = 0 \mod \Lambda_\Omega$. Let $P = \pi(z_P)$ and $Q = \pi(z_Q)$. Let*

$$L(z_P, z_Q) = \frac{\theta\begin{bmatrix} a \\ b \end{bmatrix}(l.z_P + z_Q, \Omega) \;\; \theta\begin{bmatrix} a \\ b \end{bmatrix}(0, \Omega)}{\theta\begin{bmatrix} a \\ b \end{bmatrix}(z_Q, \Omega) \;\; \theta\begin{bmatrix} a \\ b \end{bmatrix}(l.z_P, \Omega)},$$

$$R(z_P, z_Q) = \frac{\theta\begin{bmatrix} a \\ b \end{bmatrix}(l.z_Q + z_P, \Omega) \;\; \theta\begin{bmatrix} a \\ b \end{bmatrix}(0, \Omega)}{\theta\begin{bmatrix} a \\ b \end{bmatrix}(z_P, \Omega) \;\; \theta\begin{bmatrix} a \\ b \end{bmatrix}(l.z_Q, \Omega)}.$$

If $L(z_P, z_Q)$ and $R(z_P, z_Q)$ are well defined and non null, then

$$e_W(P,Q) = L(z_P, z_Q)^{-1}.R(z_P, z_Q) = e_\Omega(z_P, z_Q)^l.$$

The algorithm to compute the Weil pairing is composed of four calls to the function ScalarMult.

Theorem 1. *Suppose that n and l are relatively prime. For X, $Y \in A(\overline{K})$, denote by $\widetilde{X}$, $\widetilde{Y}$, $\widetilde{X+Y}$ any affine lifts of X, Y and $X+Y$. For $i \in \mathbb{Z}(\overline{n})$, let $\widetilde{X}_i$ be the ith coordinate of $\widetilde{X}$. For $\in \mathbb{N}$ and $i \in \mathbb{Z}(\overline{n})$, let*

$$f_T(\widetilde{X}, \widetilde{Y}, \widetilde{X+Y}, \widetilde{0}, l, i) = \frac{ScalarMult(\widetilde{X+Y}, \widetilde{X}, \widetilde{Y}, \widetilde{0}, l)_i}{ScalarMult(\widetilde{X}, \widetilde{X}, \widetilde{0}, \widetilde{0}, l)_i} \frac{\widetilde{0}_i}{\widetilde{Y}_i}.$$

Then for P, $Q \in A[l]$ and $i \in Z(\overline{n})$, we have

$$e_W(P,Q)^n = f_T(\widetilde{P}, \widetilde{Q}, \widetilde{P+Q}, \widetilde{0}, l, i)^{-1} f_T(\widetilde{Q}, \widetilde{P}, \widetilde{P+Q}, \widetilde{0}, l, i),$$

whenever the right hand side is well defined.

The Tate Pairing. For efficiency reasons, the pairing that will be implemented is the Tate pairing (or a variant of the Tate pairing) so we only consider the side channel attacks against the Tate pairing. Let K be a number field and suppose that A is defined over K. Recall that $l \in \mathbb{N}$ is the level of the Theta function and it is fixed once for all. In this section, we suppose that $\mu_l \subset K$ and that $A[l]$ is rational over K. Let $\overline{K}$ be the algebraic closure of K and $G = \mathrm{Gal}(\overline{K}/K)$. Let $\delta_1 : K^\star/K^{\star l} \to \mathrm{Hom}(G, \mu_l)$ (respectively $\delta_2 : A(K)/[l]A(K) \to \mathrm{Hom}(G, A[l])$) be the connecting morphism of the Galois cohomology long exact sequence associated to the Kummer exact sequence (respectively to the short exact sequence $0 \to A[l] \to A(\overline{K}) \to A(\overline{K}) \to 0$). There exists a bilinear application often referred to as the Tate pairing $e_T : A(K)/[l]A(K) \times A[l] \to K^\star/K^{\star l}$ such that for $(P,Q) \in A(K)/[l]A(K) \times A[l]$, $e_W(\delta_2(P), Q) = \delta_1(e_T(P,Q))$, where e_W is the Weil pairing over Theta functions.

Definition 2. *Let K be a number field and let A be a dimension g abelian variety over K. Let $\Omega \in \mathbb{H}_g$ be such that A is analytically isomorphic to $\mathbb{C}^g/\Lambda_\Omega$. Let a, $b \in \mathbb{Q}^g$ and l be a positive integer. Let $P \in A(K)/[l]A(K)$, $Q \in A[l](K)$ and z_P, $z_Q \in \mathbb{C}^g$ such that $\pi(z_P) = P$ and $\pi(z_Q) = Q$ where $\pi : \mathbb{C}^g \to A$ is the natural projection* [1]*. Suppose that z_P, z_Q and z_{P+Q} are chosen such that*

$$\frac{\theta\begin{bmatrix}0\\0\end{bmatrix}(z_P + z_Q, \Omega)}{\theta\begin{bmatrix}0\\0\end{bmatrix}(z_P, \Omega)} \frac{\theta\begin{bmatrix}0\\0\end{bmatrix}(0, \Omega)}{\theta\begin{bmatrix}0\\0\end{bmatrix}(z_Q, \Omega)} \in K^\star,$$

then

$$e_T(P,Q) = \frac{\theta\begin{bmatrix}0\\0\end{bmatrix}(l.z_Q + z_P, \Omega)}{\theta\begin{bmatrix}0\\0\end{bmatrix}(z_P, \Omega)} \frac{\theta\begin{bmatrix}0\\0\end{bmatrix}(0, \Omega)}{\theta\begin{bmatrix}0\\0\end{bmatrix}(l.z_Q, \Omega)}.$$

The Algorithm for Computation of Pairings over Theta Functions. Let n, $l \in \mathbb{N}$ and assume that 2 divides n and that $\gcd(n,l) = 1$. Let A be an abelian variety over $\mathbb{C}$ with period matrix Ω. We represent A as a closed subvariety of $\mathbb{P}^{n^g-1}$ by the way of level n Theta functions and suppose that this embedding is defined over K. Let $\widetilde{A}$ be the pullback of A via the natural projection $\kappa : A^{n^g} \to \mathbb{P}^{n^g-1}$. For $P \in A$, let $\widetilde{P}$ be an affine lift of P that is a point of A^{n^g} such that $\kappa(\widetilde{P}) = P$. Important ingredients of the algorithm in [17] are the Riemann addition formulas. Suppose that the Theta null point $\widetilde{0} = (\theta_i(0))_{i\in\mathbb{Z}(\overline{n})}$ is known. From [17, Theorem 1], we can construct an algorithm that takes as input $\widetilde{P} = \left(\widetilde{P_i}\right)_{i\in\mathbb{Z}(\overline{n})}$, $\widetilde{Q} = \left(\widetilde{Q_i}\right)_{i\in\mathbb{Z}(\overline{n})}$ and $\widetilde{P-Q} = \left(\widetilde{(P-Q)_i}\right)_{i\in\mathbb{Z}(\overline{n})}$ and outputs $\widetilde{P+Q} = \left(\widetilde{(P+Q)_i}\right)_{i\in\mathbb{Z}(\overline{n})}$. Let $\widetilde{P+Q} = \mathrm{PseudoAdd}(\widetilde{P}, \widetilde{Q}, \widetilde{P-Q})$. Using the Riemann addition formulas, if $n = 4$, the projective point $P+Q$ can be recovered from P and Q. As a consequence, with the knowledge of $\widetilde{P}$, $\widetilde{Q}$

[1] By abuse of notation we use P, Q to denote the corresponding points of an algebraic and analytic model of A.

and $\widetilde{P-Q}$ there is a unique affine point $\widetilde{P+Q}$ above $P+Q$ that satisfies the addition formulas from [17, Theorem 1]. The result is extended in [17] for $n=2$.

Chaining the algorithm PseudoAdd in a classical Montgomery Ladder yields an algorithm that takes as inputs $\widetilde{Q}$, $\widetilde{P+Q}$, $\widetilde{P}$, $\widetilde{0}$ and an integer l and outputs $\widetilde{P+lQ}$.

Let $\widetilde{P+lQ} = \text{ScalarMult}(\widetilde{P+Q}, \widetilde{Q}, \widetilde{P}, \widetilde{0}, l)$. In particular,

$$l\widetilde{P} = \text{ScalarMult}(\widetilde{P}, \widetilde{P}, \widetilde{0}, \widetilde{0}, l).$$

The output of the function ScalarMult is independent on the particular chain of PseudoAdd calls it uses.

Theorem 2. *Suppose that n and l are relatively prime. For X, $Y \in A(\overline{K})$, denote by $\widetilde{X}$, $\widetilde{Y}$, $\widetilde{X+Y}$ any affine lifts of X, Y and $X+Y$. For $i \in \mathbb{Z}(\overline{n})$, let $\widetilde{X}_i$ be the ith coordinate of $\widetilde{X}$. For $\in \mathbb{N}$ and $i \in \mathbb{Z}(\overline{n})$, let*

$$f_T(\widetilde{X}, \widetilde{Y}, \widetilde{X+Y}, \widetilde{0}, l, i) = \frac{\mathit{ScalarMult}(\widetilde{X+Y}, \widetilde{X}, \widetilde{Y}, \widetilde{0}, l)_i}{\mathit{ScalarMult}(\widetilde{X}, \widetilde{X}, \widetilde{0}, \widetilde{0}, l)_i} \frac{\widetilde{0}_i}{\widetilde{Y}_i}.$$

Then, for $P \in A(K)/[l]A(K)$, $Q \in A[l]$, if we suppose that $\widetilde{0}$, $\widetilde{P}$, $\widetilde{Q}$ and $\widetilde{P+Q}$ are affine lifts of 0, P, Q and $P+Q$ with coordinates in K, then we have for $i \in \mathbb{Z}(\overline{n})$,

$$e_T(P,Q)^n = f_T(\widetilde{Q}, \widetilde{P}, \widetilde{P+Q}, \widetilde{0}, l, i),$$

whenever the right hand side is well defined.

For example, let E be an elliptic curve defined by $\Omega \in \mathbb{H}_1$ and $\Omega' = \Omega/2$. Put

$$a = \theta \begin{bmatrix} 0 \\ 0 \end{bmatrix} (0, \Omega'); \quad b = \begin{bmatrix} 0 \\ 1/2 \end{bmatrix} (0, \Omega');$$

$$\mathcal{A} = \theta \begin{bmatrix} 0 \\ 0 \end{bmatrix} (0, 2\Omega'); \quad \mathcal{B} = \theta \begin{bmatrix} 1/2 \\ 0 \end{bmatrix} (0, 2\Omega').$$

The algorithm ScalarMult is composed by a doubling algorithm and a differential addition algorithm given in Figure 1.

3 Side Chanel Attacks against the Tate Pairing over Theta Function

3.1 Side Channel Attacks in Pairing Based Cryptography

The general scheme of an identity based encryption is recalled in [9]. The important point is that to decipher a message using an Identity Based Protocol,

Doubling Algorithm
Input: A point $P = (x_P : z_P)$.
Output: The double $2P = (x_{2P} : z_{2P})$
1. $x_0 = (x_P^2 + z_P^2)^2$
2. $z_0 = \frac{A^2}{B^2}(x_P^2 - z_P^2)^2$
3. $x_{2P} = x_0 + z_0$
4. $z_{2P} = \frac{a}{b}(x_0 - z_0)$
5. Return $(x_{2P} : z_{2P})$

Differential Addition Algorithm
Input: Two points $P = (x_P : z_P)$ and $Q = (x_Q, z_Q)$ on E, $R = (x_R : z_R) = P - Q$, with $x_R z_R \neq 0$.
Output: The point $P + Q = (x_{P+Q} : z_{P+Q})$
1. $x_0 = (x_P^2 + z_P^2)(x_Q^2 + z_Q^2)$
2. $z_0 = \frac{A^2}{B^2}(x_P^2 - z_P^2)(x_Q^2 - z_Q^2)$
3. $x_{P+Q} = (x_0 + z_0)/x_R$
4. $z_{P+Q} = (x_0 - z_0)/z_R$
5. Return $(x_{P+Q} : z_{P+Q})$

Fig. 1. Doubling and Differential Addition Algorithms

a computation of a pairing between a private key and a public message is performed. Side channel attacks are powerful attacks using the leakage of information during the execution of a cryptographic protocol. As soon as the algorithm involves a computation between a secret and a public data, side channel attacks can be applied in order to reveal the secret, or information about the secret. The particularity of identity based cryptography is that an attacker can know the algorithm used, the number of iterations and the exponent. The secret is only one of the arguments of the pairing. We describe here two attacks, namely the Differential Power Analysis (DPA) and the fault attack. There are other side channel attacks, but the popular ones are either a generalization of the DPA (DEMA, CPA) or fault attacks.

3.2 The Possible Targets

If we compare the efficiency of the Tate and of the Weil pairings, the former is more efficient than the later at least for the security levels considering today, when pairings are computed using a Miller's algorithm. In the case of Theta functions, the algorithmic complexity of the Tate pairing consists in two applications of the function ScalarMult, while the Weil pairing consists in four applications of this function. It is quite evident that the Tate pairing over Theta function will always be more efficient than the Weil pairing over the Theta function. So we study only the weakness of the Tate pairing considering side channel attacks. Nevertheless, the attacks described for the Tate pairing can easily be adapted to the Weil pairing. As a consequence the countermeasure proposed here must be considered also for the implementation of the Weil pairing.

The Tate pairing is composed of two applications of ScalarMult. First of all, we focus on side channel attacks against one application of ScalarMult and after that we will consider side channel attacks against the Tate pairing. The same argument can provide the result of side channel attacks against the Weil pairing, or any optimizations of the Tate pairing namely Ate, twisted Ate or optimal pairings. The function ScalarMult is a Montgomery Ladder composed by the Doubling and Differential Addition algorithms at each step. When the secret is the exponent this algorithm is an efficient countermeasure to side channel attacks. In the case of pairing based cryptography, the secret is not the exponent but one of the parameters of the Mongtomery Ladder algorithm. Consequently, the analysis considering side channel attacks against Montgomery's Ladder for the classical use in cryptography (efficient exponentiation) is no more available. We analyze the weaknesses of the algorithm to compute pairing using Theta functions. We will focus on the DPA and on the fault attack. The consideration of the DPA includes also the consideration of the Correlation Power Analysis (CPA) and the Differential Electromagnetic Attack (DEMA). Indeed the DEMA works exactly like the DPA and the CPA is an improvement of the DPA.

3.3 Differential Power Analysis Attack and Generalization

In order to simplify the explanation, we describe here only the differential power analysis (DPA) attack. As the concept is the same for all differential attacks, we include in the same family the differential power analysis (DPA) and the differential electromagnetic attack (DEMA) [4]. Further on, correlation attack [23] is just a form of DPA using the particular side-channel distinguisher i.e. Pearson correlation.

We now introduce some theoretical issues that allow the reader to understand the principle underlying the DPA attack, more details can be found in [15,18]. We consider the output of a gate whose state depends on both the plain text to be ciphered (primary inputs) and the secret key. It is called the target node. We consider now a sequence of input patterns $P_0, P_1, \ldots, P_n$ that generate the transitions $T_1(P_0 \to P_1), T_2(P_1 \to P_2), \ldots, T_n(P_{n-1} \to P_n)$ on the circuit primary inputs. A logic simulation of the circuit while monitoring the target node allows classifying these input transitions in two sets, according to a guess on the key:

- P_A, composed by the transitions that make the target node to commute from 0 to 1 and therefore that make the target gate to consume current;
- P_B, composed by the transitions that do not lead the target gate to participate to the power consumed by the circuit (i.e., transitions from 0 to 0, 1 to 1 and 1 to 0 on the target node).

Figure 2 represents the power consumption of the device when stimulated by numerous input vectors. We assume here that the guess on the secret key is correct. In other word, the simulation is performed with the key actually used in the circuit from which power consumptions are collected. Each rectangle represents the total power consumed by the circuit when a new vector is applied to the inputs. In this figure and just for clarity of explanation, the power consumption is represented by a rectangle corresponding to the average of the consumption over the transition time. The set of transitions on the circuit inputs is splitted in the two sets: in the left part there are the P_A transitions and the related consumptions while in the right part there are the P_B transitions and their corresponding consumptions. A part of the power consumption related to the transitions belonging to P_A is due to the power consumed by the target gate (shaded rectangles). Obviously, the commutation from 0 to 1 of non-target nodes also

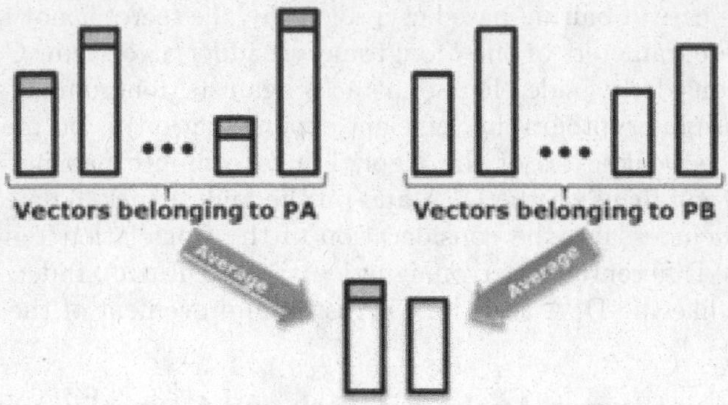

Fig. 2. Power consumption after pattern partitioning

contributes to the power consumption of the circuit but input transitions that lead to such commutations are assumed to be evenly distributed to sets P_A and P_B. If a large number of transitions are considered, mean consumptions related to sets P_A and P_B are almost equal, except for the contribution of the target node. In other words, since the two sets are classified in such a way that the set P_A always leads to a component of power consumption that is not present in the set P_B, the difference between the two mean powers computed from set P_A and set P_B must show a noticeable difference.

During a DPA attack, the target node is chosen in such a way that it depends on a small part of the key only, so that all the key guesses can be considered.

For each key guess, the two sets P_A and P_B are created according to the results of the logic simulation and the key guess under evaluation. The power mean values are calculated for each set using the simulated power traces of the circuit under attack for each transition. Finally, the differences of the mean values of the two sets are calculated. When the key guess is correct (and only in this case), P_A actually includes the input transitions that lead to a transition 0 to 1 on the target node while P_B does not include any of these transitions. The difference between the mean power obtained from P_A and P_B can be observed in this case. On the contrary, when the curves are classed in P_A or P_B independently from the actual value of the secret key, the two average curves do not present any noticeable difference. The classification process is illustrated in Figure 3 where K_x is assumed to be the correct key, the one actually used during ciphering.

3.4 DPA Attack

The computations sensitive to the DPA attack are the one involving the coordinates of the points P and Q. As a consequence, the DPA attack could only be done in the addition step. Indeed, the doubling step does not involve any computation between the coordinates of P and Q, the operations are multiplications by constant.

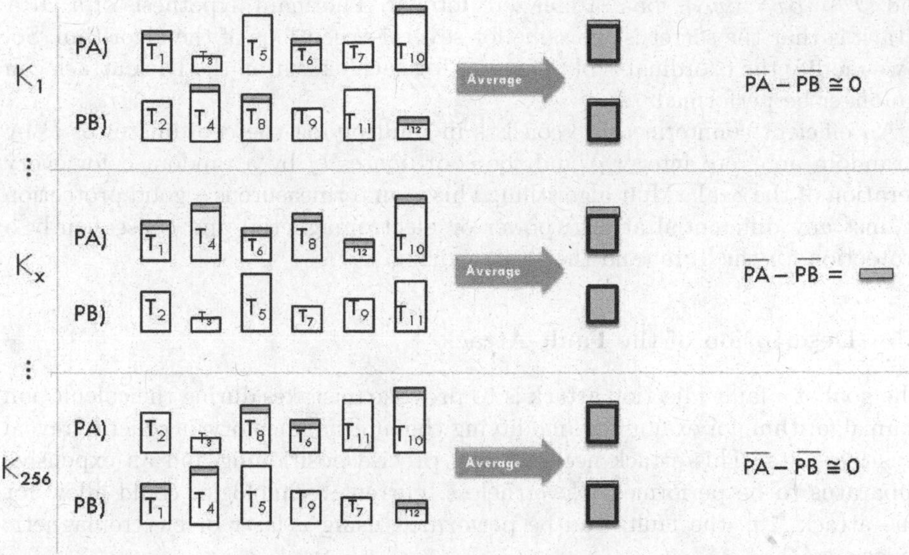

Fig. 3. Pattern classification for several key guesses

Without protection, the DPA attack is a threat against the addition step, whenever the secret is (point P or Q). According to the algorithm of ScalarMult, the argument Q of the pairing is in fact a multiple of the point Q and the point P is fixed.

The target of the DPA attack is the computation of x_0 and z_0 in the differential addition algorithm. In order to compute x_0, we have to perform a multiplication between $(x_P^2 + z_P^2)$ and $(x_Q^2 + z_Q^2)$. Suppose that the point P is public and that Q is secret, we know the value of $(x_P^2 + z_P^2)$ and the value $(x_Q^2 + z_Q^2)$ is secret. We perform the DPA attack against the multiplication $(x_P^2 + z_P^2) \times (x_Q^2 + z_Q^2)$. Assuming that the multiplication is implemented using the Schoolbook method, the guesses on the value $(x_Q^2 + z_Q^2)$ can be done by words of 32 or 64 bits and begin by the less significant bits. The result of the DPA attack against x_0 is $(x_Q^2 + z_Q^2)$.

During the computation of x_0 the Differential Addition Algorithm would give us the value $\widetilde{x_Q}^2 + \widetilde{z_Q}^2$. In parallel, another DPA attack during the computation of z_0 would give the value $\widetilde{x_Q}^2 - \widetilde{z_Q}^2$. Once we have these two values, it is easy to extract $\widetilde{x_Q}^2$ and $\widetilde{z_Q}^2$, which gives 4 possible couples for the coordinates of point Q.

As the Differential Addition Algorithm is symmetric in the coordinates of P and Q, the same attack is efficient if the point P is secret and Q is public.

Classical countermeasures presented in the case of pairing over elliptic curves can easily be adapted for the computation of pairings over Theta functions [8]. A native countermeasure is the homogeneity of the projective coordinates. Indeed, the point $P = (x_P : z_P)$ is also the point $(\lambda x_P : \lambda z_Q)$, for λ a non zero integer and $Q = (\beta x_Q, \beta y_Q)$, for β a non zero integer. The main hypothesis of a DPA attack is that the secret is the same for several executions of the algorithm. So, if we modify the coordinates of the secret for each execution the DPA attack can no longer be performed.

An efficient countermeasure consists in multiplying the coordinates of P by a random non zero integer λ and the coordinates Q by a random β for every iteration of the ScalarMult algorithm. This countermeasure is a good protection against any differential attack (power or electromagnetic) and consequently a protection for the Tate (and the Weil) pairing.

3.5 Description of the Fault Attack

The goal of a fault injection attack is to provoke mistakes during the calculation of an algorithm, for example by modifying the internal memory, in order to reveal sensitive data. This attack needs a very precise positioning and an expensive apparatus to be performed. Nevertheless, current technologies could allow for this attack [10]. the faults can be performed using a laser or electromagnetic emissions [4].

We follow the scheme of attack described in [20] and completed in [6]. We assume that the pairing is used during an Identity Based Protocol, the secret point is introduced in a smart card or an electronic device and is a parameter of the pairing. In order to find the secret, we modify the number of iterations in the Tate pairing algorithm by the following way.

First of all, we have to find the flip-flops belonging to the counter of the number of iterations (i.e. $\log_2(s)$) in the Tate pairing algorithm. This step can be done by using reverse engineering procedures. In classical architecture, the counter is divided into small pieces of 32 or 64 bits (according to the size of a word). To find it, we make one normal execution of the algorithm, without any fault. Then we choose one piece of the counter and provoke disturbances in order to modify it and consequently the number of iterations of the algorithm. For example the disturbance can be induced by a laser [1]. Nowadays lasers are thin enough to make this attack realistic [10]. Counting the clock cycles, we are able to know how many iterations the Tate pairing loop has done. Each time, we record the value of the pairing loop and the number of iterations we made.

3.6 Fault Attack

State of the Art. The principle of fault attacks in pairing based cryptography consists to force the algorithm to stop by reducing the number of iterations and by finding the results of two consecutive iterations τ and $\tau + 1$. The results of these two executions give equations that allow to find the secret. In the case of pairings over Theta function, the fault attack consists in finding one the coordinates involved during the computation of ScalarMult$(\widetilde{P+Q}, \widetilde{Q}, \widetilde{P})$. The ScalarMult algorithm is composed by the doubling and differential addition algorithms, the result of of ScalarMult are the coordinates of $\widetilde{P+lQ}$. The fault attack consists in reducing the number of iteration of ScalarMult. To do so, we can use a laser or electromagnetic emissions to locally modify the register storing l. The target of this attack is then a smart card or a FPGA. Let τ be the reduced number of iteration performed by ScalarMult. In practice τ can be recovered using the number of clock cycles made by the algorithm. Indeed, we know the binary decomposition of l, we are then able to find when the algorithm stops and how many iterations were done. Let j be the integer composed by the τ most significant bits of l, which is public. The fault attack for pairing over Theta function is easier than the classical fault attack in pairing based cryptography. We need only one fault and the result of this faulty execution to find the secret involved in the ScalarMult algorithm

The result of the pairing is the coordinates of the point $\widetilde{P+lQ}$. We can suppose that we obtain one of the two coordinates, for example the coordinate z. With the z coordinate of the result, we are able to recover the secret argument of the pairing computations.

Suppose that we can recover the coordinate z of the point $\widetilde{P+jQ}$, for $j < l$. As the points P and Q are of order l by construction, the result of the pairing itself cannot give us information. That is why we need to provoke a fault reducing the number of iterations of the ScalarMult algorithm.

Let $z_1 = z_{P+jQ}$, where j is a known integer. The equation of z_1 is the following

$$z_1 = \left[(x_j^2 + z_j^2)(x_P^2 + z_P^2) - \frac{\mathcal{A}^2}{\mathcal{B}^2}(x_j^2 - z_j^2)(x_P^2 - z_P^2) \right] \frac{1}{\overline{z}}, \tag{1}$$

where

- $P = (x_P, z_P) = (\overline{x}, \overline{z})$ (with the notations introduced above)
- $(j-1)Q = (x_j, z_j)$
- $P + jQ = (x_1, z_1)$
- $\mathcal{A}$ and $\mathcal{B}$ are constants.

We first describe the attack of the algorithm ScalarMult, before considering the fault attack against the whole Tate pairing algorithm.

If the Secret Is the Point P. Suppose that the point P is secret. The fault attack provide us z_1, the values A, B, x_j and z_j are public. All together, they verify the equation

$$\boldsymbol{\lambda} z_P = \boldsymbol{\beta}(x_P^2 + z_P^2) + \boldsymbol{\gamma}(x_P^2 - z_P^2),$$

where the data in bold $(\boldsymbol{\lambda}, \boldsymbol{\beta}, \boldsymbol{\gamma})$ are known. The coordinates x_P and z_P are the values we are looking for.

The point P is given in projective coordinates, this equality is correct for any representative of the point P, i.e. for any $\alpha \neq 0$ we have that

$$\boldsymbol{\lambda}(\alpha z_P) = \boldsymbol{\beta}((\alpha x_P)^2 + (\alpha z_P)^2) + \boldsymbol{\gamma}((\alpha x_P)^2 - (\alpha z_P)^2).$$

As the coordinates of P are such that $x_P z_P \neq 0$, we can consider that $\alpha = \frac{1}{z_P}$ and write the equation

$$\boldsymbol{\lambda} = \boldsymbol{\beta}((x'_P)^2 + 1) + \boldsymbol{\gamma}((x'_P)^2 - 1),$$

which leads to

$$(x'_P)^2 = \frac{\boldsymbol{\lambda} - \boldsymbol{\beta} + \boldsymbol{\gamma}}{\boldsymbol{\beta} - \boldsymbol{\gamma}}.$$

Up to the sign, we find one coordinate of a representative of the point P and from that point we can find the secret.

if the secret is the point Q. The formulas are symmetric in the coordinates of P and jQ. Following the same scheme, we obtain z_1 for j not equal to the order of Q and that gives the coordinates of a representative of jQ, knowing j. To find the coordinates of Q, we just have to compute the inverse of j mod (l) and after that we can recover the coordinates of the point Q.

The condition to perform the fault attack when Q is secret is to stop the computation before $j = l$, as Q is a point of order l. This is a simplification of the fault attack against the pairing considering Miller's algorithm, because we only need one faulty execution of ScalarMult.

Considering the computation of the Tate Pairing. Recall that the algorithm to compute the Tate pairing is

$$e_T = \frac{\text{ScalarMult}(\widetilde{P+Q}, \widetilde{Q}, \widetilde{P}, l)_i}{\text{ScalarMult}(\widetilde{Q}, \widetilde{Q}, \widetilde{0}, l)_i} \frac{\widetilde{0}_i}{\widetilde{P}_i}.$$

The attacks described above for ScalarMult can be directly adapted to the Tate pairing (and also to the Weil pairing). For efficiency reasons, the computation of ScalarMult$(\widetilde{P+Q}, \widetilde{Q}, \widetilde{P}, l)_i$ and ScalarMult$(\widetilde{Q}, \widetilde{Q}, \widetilde{0}, l)_i$ would certainly be implemented in parallel. As a consequence, the fault attack forces the algorithm to stop after the same number of iterations and the result

$$\text{ScalarMult}(\widetilde{P+Q}, \widetilde{Q}, \widetilde{P}, j)_i \text{ and ScalarMult}(\widetilde{Q}, \widetilde{Q}, \widetilde{0}, j)_i,$$

for the same integer j. For both cases, either P secret or either Q, the homogeneity of projective coordinates is a trapdoor that gives information about the secret. Let P be the secret point and Q be public, then the coordinate $\widetilde{P}_i$ is also secret, but the homogeneity of the projective coordinates allows us to consider that for example the z coordinate is set to 1, exactly like in the attack described above. We just have to be careful to set the same coordinate to 1 in both calls to ScalarMult, the z one for example. The Equation (1) would give a slightly different system but linear and easily solvable. The method is the same if the point Q is secret.

Countermeasure to the Fault Attack. Considering that the fault attack uses the homogeneity of the coordinates, the countermeasure to the DPA attack is clearly not sufficient. We have to present another countermeasure and this countermeasure must protect the pairing algorithm from the fault and the DPA attacks. So, we have to modify the coordinates of the point P and Q for every pairing computation. A solution would be to use the bilinearity of the pairing [8]. Indeed, if we compute the Tate pairing between the points P and Q, the bilinearity induces that

$$e_T(P, Q) = e_T\left(\delta P, (\delta^{-1} \mod (l))Q\right),$$

for a non zero integer δ. The cost of this countermeasure consists in two exponentiations over t he variety $A(K)$.

4 Conclusion

We analyze the weaknesses of the pairings over Theta function with respect to side channel attacks. We consider the differential power analysis and the fault attack. The scheme of the differential power analysis embraces the differential electromagnetic attack and the correlation power analysis. The ScalarMult algorithm is sensitive to the DPA attack, but the homogeneity of the projective coordinates provides a native countermeasure. Unfortunately, the homogeneity is a trapdoor for the fault attack. The fault attack against pairing over Theta functions is easier than in the case of pairings using the Miller's algorithm. We only need one fault to recover the secret. As the homogeneity of the coordinates is no longer a countermeasure, we present an alternative countermeasure. This countermeasure relies on the bilinearity of pairings and is efficient for all side channel attacks.

Acknowledgment. The author wishes to thank the anonymous referees for their helpful remarks and comments.

References

1. Anderson, R., Kuhn, M.: Tamper resistance: a cautionary note. In: WOEC 1996: Proceedings of the Second USENIX Workshop on Electronic Commerce, pp. 1–11. USENIX Association, Berkeley (1996)
2. Cheung, R.C.C., Duquesne, S., Fan, J., Guillermin, N., Verbauwhede, I., Yao, G.X.: Fpga implementation of pairings using residue number system and lazy reduction. In: Preneel, B., Takagi, T. (eds.) CHES 2011. LNCS, vol. 6917, pp. 421–441. Springer, Heidelberg (2011)
3. Costello, C., Lange, T., Naehrig, M.: Faster pairing computations on curves with high-degree twists. In: Nguyen, P.Q., Pointcheval, D. (eds.) PKC 2010. LNCS, vol. 6056, pp. 224–242. Springer, Heidelberg (2010)
4. De Mulder, E., Örs, S.B., Preneel, B., Verbauwhede, I.: Differential power and electromagnetic attacks on a FPGA implementation of elliptic curve cryptosystems. Comput. Electr. Eng. 33(5-6), 367–382 (2007)

5. Duquesne, S., Fouotsa, E.: Tate pairing computation on jacobi's elliptic curves. In: Abdalla, M., Lange, T. (eds.) Pairing 2012. LNCS, vol. 7708, pp. 254–269. Springer, Heidelberg (2013)
6. El Mrabet, N.: What about vulnerability to a fault attack of the miller's algorithm during an identity based protocol? In: Park, J.H., Chen, H.-H., Atiquzzaman, M., Lee, C., Kim, T.-h., Yeo, S.-S. (eds.) ISA 2009. LNCS, vol. 5576, pp. 122–134. Springer, Heidelberg (2009)
7. El Mrabet, N., Di Natale, G., Flottes, M.L.: A practical differential power analysis attack against the miller algorithm. In: PRIME 2009 - 5th Conference on Ph.D. Research in Microelectronics and Electronics, Circuits and Systems Magazine. IEEE Xplore (2009)
8. El Mrabet, N., Page, D., Vercauteren, F.: Fault attacks on pairing based cryptography: A state of the art. In: Joye, M., Tunstall, M. (eds.) Fault Analysis in Cryptography. Information Security and Cryptography, pp. 221–236. Springer (2012)
9. Galbraith, S.: Pairings in Advances in Elliptic Curve Cryptography. London Mathematical Society Lecture Note Series, vol. 317. Cambridge University Press (2005)
10. Habing, D.: The use of lasers to simulate radiation-induced transients in semiconductor devices and circuits. IEEE Transactions on Nuclear Science 12(5), 91–100 (1965)
11. Hess, F.: Pairing Lattices. In: Galbraith, S.D., Paterson, K.G. (eds.) Pairing 2008. LNCS, vol. 5209, pp. 18–38. Springer, Heidelberg (2008)
12. Hess, F., Smart, N., Vercauteren, F.: The Eta Pairing Revisited, vol. 52, pp. 4595–4602 (2006)
13. Ionica, S., Joux, A.: Another approach to pairing computation in Edwards coordinates. In: Chowdhury, D.R., Rijmen, V., Das, A. (eds.) INDOCRYPT 2008. LNCS, vol. 5365, pp. 400–413. Springer, Heidelberg (2008)
14. Joye, M., Tibouchi, M., Vergnaud, D.: Huff's model for elliptic curves. In: Hanrot, G., Morain, F., Thomé, E. (eds.) ANTS-IX. LNCS, vol. 6197, pp. 234–250. Springer, Heidelberg (2010)
15. Kocher, P.C., Jaffe, J., Jun, B.: Differential power analysis. In: Wiener, M. (ed.) CRYPTO 1999. LNCS, vol. 1666, pp. 388–397. Springer, Heidelberg (1999)
16. Lauter, K., Montgomery, P., Naehrig, M.: An analysis of affine coordinates for pairing computation. In: Joye, M., Miyaji, A., Otsuka, A. (eds.) Pairing 2010. LNCS, vol. 6487, pp. 1–20. Springer, Heidelberg (2010)
17. Lubicz, D., Robert, D.: Efficient pairing computation with theta functions. In: Hanrot, G., Morain, F., Thomé, E. (eds.) ANTS-IX. LNCS, vol. 6197, pp. 251–269. Springer, Heidelberg (2010)
18. Mangard, S., Oswald, E., Popp, T.: DPA book. Graz University of Technology (2007)
19. Miller, V.S.: The weil pairing, and its efficient calculation. Journal of Cryptology 17(4), 235–261 (2004)
20. Page, D., Vercauteren, F.: A fault attack on pairing-based cryptography. IEEE Trans. Computers 55, 1075–1080 (2006)
21. Vercauteren, F.: Optimal pairings. IEEE Transactions on Information Theory 56(1), 455–461 (2010)
22. Whelan, C., Scott, M.: The importance of the final exponentiation in pairings when considering fault attacks. In: Takagi, T., Okamoto, T., Okamoto, E., Okamoto, T. (eds.) Pairing 2007. LNCS, vol. 4575, pp. 225–246. Springer, Heidelberg (2007)
23. Whelan, C., Scott, M.: Side channel analysis of practical pairing implementations: Which path is more secure? In: Nguyên, P.Q. (ed.) VIETCRYPT 2006. LNCS, vol. 4341, pp. 99–114. Springer, Heidelberg (2006)

On the Efficient Generation of Generalized MNT Elliptic Curves

Georgios Fotiadis and Elisavet Konstantinou

Department of Information and Communication Systems Engineering,
University of the Aegean, 83200 Karlovassi, Samos, Greece
{gfotiadis,ekonstantinou}@aegean.gr

Abstract. Finding suitable elliptic curves for pairing-based cryptosystems is a crucial step for their actual deployment. Miyaji, Nakabayashi and Takano [12] (MNT) were the first to produce ordinary pairing-friendly elliptic curves of prime order with embedding degree $k \in \{3, 4, 6\}$. Scott and Barreto [16] as well as Galbraith et al. [10] extended this method by allowing the group order to be non-prime. The advantage of this idea is the construction of much more suitable elliptic curves, which we will call *generalized MNT curves*. A necessary step for the construction of such elliptic curves is finding the solutions of a generalized Pell equation. However, these equations are not always solvable and this fact considerably affects the efficiency of the curve construction. In this paper we discuss a way to construct generalized MNT curves through Pell equations which are always solvable and thus considerably improve the efficiency of the whole generation process. We provide analytic tables with all polynomial families that lead to non-prime pairing-friendly elliptic curves with embedding degree $k \in \{3, 4, 6\}$ and discuss the efficiency of our method through extensive experimental assessments.

Keywords: Pairing-based cryptography, MNT elliptic curves, effective polynomial families, Pell equations.

1 Introduction

Pairing-based cryptography has gained much interest during the past few years. Several pairing-based protocols have been proposed such as the well known Boneh et al.'s ID-based encryption [5] and short signatures schemes [6]. All these cryptographic schemes are based on the construction of elliptic curves that satisfy certain properties. Clearly, generating suitable elliptic curves for pairing-based cryptosystems is a very important issue in pairing-based cryptography. These curves are known as *pairing-friendly* elliptic curves [8].

Let $E/\mathbb{F}_q$ be an elliptic curve of order $\#E(\mathbb{F}_q) = n$ defined over a prime field $\mathbb{F}_q$. In most pairing-based cryptographic protocols the ideal case is to construct elliptic curves of prime order. However, such curves are rare and so the ideal case is hard to achieve in practice. To this end we may relax this condition and

T. Muntean, D. Poulakis, and R. Rolland (Eds.): CAI 2013, LNCS 8080, pp. 147–159, 2013.

allow the use of curves with $\#E(\mathbb{F}_q) = hr$ for a small cofactor $h > 1$ and r a large prime. The ρ-value is defined as $\rho = \log(q)/\log(r)$ and shows how close to the ideal case is the constructed curve. Clearly, we require the ρ-value to be as close to 1 as possible. Furthermore, let $E[r]$ denote the set of r-torsion points of $E/\mathbb{F}_q$. Then the embedding degree of $E[r]$ is the smallest positive integer $k > 1$, such that $E[r] \subseteq E(\mathbb{F}_{q^k})$, or equivalently the smallest positive integer such that $r \mid q^k - 1$, where $\mathbb{F}_{q^k}$ is a finite extension of $\mathbb{F}_q$ of degree k. According to [8], an elliptic curve E defined over a prime field $\mathbb{F}_q$ with small embedding degree and large prime order subgroup is called pairing-friendly.

A well known method to construct elliptic curves over a large prime field is the Complex Multiplication (CM) method [1]. By Hasse's theorem, $Z = 4q - t^2$ must be positive and, thus, there is a unique factorization $Z = DY^2$, with D a square free positive integer. Therefore

$$4q = t^2 + DY^2 \tag{1}$$

is satisfied for a given pair (q, t). The negative parameter $-D$ is called a *CM discriminant for the prime q*. For convenience throughout the paper, we will use (the positive integer) D to refer to the CM discriminant. Knowing the values of q and t, an elliptic curve E defined over $\mathbb{F}_q$ with $n = q + 1 - t$ number of $\mathbb{F}_q$-rational points can be constructed. The triple (q, t, n) represents the curve parameters, i.e. the order of the finite field, the Frobenius trace and the group order of $E(\mathbb{F}_q)$ respectively.

A pairing on an elliptic curve $E/\mathbb{F}_q$ is a map of the form $e : E(\mathbb{F}_q)[r] \times E(\mathbb{F}_{q^k}) \to \mathbb{F}_{q^k}^*$ which is bilinear, non-degenerate and efficiently computable. As mentioned in [16] the most commonly used pairings are the Weil and Tate pairings [2,9]. In order to use pairings in cryptography, we must guarantee that the discrete logarithm problem (DLP) in both $E(\mathbb{F}_q)[r]$ and $\mathbb{F}_{q^k}^*$ is computationally infeasible. Thus the embedding degree must be chosen to be large enough in order to keep the DLP in $\mathbb{F}_{q^k}^*$ as hard as possible, but also k must be small enough for the efficient arithmetic in $\mathbb{F}_{q^k}^*$. As stated in [16], a good choice for an 80-bit security level is $\log r \approx 160$ and $k \log q \approx 1024$ bits, so that the cryptosystem can resist attacks both in elliptic curve groups and in finite fields.

Miyaji, Nakabayashi and Takano in 2001 [12] were the first who proposed a method (the so called MNT method) for the construction of prime order pairing-friendly elliptic curves with embedding degrees $k \in \{3, 4, 6\}$. Using the CM equation (1) and representing the elliptic curve parameters (q, t, n) as polynomials in $\mathbb{Z}[x]$, they created three Pell-type equations, one for each $k \in \{3, 4, 6\}$. The solutions of these equations lead to potential suitable curve parameters (q, t, n). Scott and Barreto [16] extended the idea of Miyaji et al. by allowing the group order to contain a large prime factor r and a positive small integer $h > 1$ called *cofactor*. In particular, they describe an explicit algorithm that constructs more Pell-type equations for $h > 1$, whose solutions lead to the generation of much more suitable elliptic curves when $k \in \{3, 4, 6\}$. Galbraith, McKee and Valença [10] also extended the MNT method for $k \in \{3, 4, 6\}$ by using non-prime elliptic curves. The difference of their work from [16] is that Galbraith et al. represent the curve parameters (q, t, r) as polynomial families $(q(x), t(x), r(x))$. In their paper they give all polynomial families for $h \in \{2, 3, 4, 5\}$ when $k \in \{3, 4, 6\}$. In [7], Duan, Cui and Wah Chan present a general algorithm for the construction of pairing friendly elliptic curves with arbitrary embedding degree and similarly to [10] they represent the curve parameters as polynomial families. In their method they also construct Pell-type equations from which they obtain suitable curve parameters. Furthermore they introduce the term of *effective polynomial families* by inducing some restrictions on the choice of polynomials $(q(x), t(x), r(x))$.

In this paper we further investigate the construction of generalized MNT elliptic curves with embedding degree $k \in \{3, 4, 6\}$ by using quadratic families that have better chances in producing suitable elliptic curve parameters. In particular we extend the idea of effective polynomial families, first introduced in [7], and enhance them with the ability to lead to generalized Pell equations which are always solvable. The solutions of these Pell equations can be tested for suitability in more than one quadratic families. This observation increases the chances of finding suitable parameters and speeds up the method considerably. While previous works in [10,16] study cases where $h \leq 5$ we extend the search to families with larger cofactors $h > 5$, but not too large since we wish to keep the ρ-value as close to one as possible. The advantage of our method is that we avoid solving Pell equations leading to a small number of suitable curve parameters. We also present experimental evidence that our method can considerably speed up the generation of generalized MNT elliptic curves.

The paper is organized as follows. In Section 2 we present previous work for the generation of MNT elliptic curves with embedding degree $k \in \{3, 4, 6\}$. In Section 3 we describe our method for the construction of generalized MNT curves. In Section 4 we present our experimental results and we conclude the paper in Section 5.

2 Previous Work

In this section we give a brief overview of previous work concerning the generation of pairing-friendly elliptic curves with embedding degree $k \in \{3, 4, 6\}$. All methods share a common characteristic: in order to generate the curve parameters, they use the solutions of some Pell-type equations. These equations are of the form

$$X^2 - SDY^2 = m \tag{2}$$

where $S, m \in \mathbb{Z}$ and $S > 0$. The integer D represents the CM discriminant and it is positive and square-free. If a Pell equation of this form is solvable, then there is an infinite number of integral pairs (X_i, Y_i) satisfying it. For more detailed analysis on the theory of Pell equations the interested reader can consult [13]. Throughout the paper we will consider elliptic curves E defined over a finite field $\mathbb{F}_q$ where q is a large prime and $\#E(\mathbb{F}_q) = n = hr$ for some large prime r and a cofactor $h \geq 1$.

Miyaji, Nakabayashi and Takano were the first to describe a method for producing ordinary pairing-friendly elliptic curves of prime order with embedding

Table 1. MNT Families

k	q(x)	t(x)	n(x)	Pell Equation	Suitable X
3	$12x^2 - 1$	$\pm 6x - 1$	$12x^2 \pm 6x + 1$	$X^2 - 3DY^2 = 24$	$X = 6x \pm 3$
4	$x^2 + x + 1$	$-x, x + 1$	$x^2 + 2x + 2, x^2 + 1$	$X^2 - 3DY^2 = -8$	$X = 3x + 2, 3x + 1$
6	$4x^2 + 1$	$\pm 2x + 1$	$4x^2 \pm 2x + 1$	$X^2 - 3DY^2 = -8$	$X = 6x \mp 1$

degree $k \in \{3, 4, 6\}$ (e.g. $\#E(\mathbb{F}_q) = n$ is a large prime number). In their work they represent the values (q, t) as polynomials $q(x), t(x) \in \mathbb{Z}[x]$, such that the polynomial $n(x) = q(x) + 1 - t(x)$ divides $\Phi_k(q(x))$, where $\Phi_k(x)$ is the kth-cyclotomic polynomial for $k \in \{3, 4, 6\}$. When the polynomial $q(x)$ is quadratic, we will refer to the families $(q(x), t(x), n(x))$ as *quadratic polynomial families.*

The quadratic polynomial families of Miyaji et al. are presented in Table 1 and are known as the MNT families. For any pair $(q(x), t(x))$ of Table 1 substitute them into the CM equation (Eq. 1) to get

$$4q(x) - t^2(x) = DY^2 \tag{3}$$

Multiplying by a constant factor and completing the squares yields to the Pell-type equations of Table 1. We refer to these equations as the MNT equations. Suppose that the integral pair (X_i, Y_i) represents a solution of an equation in Table 1, for some $i \in \mathbb{N}$. Then check if there is an integer x_0 such that X_i is suitable, i.e. if it is written in the form given in the last column of Table 1. If such a x_0 exists, substitute x_0 into the corresponding polynomials $q(x), t(x)$ and $r(x)$ and check if $q(x_0)$ is prime, $|t(x_0)| \leq 2\sqrt{q(x_0)}$ and $n(x_0)$ is also prime. If these conditions hold, the triple $(q(x_0), t(x_0), n(x_0))$ represents the suitable elliptic curve parameters. An implementation of the MNT method can be found in [11].

In [16], Scott and Barreto argue that by using the MNT method we can find few curves for actual deployment and furthermore these are the only curves available if we insist on constructing prime order pairing-friendly elliptic curves with $k \in \{3, 4, 6\}$. To overcome this problem, they generalized the method by allowing the use of curves with nearly prime order, i.e. $\#E(\mathbb{F}_q) = n = hr$ where r is a large prime and $h > 1$. Note that in this case the field size q satisfies the relation $q = hr + t - 1$. The advantage of this idea is the construction of more Pell-type equations leading to the generation of much more suitable curve parameters.

Since the group $E(\mathbb{F}_q)$ has a subgroup of prime order r and k is the embedding degree of this subgroup, we must have $r \mid q^k - 1$ and $r \nmid q^i - 1$ for any $i \in \{1, \ldots, k-1\}$, according to the definition of the embedding degree. This condition is equivalent to $r \mid \Phi_k(t-1)$ and $r \nmid \Phi_i(t-1)$ for any $i \in \{1, \ldots, k-1\}$, as shown in Lemma 1 in [3]. Thus we may assume that $\Phi_k(t-1) = ar$ for some positive integer a. Now substitute q, $x = t - 1$ and $r = \Phi_k(t-1)/a$ into Eq. (1) to obtain the equivalent equation

$$DY^2 = 4h\frac{\Phi_k(x)}{a} - (x-1)^2. \tag{4}$$

By setting $x = (X - a_k)/(4h - a)$, $\lambda = -2\lfloor k/2 \rfloor + 4$, $a_k = \lambda h + a$ and $f_k = a_k^2 - (4h-a)^2$ Eq. (4) is transformed into

$$X^2 - a(4h-a)DY^2 = f_k. \quad (5)$$

This equation has the form of Eq. (2) where D is the CM discriminant. Thus $a(4h-a) > 0$ forcing $a < 4h$. If the above Pell equation is solvable for some values D, h and a with solution an integral pair (X_i, Y_i), then it is checked if $x_0 = (X_i - a_k)/(4h-a)$ is integer. If this is the case, check if $q = hr + x_0$ is prime and $r = \Phi_k(x_0)/a$ is also prime. As mentioned in [16], we may further relax the condition on the group order by allowing r to contain itself a large prime factor, i.e. $r = ms$, for some $m \geq 1$ and s a large prime. If both conditions hold, the integers (q, t, r) are suitable elliptic curve parameters.

Galbraith, McKee and Valença [10] (GMV) also generalize the MNT method by using non-prime elliptic curves. In their work they present a complete characterization of all polynomial families $(q(x), t(x), r(x))$ with cofactors $h \in \{2, 3, 4, 5\}$ for cases where $k \in \{3, 4, 6\}$. Their polynomial families appear in [10] and lead to the same Pell equations as in the case of Scott and Barreto method. In order to find suitable curve parameters for a fixed embedding degree k and a cofactor h, the GMV method proceeds as the original MNT method.

In [7], Duan, Cui and Wah Chan present an alternative way for producing pairing-friendly elliptic curves with arbitrary embedding degree k. Following the same approach as [10], they represent the curve parameters as polynomials $q(x), t(x), r(x) \in \mathbb{Z}[x]$. Furthermore they introduce the concept of *effective polynomial families*. According to their definition a polynomial family $(q(x), t(x), r(x))$ is called effective if the polynomial $f(x) = 4q(x) - t^2(x)$ can be factorized with one square polynomial, or it is quadratic and factorable, or it only contains terms with smaller degree compared to $q(x)$. An example for the first case is studied by Barreto and Naehrig in [4] for $k = 12$. Duan et al. argue that an effective polynomial family has better chances in producing suitable elliptic curve parameters.

Although the method of Duan et al. is suitable for any k we focus on the case where $k \in \{3, 4, 6\}$. If we substitute $q(x) = hr(x) + 1 - t(x)$ in Eq. (3) we have that

$$f(x) = DY^2 = 4hr(x) - (t(x) - 2)^2. \quad (6)$$

Then, we choose a quadratic polynomial $r(x)$ and since we wish r to be prime, the polynomial $r(x)$ must be irreducible over $\mathbb{Z}[x]$. A linear trace polynomial $t(x)$ must also be chosen, such that $r(x) \mid \Phi_k(t(x) - 1)$. Knowing $r(x)$ and $t(x)$ we may compute $f(x)$ and $q(x)$. Since $\deg r(x) = 2$, the polynomial $f(x)$ is quadratic and a generalized Pell equation should be solved. Using the solutions of these equations we may search for suitable curve parameters in the usual way.

3 The Proposed Method

We focus on the generation of pairing-friendly elliptic curves with embedding degree $k \in \{3, 4, 6\}$ and we determine a way to construct quadratic polynomial families that have better chances in producing suitable elliptic curve parameters. To this end we adopt the remarks from the work of Duan, Cui and Wah Chan [7] about effective polynomial families. In our study we will consider effective polynomial families where the polynomial $f(x) = 4q(x) - t^2(x)$ is quadratic and factorable. We present a complete characterization of all such polynomial families and we argue that these families lead to a special kind of Pell equations which are always solvable and this fact considerably improves the efficiency of the whole generation method. We also extend the ideas presented in the previous section by allowing the cofactor to take values larger than the ones studied by Scott and Barreto and Galbraith et al. i.e. $h > 5$. We begin our study by analyzing the case $k = 6$, while the same ideas hold for the other two cases $k \in \{3, 4\}$.

3.1 The Case of k = 6

Suppose that $q(x), t(x), r(x) \in \mathbb{Z}[x]$ is a polynomial representation for the field size, the trace polynomial and the subgroup order respectively. Let a be a positive integer and suppose that the trace polynomial is linear of the form $t(x) = ax + b$ for some $b \in \mathbb{Z}$. Substitute $t(x) - 1$ into $\Phi_6(x)$ to obtain

$$\Phi_6(t(x) - 1) = a^2x^2 + a(2b - 3)x + b^2 - 3b + 3. \tag{7}$$

Since $\Phi_6(t(x) - 1)$ must be divisible by $r(x)$, we may set

$$r(x) = ax^2 + (2b - 3)x + \frac{b^2 - 3b + 3}{a}. \tag{8}$$

and thus a must be chosen such that the congruence $b^2 - 3b + 3 \equiv 0 \bmod a$ is satisfied for some $b \in \mathbb{Z}$. The polynomial $r(x)$ is irreducible over $\mathbb{Z}[x]$, since its discriminant is equal to $\Delta_r = -3 < 0$. Because $r(x)$ represents the order of a subgroup of $E(\mathbb{F}_q)$, it has to represent primes and therefore the condition that $r(x)$ is irreducible over $\mathbb{Z}[x]$ is essential. We may then assume that the order of $E(\mathbb{F}_q)$ is given by a small integer cofactor h times the polynomial $r(x)$, i.e. $\#E(\mathbb{F}_q) = hr(x)$. Now substitute $r(x)$ into $q(x) = hr(x) + t(x) - 1$ to obtain the corresponding field polynomial

$$q(x) = ahx^2 + (2bh - 3h + a)x + \frac{b^2h - 3bh + 3h + ab - a}{a}. \tag{9}$$

Note that $(b^2h - 3bh + 3h + ab - a)/a \in \mathbb{Z}$, since we have chosen $a \mid (b^2 - 3b + 3)$. Furthermore the field size must be prime and thus the polynomial $q(x)$ must be irreducible over $\mathbb{Z}[x]$. This means that the integer $\Delta_q = (a - h)^2 - 4h^2$ must not be a perfect square and also the coefficients of $q(x)$ must not have a common factor. Now substitute $q(x)$ and $t(x)$ into Eq. (3) represented in polynomial field and set $f(x) = 4q(x) - t^2(x)$. We obtain the quadratic polynomial

$$f(x) = a(4h - a)x^2 + 2\big((4h - a)b + 2a - 6h\big)x$$

Table 2. Some effective polynomial families for $\mathbf{k = 6}$

h	t(x)	q(x)	r(x)	Pell Equation
4	$13x + 5$	$52x^2 + 41x + 8$	$13x^2 + 7x + 1$	$(39x + 17)^2 - 39DY^2 = 4^2$
	$13x + 11$	$52x^2 + 89x + 38$	$13x^2 + 19x + 7$	$(39x + 35)^2 - 39DY^2 = 4^2$
9	$31x + 7$	$279x^2 + 130x + 15$	$31x^2 + 11x + 1$	$(155x + 43)^2 - 155DY^2 = 12^2$
	$31x + 27$	$279x^2 + 490x + 215$	$31x^2 + 51x + 21$	$(155x + 143)^2 - 155DY^2 = 12^2$
12	$39x + 18$	$468x^2 + 435x + 101$	$39x^2 + 33x + 7$	$(117x + 56)^2 - 39DY^2 = 4^2$
	$39x + 24$	$468x^2 + 579x + 179$	$39x^2 + 45x + 13$	$(117x + 74)^2 - 39DY^2 = 4^2$
16	$49x + 20$	$784x^2 + 641x + 131$	$49x^2 + 37x + 7$	$(735x + 302)^2 - 735DY^2 = 8^2$
	$49x + 32$	$784x^2 + 1025x + 335$	$49x^2 + 61x + 19$	$(735x + 482)^2 - 735DY^2 = 8^2$
25	$79x + 25$	$1975x^2 + 1254x + 199$	$79x^2 + 47x + 7$	$(1659x + 533)^2 - 1659DY^2 = 20^2$
	$79x + 57$	$1975x^2 + 2854x + 1031$	$79x^2 + 111x + 39$	$(1659x + 1205)^2 - 1659DY^2 = 20^2$
25	$91x + 11$	$2275x^2 + 566x + 35$	$91x^2 + 19x + 1$	$(819x + 131)^2 - 819DY^2 = 40^2$
	$91x + 18$	$2275x^2 + 916x + 92$	$91x^2 + 33x + 3$	$(819x + 194)^2 - 819DY^2 = 40^2$
	$91x + 76$	$2275x^2 + 3816x + 1600$	$91x^2 + 149x + 61$	$(819x + 716)^2 - 819DY^2 = 40^2$
	$91x + 83$	$2275x^2 + 4166x + 1907$	$91x^2 + 163x + 73$	$(819x + 779)^2 - 819DY^2 = 40^2$
36	$109x + 47$	$3924x^2 + 3385x + 730$	$109x^2 + 91x + 19$	$(3815x + 1647)^2 - 3815DY^2 = 12^2$
	$109x + 65$	$3924x^2 + 4681x + 1396$	$109x^2 + 127x + 37$	$(3815x + 2277)^2 - 3815DY^2 = 12^2$

$$+\frac{(4h-a)b^2 + 2(2a-6h)b + 12h - 4a}{a}.$$

Since $\deg f(x) = 2$, this will lead us to a generalized Pell equation. Following the definition of Duan et al. when $f(x)$ is factorable over $\mathbb{Z}[x]$ we have better chances in finding suitable pairing-friendly elliptic curves and in this case the triple $(q(x), t(x), r(x))$ is an effective polynomial family. In particular suppose that the above polynomial $f(x)$ is factorable over $\mathbb{Z}[x]$. Then the integer $\Delta_f = 16h(a-3h)$ must be positive and perfect square. Moreover since $\Delta_f > 0$ we get that $a > 3h$. Multiplying the relation $f(x) = 4q(x) - t^2(x) = DY^2$ by $a(4h-a)$, completing the squares and setting $X = a(4h-a)x + (4h-a)b + 2a - 6h$ we obtain an equation of the form

$$X^2 - a(4h-a)DY^2 = \left(2\sqrt{h(a-3h)}\right)^2. \quad (10)$$

This is a generalized Pell equation and in fact it is the same as the one found by Scott and Barreto, since $f_6 = 4h(a-3h)$. The difference is that we consider these equations only when f_6 is a perfect square. Furthermore, combining the two inequalities for a we conclude that an equation of the form of Eq. (10) is possible, if a is chosen in the range $3h < a < 4h$.

Conversely, suppose that the polynomial $f(x)$ is quadratic of the form $f(x) = ax^2 + bx + c \in \mathbb{Z}[x]$ that leads to a generalized Pell equation of the form $X^2 - SDY^2 = m$ with m a perfect square. Multiply $f(x)$ by $4a$ and complete the squares to obtain the Pell equation $(2ax+b)^2 - aD(2Y)^2 = b^2 - 4ac$ where $S = a$ and $m = b^2 - 4ac$. Since m is a perfect square, we have that the integer $b^2 - 4ac$ must be a perfect square which in turn means that $f(x)$ is factorable over $\mathbb{Z}[x]$. Hence we have shown that in order to get a generalized Pell equation of the form of Eq. (10) we must have $f(x)$ factorable and thus $h(a-3h)$ must be a perfect square. The above discussion actually indicates that these are the only Pell equations of this form for $k=6$. We conclude that *all* effective polynomial families for $k=6$ have the next parametric polynomial representation

$$t(x) = ax + b$$
$$r(x) = ax^2 + (2b-3)x + \frac{b^2 - 3b + 3}{a}$$
$$q(x) = ahx^2 + (2bh - 3h + a)x + \frac{b^2h - 3bh + 3h + ab - a}{a}$$

where the following conditions must be satisfied: (i) the integer $h(a-3h)$ is a perfect square, (ii) the congruence $b^2 - 3b + 3 \equiv 0 \bmod a$ is solvable, (iii) the integer $(a-h)^2 - 4h^2$ is not a perfect square and (iv) the coefficients of $q(x)$ have no common factor. The last two conditions guarantee that $q(x)$ has no constant or linear factors. Some examples of Pell equations of the form of Eq. (10), obtained by effective polynomial families are given in Table 2.

Pell equations of the form of Eq. (10) are considered as a special case and this is because they have a very usefull advantage compared to others. In particular consider the standard Pell equation

$$U^2 - a(4h-a)DV^2 = 1. \tag{11}$$

By Theorem 4.1 [14] Eq. (11) is always solvable for every positive integer D, such that $a(4h-a)D$ is not a perfect square. Suppose that the pairs (U_i, V_i) define a sequence of solutions for Eq. (11), with $i \in \mathbb{N}$. Then the pairs $(X_i, Y_i) = (2\sqrt{h(a-3h)}U_i, 2\sqrt{h(a-3h)}V_i)$ represent the corresdonding solutions of the generalized Pell equation (10). Thus there is always at least one class of solutions for Eq. (10) arising from the units in the quadratic field $\mathbb{Q}(\sqrt{a(4h-a)D})$. Of course in most cases there are more than one classes of solutions. This is a very important observation because the more integer solutions we have to test, the more possibilities we have to generate suitable curve parameters. Once a solution (X_i, Y_i) of the appropriate size is obtained, we follow the standard MNT method in order to construct the curve parameters. More precisely check if there is a $x_0 \in \mathbb{Z}$ such that X_i is written as $X_i = a(4h-a)x_0 + (4h-a)b + 2a - 6h$, for some $b \in \mathbb{Z}$ satisfying the congruence $b^2 - 3b + 3 \equiv 0 \bmod a$. If such a x_0 exists, substitute into $q(x)$ and $r(x)$ and check if $q(x_0)$ is prime and $r(x_0)$ is prime or nearly prime.

The above procedure generalizes the work of Duan et al. [7] since it defines a parametric representation of all effective polynomial families $(q(x), t(x), r(x))$ such that $f(x)$ is quadratic and factorable. This analysis also shows that for a chosen pair (a, h) such that $h(a-3h)$ is a perfect square and $b^2 - 3b + 3 \equiv 0 \bmod a$ is solvable there are more than one effective polynomial families and the number of these families depends on the number of different $b \in \mathbb{Z}_a$ satisfying the above congruence. All these different families lead to the same generalized Pell equation. For example the effective polynomial family proposed in [7] for

$k = 6$, $h = 9$ and $t(x) = 31x + 7$ is not the only one. In Table 2 we have shown that there is a second family for $t(x) = 31x + 27$. Thus in our case we solve this generalized Pell equation only once and we are searching for suitable values $q(x_0)$ and $r(x_0)$ for all effective polynomial families leading to this Pell equation. Following the strategy of Duan et al., the same Pell equation may be solved more than once which induces a considerable delay in the execution time.

3.2 The Case of k = 3, 4

For the cases where $k = 3, 4$ we follow the same arguments as in the case of $k = 6$. In particular when $k = 3$ we may represent the quadratic families $(q(x), t(x), r(x))$ by the parametrization

$$\begin{aligned} t(x) &= ax + b \\ r(x) &= ax^2 + (2b - 1)x + \frac{b^2 - b + 1}{a} \\ q(x) &= ahx^2 + (2bh - h + 1)x + \frac{b^2h - bh + h + ab - a}{a} \end{aligned}$$

where the following conditions are satisfied: (1) the integer $48h(a-h)$ is a perfect square, (2) the congruence $b^2 - b + 1 \equiv 0 \bmod a$ is solvable, (3) the integer $(a+h)^2 - 4h^2$ is not a perfect square and (4) the coefficients of $q(x)$ are coprime. Furthermore a and h must also satisfy the relations $4h - a > 0$ and $a - h > 0$ and thus a lies in the range $h < a < 4h$. Multiplying the relation $f(x) = 4q(x) - t^2(x)$ by $a(4h - a)$, completing the squares and setting $X = a(4h - a)x + (4h - a)b + 2a - 2h$ we conclude to the special Pell equation

$$X^2 - a(4h - a)DY^2 = \left(2\sqrt{3h(a - h)}\right)^2. \quad (12)$$

In the same way if $k = 4$ then there is a parametrization of the quadratic families $(q(x), t(x), r(x))$ as

$$\begin{aligned} t(x) &= ax + b \\ r(x) &= ax^2 + 2(b - 1)x + \frac{b^2 - 2b + 2}{a} \\ q(x) &= ahx^2 + (2bh - 2h + a)x + \frac{b^2h - 2bh + 2h + ab - a}{a} \end{aligned}$$

where the following conditions are satisfied: (1) the integer $32h(a - 2h)$ is a perfect square, (2) the congruence $b^2 - 2b + 2 \equiv 0 \bmod a$ is solvable, (3) the integer $a^2 - 4h^2$ is not a perfect square and (4) the coefficients of $q(x)$ have no common factor. Multiplying the relation $f(x) = 4q(x) - t^2(x)$ by $a(4h - a)$, completing the squares and setting $X = a(4h - a)x + (4h - a)b + 2a - 4h$ yields the special Pell equation

$$X^2 - a(4h - a)DY^2 = \left(2\sqrt{2h(a - 2h)}\right)^2. \quad (13)$$

If we wish to find suitable curve parameters in both cases we proceed in the usual way. Note here that the number of effective polynomial families decreases as the value of k increases in $\{3, 4, 6\}$. The reason is that the choices for a are decreased. In particular when $k = 3$ the integer a is chosen in the range $(h, 4h)$ while if $k = 6$, the integer a lies in $(3h, 4h)$.

4 Experimental Results

The most crucial step in the above procedure is solving a generalized Pell equation of the form of Eq. (2). A well known method used to solve any kind of Pell equations is the LMM algorithm [14,15]. Alternative ways are also presented in [15]. One of these methods finds all solutions of an equation of the form (2) by computing the simple continued fraction expansion of the quadratic irrational $\sqrt{SD}$, but it is only suitable for values of the CM discriminant D such that $m^2/S < D$. This method is also implemented by Karabina and Teske in [11] for the original MNT equations. When $m^2/S > D$ this procedure finds only some of the solutions for some D. Thus in our implementation we might lost a few suitable parameters. For more precise results, one should implement the LMM algorithm when $m^2/S > D$.

Table 3 presents the number of suitable curve parameters obtained by effective polynomial families for certain choices of h when $k \in \{3, 4, 6\}$. The criteria for suitability are the same as those in [16]. In particular the field size q is chosen such that $768 \leq k \log q \leq 1536$ and the group order r is chosen to be a product $r = ms$ for some prime s with $\log s > 128$ bits.

For example when $k = 6$ we are looking for primes q such that $128 \leq \log q \leq 256$ bits. In this case the most lucky families appear when $h = 4$ and $a = 13$ where we found 384 suitable triples (q, t, r). When $k = 3$ the field size q must be chosen between the sizes $256 \leq \log q \leq 512$. The most lucky case appears when $h = 4$ and $a = 7$ where we found 392 suitable curve parameters. When $k = 4$ the best results appear when $h = 18$ and $a = 37$ where we found 60 suitable triples (q, t, r) with $192 \leq \log q \leq 384$.

Table 3. Suitable parameters for $\mathbf{k} \in \{\mathbf{3}, \mathbf{4}, \mathbf{6}\}$ and $\mathbf{h} > \mathbf{1}$ from effective polynomial families ($\mathbf{768} \leq \mathbf{k} \log \mathbf{q} \leq \mathbf{1536}$, $\log \mathbf{s} > \mathbf{128}$ and $\mathbf{D} < \mathbf{10^5}$)

k = 3			k = 4			k = 6		
Cofactor h	a	Suitable (q, t, r)	Cofactor h	a	Suitable (q, t, r)	Cofactor h	a	Suitable (q, t, r)
4	7	392	8	17	52	4	13	384
12	21	46	8	25	19	9	31	13
12	37	45	16	34	52	12	39	72
16	19	57	16	50	19	16	49	37
16	43	10	18	37	60	25	79	7
36	111	36	18	61	23	25	91	17
48	49	33	32	65	53	36	109	40

Table 4. Time required for the generation of suitable triples $(\mathbf{q}, \mathbf{t}, \mathbf{r})$ when $\mathbf{k} \in \{3, 4, 6\}$ $(768 \leq \mathbf{k} \log \mathbf{q} \leq 1536$ and $\log \mathbf{s} > 128)$

	k = 3			k = 4			k = 6		
Triples	h	SB Method (sec)	Effective Families (sec)	h	SB Method (sec)	Effective Families (sec)	h	SB Method (sec)	Effective Families (sec)
	4		**a = 7**	**8**		**a = 17**	**4**		**a = 13**
1		9.01	18.03		20.43	3.38		0.26	0.82
5		212.38	50.87		733.42	25.29		36.94	9.24
10		739.51	76.68		2717.25	100.61		377.71	11.69
20		1172.41	208.80		3670.81	383.85		1809.48	23.84
30		1641.05	310.70		6053.82	962.65		1874.23	45.34
	12		**a = 21**	**16**		**a = 34**	**9**		**a = 31**
1		8.68	9.26		4.82	3.45		12.49	18.58
5		279.08	132.92		240.88	27.50		72.48	226.62
10		931.51	1635.00		1112.30	121.02		3773.71	3176.45
	16		**a = 19**	**18**		**a = 37**	**12**		**a = 39**
1		2.60	12.62		21.40	44.23		111.06	0.19
5		1303.61	70.88		92.37	255.47		3118.85	50.67
10		3135.41	165.36		3869.22	638.52		6537.99	275.64
	48		**a = 49**	**32**		**a = 65**	**16**		**a = 49**
1		1.68	0.51		307.74	5.65		0.94	11.96
5		157.48	99.35		5899.03	44.15		298.87	35.30
10		6386.99	1170.49		13121.12	199.86		1168.44	141.01

According to our earlier analysis we expect that the number of suitable parameters obtained from effective polynomial families is larger than the number of parameters from non-effective ones. Thus we argue that one may use only the effective polynomial families for finding suitable triples (q, t, r) when $k \in \{3, 4, 6\}$. In order to show the efficiency of our method, we implemented the algorithm proposed by Scott and Barreto and compared the time required for the construction of a fixed number of suitable parameters with their method and our proposal. The results appear in Table 4. In the case where there are more than one effective polynomial families, we studied only the first one, i.e. the first $a \in \mathbb{Z}$ leading to an effective polynomial family.

In almost all cases, we observe that our method is faster than the method of Scott and Barreto, especially as the number of the desired suitable triples (q, t, r) increases. This is because the Pell equations from non-effective polynomial families are not always solvable and thus there might be a large distance between the suitable values of D. For example consider the case where $k = 6$ and $h = 4$. If we wish to construct only one elliptic curve (e.g. one triple (q, t, r)), the algorithm of Scott and Barreto requires 0.26 seconds, while our method needs 0.82 seconds. If we wish to construct 5 elliptic curves (or the first 5 triples), Scott and Barreto method requires 36.94 seconds, while our method needs only 9.24. For a required number of 10 parameters, the difference is more clear. Taking the number of suitable parameters even further, say 20 or 30 the method of Scott

Table 5. Time required for the generation of suitable triples $(\mathbf{q},\mathbf{t},\mathbf{r})$ when $\mathbf{k}=\mathbf{6}$ and $\mathbf{3072}\leq\mathbf{k}\log\mathbf{q}\leq\mathbf{4608}$

Triples	h	SB Method (sec)	Effective Families (sec)	h	SB Method (sec)	Effective Families (sec)	h	SB Method (sec)	Effective Families (sec)
	4		a = 13	12		a = 39	16		a = 49
1		3.68	9.60		1278.23	1.18		15.13	35.75
5		1004.76	106.19		8703.31	228.67		1465.73	356.10

and Barreto needs to solve more than one Pell equations. This fact provides a considerable delay in the whole procedure. The same remarks hold also for the case $k=3$ and $h=4$. Furthermore since the density of the values of D is larger in our case, we expect that for a fixed number of suitable triples the values of the discriminants will be smaller in the case of effective polynomial families than in the case of Scott and Barreto method. For example when $k=6$ and $h=4$ the first 30 suitable triples appear for values of $D\leq 2221$, while in the case of Scott and Barreto the same number of triples were found for $D\leq 97282$. In order to achieve higher security levels, we may increase the size of the prime q. In this case we observe the same behaviour as in Table 4. Some indicative results are presented in Table 5.

5 Conclusion

According to Scott and Barreto [16] the construction of generalized MNT elliptic curves is based on solving several Pell-type equations of the form of Eq. (2). For certain choices of cofactor h, some of these equations have more chances than others in producing suitable elliptic curve parameters. In particular the most lucky quadratic polynomial families $(q(x),t(x),r(x))$ are those for which $f(x)=4q(x)-t^2(x)$ is quadratic and factorable. The Pell equations obtained by such families have the advantage that they are always solvable for every positive and square-free integer D and thus the more solutions we have to test for suitability, the higher is the probability to get suitable curve parameters. This observation also implies that this special kind of Pell equations provides even more flexibility on the CM discriminant, since there are no congruential restrictions on D. In this work we isolate these equations and introduce a procedure that uses only these special equations to construct the desired generalized MNT elliptic curves.Based on our experimental assessments, we argue that our method can considerably speed up the algorithm proposed in [16]. This is theoretically explained (mainly) from the fact that we manage to avoid the solution of "unlucky" Pell equations.

References

1. Atkin, A.O.L., Morain, F.: Elliptic Curves and Primality Proving. Mathematics of Computation 61, 29–68 (1993)
2. Barreto, P.S.L.M., Kim, H.Y., Lynn, B., Scott, M.: Efficient Algorithms for Pairing-Based Cryptosystems. In: Yung, M. (ed.) CRYPTO 2002. LNCS, vol. 2442, pp. 354–368. Springer, Heidelberg (2002)
3. Barreto, P.S.L.M., Lynn, B., Scott, M.: Constructing Elliptic Curves with Prescribed Embedding Degrees. In: Cimato, S., Galdi, C., Persiano, G. (eds.) SCN 2002. LNCS, vol. 2576, pp. 257–267. Springer, Heidelberg (2003)
4. Barreto, P.S.L.M., Naehrig, M.: Pairing-Friendly Elliptic Curves of Prime Order. In: Preneel, B., Tavares, S. (eds.) SAC 2005. LNCS, vol. 3897, pp. 319–331. Springer, Heidelberg (2006)
5. Boneh, D., Franklin, M.: Identity-Based Encryption from the Weil Pairing. SIAM Journal of Computing 32(3), 586–615 (2003)
6. Boneh, D., Lynn, B., Shacham, H.: Short Signatures from the Weil Pairing. In: Boyd, C. (ed.) ASIACRYPT 2001. LNCS, vol. 2248, pp. 514–532. Springer, Heidelberg (2001)
7. Duan, P., Cui, S., Wah Chan, C.: Finding More Non-Supersingular Elliptic Curves for Pairing-Based Cryptosystems. International Journal of Information Technology 2(2), 157–163 (2005)
8. Freeman, D., Scott, M., Teske, E.: A Taxonomy of Pairing-Friendly Elliptic Curves. Journal of Cryptology 23, 224–280 (2010)
9. Galbraith, S.D., Harrison, K., Soldera, D.: Implementing the Tate Pairing. In: Fieker, C., Kohel, D.R. (eds.) ANTS 2002. LNCS, vol. 2369, pp. 324–337. Springer, Heidelberg (2002)
10. Galbraith, S.D., McKee, J., Valença, P.: Ordinary Abelian Varieties Having Small Embedding Degree. Finite Fields and Their Applications 13(4), 800–814 (2007)
11. Karabina, K., Teske, E.: On Prime-Order Elliptic Curves with Embedding Degrees $k = 3, 4$, and 6. In: van der Poorten, A.J., Stein, A. (eds.) ANTS-VIII 2008. LNCS, vol. 5011, pp. 102–117. Springer, Heidelberg (2008)
12. Miyaji, A., Nakabayashi, M., Takano, S.: New Explicit Conditions of Elliptic Curve Traces for FR-Reduction. IEICE Transactions Fundamentals E84-A(5), 1234–1243 (2001)
13. Mollin, R.A.: Fundamental Number Theory with Applications. CRC Press, Boca Raton (1998)
14. Mollin, R.A.: Simple Continued Fraction Solutions for Diophantine Equations. Expositiones Mathematicae 19, 55–73 (2001)
15. Robertson, J.P.: Solving the Generalized Pell Equation $x^2 - Dy^2 = N$ (2004), `http://hometown.aol.com/jpr2718/`
16. Scott, M., Barreto, P.S.L.M.: Generating more MNT Elliptic Curves. Designs, Codes and Cryptography 38, 209–217 (2006)

Shimura Modular Curves and Asymptotic Symmetric Tensor Rank of Multiplication in any Finite Field

Stéphane Ballet[1], Jean Chaumine[2], and Julia Pieltant[3]

[1] Aix-Marseille Université, CNRS IML FRE 3529
Case 930, 13288 Marseille Cedex 9, France
`stephane.ballet@univ-amu.fr`

[2] Université de la Polynésie Française, GAATI EA 3893
B.P. 6570, 98702 Faa'a, Tahiti, France
`jean.chaumine@upf.pf`

[3] INRIA Saclay, LIX, École Polytechnique, 91128 Palaiseau Cedex, France
`pieltant@lix.polytechnique.fr`

Abstract. We obtain new asymptotical bounds for the symmetric tensor rank of multiplication in any finite extension of any finite field $\mathbb{F}_q$. In this aim, we use the symmetric Chudnovsky-type generalized algorithm applied on a family of Shimura modular curves defined over $\mathbb{F}_{q^2}$ attaining the Drinfeld-Vlăduţ bound and on the descent of this family over the definition field $\mathbb{F}_q$.

Keywords: Algebraic function field, tower of function fields, tensor rank, algorithm, finite field, modular curve, Shimura curve.

1 Introduction

1.1 General Context

The determination of the tensor rank of multiplication in finite fields is a problem which has been widely studied over the past decades both for its theoretical and practical importance. Besides it allows one to obtain multiplication algorithms with a low bilinear complexity, which determination is of crucial significance in cryptography, it has also its own interest in algebraic complexity theory. The pioneer work of D.V. and G.V. Chudnovsky [15] resulted in the design of a Karatsuba-like algorithm where the interpolation is done on points of algebraic curves with a sufficient number of rational points over the ground field. Following these footsteps, several improvements and generalizations of this algorithm leading to ever sharper bounds have been proposed since by various authors [9,1,14,19], and have required to investigate and combine different techniques and objects from algebraic geometry such as evaluations on places of arbitrary degree, generalized evaluations, towers of algebraic function fields... Furthermore, a lot of connexions with other topics have been made : Shparlinski, Tsfasman and Vlăduţ [21] have first developed a correspondence between decompositions of the tensor of multiplication and a family of linear codes with good

T. Muntean, D. Poulakis, and R. Rolland (Eds.): CAI 2013, LNCS 8080, pp. 160–172, 2013.

parameters that they called *(exact) supercodes*. These codes, renamed *multiplication friendly codes*, had recently be more extensively studied and exploited by Cascudo, Cramer, Xing and Yang [13] to obtain good asymptotic results on the tensor rank. Moreover they combined their notion of multiplication friendly codes with two newly introduced primitives for function fields over finite fields [11], namely the torsion limit and systems of Riemann-Roch equations, to get news results not only on asymptotic tensor rank but also on linear secret sharing systems and frameproof codes. This stresses that the tensor rank determination problem has just as many mathematical interests as consequences and applications in various domains of computer science.

1.2 Tensor Rank of Multiplication

Let $q = p^s$ be a prime power, $\mathbb{F}_q$ be the finite field with q elements and $\mathbb{F}_{q^n}$ be the degree n extension of $\mathbb{F}_q$. The multiplication of two elements of $\mathbb{F}_{q^n}$ is an $\mathbb{F}_q$-bilinear application from $\mathbb{F}_{q^n} \times \mathbb{F}_{q^n}$ onto $\mathbb{F}_{q^n}$. Then it can be considered as an $\mathbb{F}_q$-linear application from the tensor product $\mathbb{F}_{q^n} \otimes_{\mathbb{F}_q} \mathbb{F}_{q^n}$ onto $\mathbb{F}_{q^n}$. Consequently it can be also considered as an element T of $(\mathbb{F}_{q^n} \otimes_{\mathbb{F}_q} \mathbb{F}_{q^n})^\star \otimes_{\mathbb{F}_q} \mathbb{F}_{q^n}$, namely an element of $\mathbb{F}_{q^n}{}^\star \otimes_{\mathbb{F}_q} \mathbb{F}_{q^n}{}^\star \otimes_{\mathbb{F}_q} \mathbb{F}_{q^n}$. More precisely, when T is written

$$T = \sum_{i=1}^{r} x_i^\star \otimes y_i^\star \otimes c_i, \tag{1}$$

where the r elements $x_i^\star$ and the r elements $y_i^\star$ are in the dual $\mathbb{F}_{q^n}{}^\star$ of $\mathbb{F}_{q^n}$ and the r elements c_i are in $\mathbb{F}_{q^n}$, the following holds for any $x, y \in \mathbb{F}_{q^n}$:

$$x \cdot y = \sum_{i=1}^{r} x_i^\star(x) y_i^\star(y) c_i.$$

Unfortunately, the decomposition (1) is not unique.

Definition 1. *The minimal number of summands in a decomposition of the tensor T of the multiplication is called the bilinear complexity of the multiplication and is denoted by $\mu_q(n)$:*

$$\mu_q(n) = \min \left\{ r \;\middle|\; T = \sum_{i=1}^{r} x_i^\star \otimes y_i^\star \otimes c_i \right\}.$$

However, the tensor T admits also a symmetric decomposition:

$$T = \sum_{i=1}^{r} x_i^\star \otimes x_i^\star \otimes c_i. \tag{2}$$

Definition 2. *The minimal number of summands in a symmetric decomposition of the tensor T of the multiplication is called the symmetric bilinear complexity of the multiplication and is denoted by $\mu_q^{\mathrm{sym}}(n)$:*

$$\mu_q^{\mathrm{sym}}(n) = \min \left\{ r \;\middle|\; T = \sum_{i=1}^{r} x_i^\star \otimes x_i^\star \otimes c_i \right\}.$$

One easily gets that $\mu_q(n) \leq \mu_q^{\mathrm{sym}}(n)$. We know some cases where $\mu_q(n) = \mu_q^{\mathrm{sym}}(n)$ but to the best of our knowledge, no example is known where we can prove that $\mu_q(n) < \mu_q^{\mathrm{sym}}(n)$. However, better upper bounds have been established in the asymmetric case and this may suggest that in general the asymmetric bilinear complexity of the multiplication and the symmetric one are distinct. In any case, at the moment, we must consider separately these two quantities. Remark that from an algorithmic point on view, as well as for some specific applications, a symmetric bilinear algorithm can be more interesting than an asymmetric one, unless if *a priori*, the constant factor in the bilinear complexity estimation is a little worse. In this note we study the asymptotic behavior of the symmetric bilinear complexity of the multiplication. More precisely we study the two following quantities:

$$M_q^{\mathrm{sym}} = \limsup_{k \to \infty} \frac{\mu_q^{\mathrm{sym}}(k)}{k}, \tag{3}$$

$$m_q^{\mathrm{sym}} = \liminf_{k \to \infty} \frac{\mu_q^{\mathrm{sym}}(k)}{k}. \tag{4}$$

1.3 Known Results

The bilinear complexity $\mu_q(n)$ of the multiplication in the n-degree extension of a finite field $\mathbb{F}_q$ is known for certain values of n. In particular, S. Winograd [24] and H. de Groote [16] have shown that this complexity is $\geq 2n-1$, with equality holding if and only if $n \leq \frac{1}{2}q + 1$. Using the principle of the D.V. and G.V. Chudnovsky algorithm [15] applied to elliptic curves, M.A. Shokrollahi has shown in [20] that the symmetric bilinear complexity of multiplication is equal to $2n$ for $\frac{1}{2}q + 1 < n < \frac{1}{2}(q + 1 + \epsilon(q))$ where ϵ is the function defined by:

$$\epsilon(q) = \begin{cases} \text{greatest integer} \leq 2\sqrt{q} \text{ prime to } q, & \text{if } q \text{ is not a perfect square} \\ 2\sqrt{q}, & \text{if } q \text{ is a perfect square.} \end{cases}$$

Moreover, U. Baum and M.A. Shokrollahi have succeeded in [10] to construct effective optimal algorithms of type Chudnovsky in the elliptic case.

Recently in [3], [4], [9], [8], [7], [6] and [5] the study made by M.A. Shokrollahi has been generalized to algebraic function fields of genus g.

Let us recall that the original algorithm of D.V. and G.V. Chudnovsky introduced in [15] is symmetric by definition and leads to the following theorem:

Theorem 1. *Let $q = p^r$ be a power of the prime p. The symmetric tensor rank $\mu_q^{\mathrm{sym}}(n)$ of multiplication in any finite field $\mathbb{F}_{q^n}$ is linear with respect to the extension degree; more precisely, there exists a constant C_q such that:*

$$\mu_q^{\mathrm{sym}}(n) \leq C_q n.$$

General forms for C_q have been established since, depending on the cases where q is a prime or a prime power, a square or not... In order to obtain these good estimates for the constant C_q, S. Ballet has given in [3] some easy to verify conditions allowing the use of the D.V. and G.V. Chudnovsky algorithm. Then S. Ballet and R. Rolland have generalized in [9] the algorithm using places of degree one and two. The best finalized version of this algorithm in this direction is a generalization introduced by N. Arnaud in [1] and developed later by M. Cenk and F. Özbudak in [14]. This generalization uses several coefficients, instead of just the first one in the local expansion at each place on which we perform evaluations. Recently, Randriambolona introduced in [19] a new generalization of the algorithm, which allows asymmetry in the construction.

From the results of [3] and the generalized symmetric algorithm, we obtain (cf. [3], [9]):

Theorem 2. *Let q be a prime power and let $n > 1$ be an integer. Let $F/\mathbb{F}_q$ be an algebraic function field of genus g and N_k be the number of places of degree k in $F/\mathbb{F}_q$. If $F/\mathbb{F}_q$ is such that $2g+1 \leq q^{\frac{n-1}{2}}(q^{\frac{1}{2}}-1)$ then:*

1) if $N_1 > 2n+2g-2$, then

$$\mu_q^{\mathrm{sym}}(n) \leq 2n+g-1,$$

2) if there is a non-special divisor of degree $g-1$ and $N_1+2N_2 > 2n+2g-2$, then

$$\mu_q^{\mathrm{sym}}(n) \leq 3n+3g,$$

3) if $N_1+2N_2 > 2n+4g-2$, then

$$\mu_q^{\mathrm{sym}}(n) \leq 3n+6g.$$

Theorem 3. *Let q be a square ≥ 25. Then*

$$m_q^{\mathrm{sym}} \leq 2\left(1+\frac{1}{\sqrt{q}-3}\right).$$

Moreover, let us recall a very useful lemma due to D.V. and G.V. Chudnovsky [15] and Shparlinski, Tsfasman and Vlăduţ [21, Lemma 1.2 and Corollary 1.3].

Lemma 1. *For any prime power q and for all positive integers n and m, one has*

$$\mu_q(m) \leq \mu_q(mn) \leq \mu_q(n)\cdot\mu_{q^n}(m),$$
$$m_q \leq m_{q^n}\cdot\mu_q(n)/n,$$
$$M_q \leq M_{q^n}\cdot\mu_q(n).$$

Note that these inequalities are also true in the symmetric case. Recall the following definitions that will be useful in the sequel. Let $F/\mathbb{F}_q$ be a function field over the finite field $\mathbb{F}_q$ and $N_1(F)$ be the number of places of degree one of $F/\mathbb{F}_q$. Let us define:

$$N_q(g) = \max\left\{N_1(F) \,\middle|\, F \text{ is a function field over } \mathbb{F}_q \text{ of genus } g\right\}$$

and

$$A(q) = \limsup_{g \to +\infty} \frac{N_q(g)}{g}.$$

We know that (Drinfeld-Vlăduţ bound):

$$A(q) \leq q^{\frac{1}{2}} - 1,$$

the bound being reached if and only if q is a square.

2 New Upper Bounds for m_q^{sym} and M_q^{sym}

In this section, we give upper bounds for the asymptotical quantities M_q^{sym} and m_q^{sym} which are defined respectively by (3) and (4). As was noted in [11, p. 694] and more precisely in [12, Section 5] (cf. also [18]), Theorems 3.1 and 3.9 in [21] are not completely correct. We are going to repair that in the following two propositions.

Proposition 1. *Let q be a prime power such that $A(q) > 2$. Then*

$$m_q^{\mathrm{sym}} \leq 2\left(1 + \frac{1}{A(q) - 2}\right).$$

Proof. Let $\{F_s/\mathbb{F}_q\}_s$ be a sequence of algebraic function fields defined over $\mathbb{F}_q$. Let us denote by g_s the genus of $F_s/\mathbb{F}_q$ and by $N_1(s)$ the number of places of degree 1 of $F_s/\mathbb{F}_q$. Suppose that the sequence $(F_s/\mathbb{F}_q)_s$ was chosen such that:

1. $\lim_{s \to +\infty} g_s = +\infty$,
2. $\lim_{s \to +\infty} \frac{N_1(s)}{g_s} = A(q)$.

Let ϵ be any real number such that $0 < \epsilon < \frac{A(q)}{2} - 1$. Let us define the following integer

$$n_s = \left\lfloor \frac{N_1(s) - 2g_s(1+\epsilon)}{2} \right\rfloor.$$

Let us remark that

$$N_1(s) = g_s A(q) + o(g_s),$$

$$\text{so } N_1(s) - 2(1+\epsilon)g_s = g_s\left(A(q) - 2(1+\epsilon)\right) + o(g_s).$$

Then the following holds:

1. there exists an integer s_0 such that for any $s \geq s_0$ the integer n_s is strictly positive,
2. for any real number c such that $0 < c < A(q) - 2(1+\epsilon)$ there exists an integer s_1 such that for any integer $s \geq s_1$ the following holds: $n_s \geq \frac{c}{2} g_s$, hence n_s tends to $+\infty$,
3. there exists an integer s_2 such that for any integer $s \geq s_2$ the following holds: $2g_s + 1 \leq q^{\frac{n_s - 1}{2}}\left(q^{\frac{1}{2}} - 1\right)$ and consequently there exists a place of degree n_s (cf. [22, Corollary 5.2.10 (c) p. 207]),

4. the following inequality holds: $N_1(s) > 2n_s + 2g_s - 2$ and consequently, using Theorem 2 we conclude that $\mu_q^{\mathrm{sym}}(n_s) \le 2n_s + g_s - 1$.

Consequently,

$$\frac{\mu_q^{\mathrm{sym}}(n_s)}{n_s} \le 2 + \frac{g_s - 1}{n_s},$$

so

$$m_q^{\mathrm{sym}} \le 2 + \lim_{s \to +\infty} \frac{2g_s - 2}{N_1(s) - 2(1+\epsilon)g_s - 2} \le 2\left(1 + \frac{1}{A(q) - 2(1+\epsilon)}\right).$$

This inequality holding for any $\epsilon > 0$ sufficiently small, we then obtain the result. □

Corollary 1. *Let $q = p^m$ be a prime power such that $q \ge 4$. Then*

$$m_{q^2}^{\mathrm{sym}} \le 2\left(1 + \frac{1}{q-3}\right).$$

Note that this corollary lightly improves Theorem 3. Now in the case of arbitrary q, we obtain:

Corollary 2. *For any $q = p^m > 3$,*

$$m_q^{\mathrm{sym}} \le 3\left(1 + \frac{1}{q-3}\right).$$

Proof. For any $q = p^m > 3$, we have $q^2 = p^{2m} \ge 16$ and thus Corollary 1 gives $m_{q^2}^{\mathrm{sym}} \le 2\left(1 + \frac{1}{q-3}\right)$. Then, by Lemma 1, we have

$$m_q^{\mathrm{sym}} \le m_{q^2}^{\mathrm{sym}} \cdot \mu_q^{\mathrm{sym}}(2)/2$$

which gives the result since $\mu_q^{\mathrm{sym}}(2) = 3$ for any q. □

Now, we are going to show that for M_q^{sym} the same upper bound as for m_q^{sym} can be proved though only in the case of q being an even power of a prime. However, we are going to prove that in the case of q being an odd power of a prime, the difference between the two bounds is very slight.

Proposition 2. *Let $q = p^m$ be a prime power such that $q \ge 4$. Then*

$$M_{q^2}^{\mathrm{sym}} \le 2\left(1 + \frac{1}{q-3}\right).$$

Proof. Let $q = p^m$ be a prime power such that $q \ge 4$. Let us consider two cases. First, we suppose that $q = p$. Moreover, firstly, let us consider the characteristic p such that $p \ne 11$. Then it is known ([23] and [21]) that the curve $X_k = X_0(11\ell_k)$, where ℓ_k is the kth prime number, has a genus $g_k = \ell_k$ and satisfies $N_1(X_k(\mathbb{F}_{q^2})) \geq (q-1)(g_k+1)$ where $N_1(X_k(\mathbb{F}_{q^2}))$ denotes the number of rational points over $\mathbb{F}_{q^2}$ of the curve X_k. Let us consider a sufficiently large n. There exist two consecutive prime numbers ℓ_k and ℓ_{k+1} such that $(p-1)(\ell_{k+1}+1) > 2n + 2\ell_{k+1} - 2$ and $(p-1)(\ell_k+1) \leq 2n + 2\ell_k - 2$. Let us consider the algebraic function field $F_{k+1}/\mathbb{F}_{p^2}$ associated to the curve X_{k+1} of genus ℓ_{k+1} defined over $\mathbb{F}_{p^2}$. Let $N_i(F_k/\mathbb{F}_{p^2})$ be the number of places of degree i of $F_k/\mathbb{F}_{p^2}$. Then we get $N_1(F_{k+1}/\mathbb{F}_{p^2}) \geq (p-1)(\ell_{k+1}+1) > 2n + 2\ell_{k+1} - 2$. Moreover, it is known that $N_n(F_{k+1}/\mathbb{F}_{p^2}) > 0$ for any integer n sufficiently large. We also know that $\ell_{k+1} - \ell_k \leq \ell_k^{0,525}$ for any integer $k \geq k_0$ where k_0 can be effectively determined by [2]. Then there exists a real number $\epsilon > 0$ such that $\ell_{k+1} - \ell_k = \epsilon \ell_k \leq \ell_k^{0,525}$ namely $\ell_{k+1} \leq (1+\epsilon)\ell_k$. It is sufficient to choose ϵ such that $\epsilon \ell_k^{0,475} \leq 1$. Consequently, for any integer n sufficiently large, this algebraic function field $F_{k+1}/\mathbb{F}_{p^2}$ satisfies Theorem 2, and so $\mu_{p^2}^{\mathrm{sym}}(n) \leq 2n + \ell_{k+1} - 1 \leq 2n + (1+\epsilon)\ell_k - 1$ with $\ell_k \leq \frac{2n}{p-3} - \frac{p+1}{p-3}$. Thus, as $n \to +\infty$ then $\ell_k \to +\infty$ and $\epsilon \to 0$, so we obtain $M_{p^2}^{\mathrm{sym}} \leq 2\left(1 + \frac{1}{p-3}\right)$. Note that for $p = 11$, Proposition 4.1.20 in [23] enables us to obtain $g_k = \ell_k + O(1)$.

Now, let us study the more difficult case where $q = p^m$ with $m > 1$. We use the Shimura curves as in [21]. Recall the construction of this good family. Let L be a totally real abelian over $\mathbb{Q}$ number field of degree m in which p is inert, thus the residue class field $\mathcal{O}_L/(p)$ of p, where $\mathcal{O}_L$ denotes the ring of integers of L, is isomorphic to the finite field $\mathbb{F}_q$. Let $\wp$ be a prime of L which does not divide p and let B be a quaternion algebra for which

$$B \otimes_{\mathbb{Q}} \mathbb{R} = \mathrm{M}_2(\mathbb{R}) \otimes \mathbb{H} \otimes \cdots \otimes \mathbb{H}$$

where $\mathbb{H}$ is the skew field of Hamilton quaternions. Let B be also unramified at any finite place if $(m-1)$ is even; let B be also unramified outside infinity and $\wp$ if $(m-1)$ is odd. Then, over L one can define the Shimura curve by its complex points $X_\Gamma(\mathbb{C}) = \Gamma \setminus \mathfrak{h}$, where $\mathfrak{h}$ is the Poincaré upper half-plane and Γ is the group of units of a maximal order $\mathcal{O}$ of B with totally positive norm modulo its center. Hence, the considered Shimura curve admits an integral model over L and it is well known that its reduction $X_{\Gamma,p}(\mathbb{F}_{p^{2m}})$ modulo p is good and is defined over the residue class field $\mathcal{O}_L/(p)$ of p, which is isomorphic to $\mathbb{F}_q$ since p is inert in L. Moreover, by [17], the number $N_1(X_{\Gamma,p}(\mathbb{F}_{q^2}))$ of $\mathbb{F}_{q^2}$-points of $X_{\Gamma,p}$ is such that $N_1(X_{\Gamma,p}(\mathbb{F}_{q^2})) \geq (q-1)(g+1)$, where g denotes the genus of $X_{\Gamma,p}(\mathbb{F}_{q^2})$. Let now ℓ be a prime which is greater than the maximum order of stabilizers Γ_z, where $z \in \mathfrak{h}$ is a fixed point of Γ and let $\wp \nmid \ell$. Let $\Gamma_0(\ell)_\ell$ be the following subgroup of $\mathrm{GL}_2(\mathbb{Z}_\ell)$:

$$\Gamma_0(\ell)_\ell = \left\{ \begin{pmatrix} a & b \\ c & d \end{pmatrix} \in \mathrm{GL}_2(\mathbb{Z}_\ell)\,;\, c \equiv 0 \pmod{\ell} \right\}.$$

Suppose that ℓ splits completely in L. Then there exists an embedding $L \to \mathbb{Q}_\ell$ where $\mathbb{Q}_\ell$ denotes the usual ℓ-adic field, and since $B \otimes_{\mathbb{Q}} \mathbb{Q}_\ell = \mathrm{M}_2(\mathbb{Q}_\ell)$, we have a natural map:

$$\phi_\ell : \Gamma \to \mathrm{GL}_2(\mathbb{Z}_\ell).$$

Let Γ_ℓ be the inverse image of $\Gamma_0(\ell)_\ell$ in Γ under ϕ_ℓ. Then Γ_ℓ is a subgroup of Γ of index ℓ. We consider the Shimura curve X_ℓ with

$$X_\ell(\mathbb{C}) = \Gamma_\ell \setminus \mathfrak{h}.$$

It admits an integral model over L and so can be defined over L. Hence, its reduction $X_{\ell,p}$ modulo p is good and it is defined over the residue class field $\mathcal{O}_L/(p)$ of p, which is isomorphic to $\mathbb{F}_q$ since p is inert in L. Moreover the supersingular $\mathbb{F}_p$-points of $X_{\Gamma,p}$ split completely in the natural projection

$$\pi_\ell : X_{\ell,p} \rightarrow X_{\Gamma,p}.$$

Thus, the number of rational points of $X_{\ell,p}(\mathbb{F}_{q^2})$ verifies:

$$N_1(X_{\ell,p}(\mathbb{F}_{q^2})) \geq \ell(q-1)(g+1).$$

Moreover, since ℓ is greater than the maximum order of a fixed point of Γ on $\mathfrak{h}$, the projection π_ℓ is unramified and thus by Hurwitz formula,

$$g_\ell = 1 + \ell(g-1)$$

where g_ℓ is the genus of X_ℓ (and also of $X_{\ell,p}$).

Note that since the field L is abelian over $\mathbb{Q}$, there exists an integer N such that the field L is contained in a cyclotomic extension $\mathbb{Q}(\zeta_N)$ where ζ_N denotes a primitive root of unity with minimal polynomial Φ_N. Let us consider the reduction $\Phi_{N,\ell}$ of Φ_N modulo the prime ℓ. Then, the prime ℓ is totally split in the integer ring of L if and only if the polynomial $\Phi_{N,\ell}$ is totally split in $\mathbb{F}_\ell = \mathbb{Z}/\ell\mathbb{Z}$ i.e. if and only if $\mathbb{F}_\ell$ contains the Nth roots of unity which is equivalent to $N \mid \ell - 1$. Hence, any prime ℓ such that $\ell \equiv 1 \pmod{N}$ is totally split in $\mathbb{Q}(\zeta_N)$ and then in L. Since ℓ runs over primes in an arithmetical progression, the ratio of two consecutive prime numbers $\ell \equiv 1 \pmod{N}$ tends to one.

Then for any real number $\epsilon > 0$, there exists an integer k_0 such that for any integer $k \geq k_0$, $\ell_{k+1} \leq (1+\epsilon)\ell_k$ where ℓ_k and ℓ_{k+1} are two consecutive prime numbers congruent to one modulo N. Then there exists an integer n_ϵ such that for any integer $n \geq n_\epsilon$, the integer k such that the two following inequalities hold

$$\ell_{k+1}(q-1)(g+1) > 2n + 2g_{\ell_{k+1}} - 2$$

and

$$\ell_k(q-1)(g+1) \leq 2n + 2g_{\ell_k} - 2,$$

satisfies $k \geq k_0$; where $g_{\ell_i} = 1 + \ell_i(g-1)$ for any integer i.

Let us consider the algebraic function field $F_k/\mathbb{F}_{q^2}$ defined over the finite field $\mathbb{F}_{q^2}$ associated to the Shimura curve X_{ℓ_k} of genus g_{ℓ_k}. Let $N_i(F_k/\mathbb{F}_{q^2})$ be the number of places of degree i of $F_k/\mathbb{F}_{q^2}$. Then

$$N_1(F_{k+1}/\mathbb{F}_{q^2}) \geq \ell_{k+1}(q-1)(g+1) > 2n + 2g_{\ell_{k+1}} - 2$$

where g is the genus of the Shimura curve $X_{\Gamma,p}(\mathbb{F}_{q^2})$. Moreover, it is known that there exists an integer n_0 such that for any integer $n \geq n_0$, $N_n(F_{k+1}/\mathbb{F}_{q^2}) > 0$.

Consequently, for any integer $n \geq \max(n_\epsilon, n_0)$ this algebraic function field $F_{k+1}/\mathbb{F}_{q^2}$ satisfies Theorem 2 and so

$$\mu_{q^2}^{\mathrm{sym}}(n) \leq 2n + g_{\ell_{k+1}} - 1 \leq 2n + \ell_{k+1}(g-1) \leq 2n + (1+\epsilon)\ell_k(g-1)$$

with $\ell_k < \frac{2n}{(q-1)(g+1)-2(g-1)}$. Thus, for any real number $\epsilon > 0$ and for any $n \geq \max(n_\epsilon, n_0)$, we obtain $\mu_{q^2}^{\mathrm{sym}}(n) \leq 2n + \frac{2n(1+\epsilon)(g-1)}{(q-1)(g+1)-2(g-1)}$ which gives $M_{q^2}^{\mathrm{sym}} \leq 2\left(1 + \frac{1}{q-3}\right)$. □

Proposition 3. *Let $q = p^m$ be a prime power with odd m such that $q \geq 5$. Then*

$$M_q^{\mathrm{sym}} \leq 3\left(1 + \frac{2}{q-3}\right).$$

Proof. It is sufficient to consider the same families of curves than in Proposition 2. These families of curves $\{X_k\}$ are defined over the residue class field of p which is isomorphic to $\mathbb{F}_q$. Hence, we can consider the associated algebraic function fields $F_k/\mathbb{F}_q$ defined over $\mathbb{F}_q$. If $q = p$, we have $N_1(F_{k+1}/\mathbb{F}_{p^2}) = N_1(F_{k+1}/\mathbb{F}_p) + 2N_2(F_{k+1}/\mathbb{F}_p) \geq (p-1)(\ell_{k+1}+1) > 2n + 2\ell_{k+1} - 2$ since $F_{k+1}/\mathbb{F}_{p^2} = F_{k+1}/\mathbb{F}_p \otimes_{\mathbb{F}_p} \mathbb{F}_{p^2}$. Then, for any real number $\epsilon > 0$ and for any integer n sufficiently large, we have $\mu_p^{\mathrm{sym}}(n) \leq 3n + 3g_{\ell_{k+1}} \leq 3n + 3(1+\epsilon)\ell_k$ by Theorem 2 since $N_n(F_{k+1}/\mathbb{F}_{q^2}) > 0$. Then, by using the condition $\ell_k \leq \frac{2n}{p-3} - \frac{p+1}{p-3}$, we obtain $M_p^{\mathrm{sym}} \leq 3\left(1 + \frac{2}{p-3}\right)$. If $q = p^m$ with odd m, we have $N_1(F_{k+1}/\mathbb{F}_{q^2}) = N_1(F_{k+1}/\mathbb{F}_q) + 2N_2(F_{k+1}/\mathbb{F}_q) \geq \ell_{k+1}(q-1)(g+1) > 2n + 2g_{\ell_{k+1}} - 2$ since $F_{k+1}/\mathbb{F}_{q^2} = F_{k+1}/\mathbb{F}_q \otimes_{\mathbb{F}_q} \mathbb{F}_{q^2}$. Then, for any real number $\epsilon > 0$ and for any integer n sufficiently large as in Proof of Proposition 2, we have $\mu_q^{\mathrm{sym}}(n) \leq 3n + 3g_{\ell_{k+1}} \leq 3n + 3(1+\epsilon)\ell_k(g-1)$ by Theorem 2 since $N_n(F_{k+1}/\mathbb{F}_{q^2}) > 0$. Then, by using the condition $\ell_k < \frac{2n}{(q-1)(g+1)-2(g-1)}$ we obtain $M_q^{\mathrm{sym}} \leq 3\left(1 + \frac{2}{q-3}\right)$. □

Remark 1. Note that in [13, Lemma IV.4], Elkies gives another construction of a family $\{\chi_s\}_{s=1}^{\infty}$ of Shimura curves over $\mathbb{F}_q$ satisfying for any prime power q and for any integer $t \geq 1$ the following conditions:

(i) the genus $g(F_s)$ tends to $+\infty$ as s tends to $+\infty$, where F_s stands for the function field $\mathbb{F}_q(\chi_s)$,
(ii) $\lim_{s\to+\infty} g(F_s)/g(F_{s-1}) = 1$,
(iii) $\lim_{s\to+\infty} B_{2t}(F_s)/g(F_s) = (q^t - 1)/(2t)$, where $B_{2t}(F_s)$ stands for the number of places of degree $2t$ in F_s.

However, this construction is not sufficiently explicit to enable Cascudo and al. [12] (and [13]) to derive the best bounds in all the cases (cf. Section 3). Indeed, let us recall the construction of Elkies.

Let $q = p^r$ be a prime power and put $f = rt$. Let K be a totally real number field such that $K/\mathbb{Q}$ is a Galois extension of degree f and p is totally inert in K.

Let B be a quaternion algebra over K such that the set S of non-archimedean primes of K that are ramified in B is Galois invariant. Note that B can be constructed by taking S to be either the empty set for odd f, or the set of primes lying over p for even f (see [21]).

Let $\ell \neq p$ be a rational prime outside S such that ℓ is totally inert in K (note that in [21], ℓ is chosen such that it is completely splitting). Consider the Shimura curve $X_0^B(\ell) := \Gamma_0(\ell \mathcal{O}_K)\backslash\mathfrak{h}$, where $\mathfrak{h}$ is the upper half-plane and $\Gamma_0(\ell\mathcal{O}_K)$ is the subgroup of the unit group of the maximal order of B mapping to upper triangle matrices modulo $\ell\mathcal{O}_K$. Then $X_0^B(\ell)$ is defined over the rational field $\mathbb{Q}$ and has a good reduction modulo p. Thus, the reduction of $X_0^B(\ell)$ is defined over $\mathbb{F}_p$, and therefore over $\mathbb{F}_q$ as well. This curve has at least $(p^f - 1)g_\ell$ supersingular points over $\mathbb{F}_{p^{2f}} = \mathbb{F}_{q^{2t}}$, where g_ℓ is the genus of $X_0^B(\ell)$. One knows that the ratio g_ℓ/ℓ^f tends to a fixed number a when ℓ tends to $+\infty$. Now let $\{\ell_s\}_{s=1}^{+\infty}$ be the set of consecutive primes such that ℓ_s are totally inert in K and $\ell_s \notin \mathsf{S}$. By Chebotarev's density theorem, we have $\ell_s/\ell_{s-1} \to 1$ as s tends to $+\infty$. Hence, $g_{\ell_s}/g_{\ell_{s-1}} \to 1$ as s tends to $+\infty$.

For the family of function fields $\{F_s/\mathbb{F}_q\}$ of the above Shimura curves, the number $N_{2t}(F_s)$ of $\mathbb{F}_{q^{2t}}$-rational places of F_s satisfies

$$\lim_{g(F_s)\to+\infty} \frac{N_{2t}(F_s)}{g(F_s)} = p^f - 1 = q^t - 1.$$

Moreover, (i) and (ii) are satisfied as well.

By the identity $N_{2t}(F_s) = \sum_{i|2t} iB_i(F_s)$, we get

$$\begin{aligned}\liminf_{g(F_s)\to+\infty} \frac{1}{g(F_s)}\sum_{i=1}^{2t}\frac{iB_i(F_s)}{q^t-1} &\geq \liminf_{g(F_s)\to+\infty} \frac{1}{g(F_s)}\sum_{i|2t}\frac{iB_i(F_s)}{q^t-1}\\ &= \liminf_{g(F_s)\to+\infty} \frac{N_{2t}(F_s)}{g(F_s)(q^t-1)} = 1.\end{aligned}$$

Thus, the inequality

$$\liminf_{g(F_s)\to+\infty} \frac{1}{g(F_s)}\sum_{i=1}^{2t}\frac{iB_i(F_s)}{q^t-1} \geq 1$$

is satisfied and consequently (iii) is also satisfied by [13, Lemma IV.3].

3 Comparison with the Current Best Asymptotical Bounds

In this section, we recall the results obtained in [13, Theorem IV.6 and IV.7] and [12, Theorem 5.18] which are known to give the best current estimates for M_q^{sym}, and compare these bounds to those established in Propositions 2 and 3.

3.1 Comparison with the Bounds in [13]

In [13], the authors establish the following results:

Theorem 4. *For any prime power $q \geq 2$, one has*

$$M_q^{\mathrm{sym}} \leq \mu_q^{\mathrm{sym}}(2t)\frac{q^t - 1}{t(q^t - 5)} \tag{5}$$

for any $t \geq 1$ as long as $q^t - 5 > 0$, and

$$M_{q^2}^{\mathrm{sym}} \leq \mu_{q^2}^{\mathrm{sym}}(t)\frac{2(q^t - 1)}{t(q^t - 5)} \tag{6}$$

for any $t \geq 1$ as long as $q^t - 5 > 0$.

Let us show that our results are better than those of this theorem except for some small values of q.

Bounds Over $\mathbb{F}_q$. The estimates obtained in [13, Corollary IV.8.] show that (5) gives better bounds than Proposition 3 for any $q \leq 13$. Indeed, one has:

q	5	7	8	9	11	13
M_q^{sym} [13, Cor. IV.8]	4.8	3.82	3.74	3.68	3.62	3.59
M_q^{sym} [Prop. 3]	6	4.5	4.2	4	3.75	3.6

However, as soon as $q \geq 15$, our estimate is sharper than (5). Indeed, for $q \geq 15$, Proposition 3 gives:

$$M_q^{\mathrm{sym}} \leq 3.5.$$

On the other hand, since $\mu_q^{\mathrm{sym}}(2t) \geq 4t - 1$, the best estimate that can be obtained with Bound (5) is:

$$M_q^{\mathrm{sym}} \leq (4t-1) \cdot \frac{q^t - 1}{t(q^t - 5)} = \left(4 - \frac{1}{t}\right) \cdot \left(1 + \frac{4}{q^t - 5}\right). \tag{7}$$

Thus one must have $4 - \frac{1}{t} < 3.5$ to obtain a better estimate than 3.5, which requires $t = 1$. In this case, (7) becomes:

$$M_q^{\mathrm{sym}} \leq 3\left(1 + \frac{4}{q-5}\right)$$

which is less precise than the bound of Proposition 3 for any $q \geq 15$.

Bounds Over $\mathbb{F}_{q^2}$. For $q = 4$, Proposition 2 gives $M_{q^2}^{\mathrm{sym}} \leq 4$, which is less sharp than Bound (6) applied with $t = 4$, which leads to $M_{q^2}^{\mathrm{sym}} \leq 3.56$.

However, for any $q \geq 5$, Proposition 2 gives better bounds than (6). Indeed, since $\mu_{q^2}^{\mathrm{sym}}(t) \geq 2t - 1$, the best estimate that can be obtained with (6) is:

$$M_{q^2}^{\mathrm{sym}} \leq 2(2t-1) \cdot \frac{q^t - 1}{t(q^t - 5)} = \left(4 - \frac{2}{t}\right) \cdot \left(1 + \frac{4}{q^t - 5}\right). \tag{8}$$

Since Proposition 2 gives $M_{q^2}^{\mathrm{sym}} \leq 3$ for any $q \geq 5$, it is necessary to have $4 - \frac{2}{t} < 3$ to obtain a better bound with (8), which requires $t = 1$. This is impossible for $q = 5$ since Bound (6) is undefined in this case, and for $q > 5$ and $t = 1$, (8) becomes:

$$M_{q^2}^{\mathrm{sym}} \leq 2\left(1 + \frac{4}{q-5}\right)$$

which is less sharp than the bound obtained from Proposition 2.

3.2 Comparison with the Bounds in [12]

In [12] (which is an extended version of [11]), the authors establish the following asymptotic bounds:

Theorem 5. *For a prime power q, one has*

$$M_q^{\mathrm{sym}} \leq \begin{cases} \mu_q^{\mathrm{sym}}(2t)\frac{q^t-1}{t(q^t-2-\log_q 2)}, & \text{if } 2|q \\ \mu_q^{\mathrm{sym}}(2t)\frac{q^t-1}{t(q^t-2-2\log_q 2)}, & \text{otherwise} \end{cases}$$

for any $t \geq 1$ *as long as* $q^t - 2 - \log_q 2 > 0$ *for even* q*; and* $q^t - 2 - 2\log_q 2 > 0$ *for odd* q*.*

This bound always beats the one of Proposition 3 for arbitrary q (for example, by setting $t = 1$ and $\mu_q^{\mathrm{sym}}(2t) = 4t - 1$). Nevertheless, if we focus on the case of $M_{q^2}^{\mathrm{sym}}$, then the bound of Proposition 2 is better as soon as $q > 5$ since in this case, it gives:

$$M_{q^2}^{\mathrm{sym}} < 3$$

which can not be reached with the bound of Theorem 3.2, since the best that one can get is:

$$M_q^{\mathrm{sym}} \leq \begin{cases} \left(4 - \frac{1}{t}\right)\left(1 + \frac{1+\log_q 2}{q^t-2-\log_q 2}\right), & \text{if } 2|q \\ \left(4 - \frac{1}{t}\right)\left(1 + \frac{1+2\log_q 2}{q^t-2-2\log_q 2}\right), & \text{otherwise} \end{cases}$$

which obviously can not be < 3.

References

1. Arnaud, N.: Evaluations Dérivés, multiplication dans les corps finis et codes correcteurs. PhD thesis, Université de la Méditerranée, Institut de Mathématiques de Luminy (2006)
2. Baker, R., Harman, G., Pintz, J.: The difference between consecutive primes, II. Proceedings of the London Mathematical Society 83(3), 532–562 (2001)
3. Ballet, S.: Curves with many points and multiplication complexity in any extension of $\mathbb{F}_q$. Finite Fields and Their Applications 5, 364–377 (1999)
4. Ballet, S.: Low increasing tower of algebraic function fields and bilinear complexity of multiplication in any extension of $\mathbb{F}_q$. Finite Fields and Their Applications 9, 472–478 (2003)

5. Ballet, S.: On the tensor rank of the multiplication in the finite fields. Journal of Number Theory 128, 1795–1806 (2008)
6. Ballet, S., Chaumine, J.: On the bounds of the bilinear complexity of multiplication in some finite fields. Applicable Algebra in Engineering Communication and Computing 15, 205–211 (2004)
7. Ballet, S., Le Brigand, D.: On the existence of non-special divisors of degree g and $g-1$ in algebraic function fields over $\mathbb{F}_q$. Journal on Number Theory 116, 293–310 (2006)
8. Ballet, S., Le Brigand, D., Rolland, R.: On an application of the definition field descent of a tower of function fields. In: Proceedings of the Conference Arithmetic, Geometry and Coding Theory, AGCT 2005. Séminaires et Congrès, vol. 21, pp. 187–203. Société Mathématique de France (2009)
9. Ballet, S., Rolland, R.: Multiplication algorithm in a finite field and tensor rank of the multiplication. Journal of Algebra 272(1), 173–185 (2004)
10. Baum, U., Shokrollahi, A.: An optimal algorithm for multiplcation in $\mathbb{F}_{256}/\mathbb{F}_4$. Applicable Algebra in Engineering, Communication and Computing 2(1), 15–20 (1991)
11. Cascudo, I., Cramer, R., Xing, C.: The torsion-limit for algebraic function fields and its application to arithmetic secret sharing. In: Rogaway, P. (ed.) CRYPTO 2011. LNCS, vol. 6841, pp. 685–705. Springer, Heidelberg (2011)
12. Cascudo, I., Cramer, R., Xing, C.: Torsion limits and Riemann-Roch systems for function fields and applications. ArXiv,1207.2936v1 (2012)
13. Cascudo, I., Cramer, R., Xing, C., Yang, A.: Asymptotic bound for multiplication complexity in the extensions of small finite fields. IEEE Transactions on Information Therory 58(7), 4930–4935 (2012)
14. Cenk, M., Özbudak, F.: On multiplication in finite fields. Journal of Complexity 26(2), 172–186 (2010)
15. Chudnovsky, D., Chudnovsky, G.: Algebraic complexities and algebraic curves over finite fields. Journal of Complexity 4, 285–316 (1988)
16. de Groote, H.: Characterization of division algebras of minimal rank and the structure of their algorithm varieties. SIAM Journal on Computing 12(1), 101–117 (1983)
17. Ihara, Y.: Some remarks on the number of rational points of algebraic curves over finite fields. Journal of the Faculty of Science, University of Tokyo 28, 721–724 (1981)
18. Pieltant, J.: Tours de corps de fonctions algébriques et rang de tenseur de la multiplication dans les corps finis. PhD thesis, Université d'Aix-Marseille, Institut de Mathématiques de Luminy (2012)
19. Randriambololona, H.: Bilinear complexity of algebras and the Chudnovsky-Chudnovsky interpolation method. Journal of Complexity 28, 489–517 (2012)
20. Shokhrollahi, A.: Optimal algorithms for multiplication in certain finite fields using algebraic curves. SIAM Journal on Computing 21(6), 1193–1198 (1992)
21. Shparlinski, I., Tsfasman, M., Vladut, S.: Curves with many points and multiplication in finite fields. In: Stichtenoth, H., Tsfasman, M. (eds.) Coding Theory and Algebraic Geometry, Proceedings of AGCT-3 Conference, Luminy, June 17-21, 1991. Lectures Notes in Mathematics, vol. 1518, pp. 145–169. Springer, Berlin (1992)
22. Stichtenoth, H.: Algebraic Function Fields and Codes. Lectures Notes in Mathematics, vol. 314. Springer (1993)
23. Tsfasman, M., Vladut, S.: Asymptotic properties of zeta-functions. Journal of Mathematical Sciences 84(5), 1445–1467 (1997)
24. Winograd, S.: On multiplication in algebraic extension fields. Theoretical Computer Science 8, 359–377 (1979)

Stochastic Equationality

Symeon Bozapalidis and George Rahonis

Department of Mathematics
Aristotle University of Thessaloniki
54124, Thessaloniki, Greece
{bozapali,grahonis}@math.auth.gr

Abstract. We introduce systems of equations of stochastic tree series and we consider two types of solutions, the *[IO]* and the *OI*, according to the substitutions we use to solve them. We show the existence of least *[IO]*- and *OI*-solutions whose non-zero components are proved to be stochastic tree series. A Kleene characterization holds for stochastically *OI*-equational tree series, i.e., components of least *OI*-solutions. Furthermore, we consider stochastic algebras and we state a Mezei-Wright type result relating least solutions of systems in arbitrary stochastic algebras and the term algebra.

Keywords: Stochastic tree series, systems of equations, least *[IO]*- and *OI*-solutions, Mezei-Wright result.

1 Introduction

Systems of equations of stochastic polynomials arise in several areas of Computer Science like probabilistic program verification, analysis of recursive Markov chains, multi-type branching processes, and model checking of recursive probabilistic systems (cf. [8,9,10,11]). Least solutions of such systems play a central role also in other sciences, namely in Physics, Biology and Computational Linguistics [8].

On the other hand, least *OI*-solutions of systems of equations of finite tree languages coincide, according to an important result of Mezei and Wright [15], with recognizable tree languages. Recently, in [4], most of the well-known tree transductions characterized as least *[IO]*- and *OI*-solutions of systems of equations of finite tree transformations. In the weighted setup, the aforementioned result of Mezei and Wright proved in [1] for tree series over fields, and in [3] for tree series over well ω-additive semirings. Furthermore, the relation among least *[IO]*- and *OI*-solutions of systems of equations of weighted tree transformations and weighted tree transductions was investigated in [5], and with discounting in [12]. We refer the reader to [2,3,7,13,14] for tree series, and systems of equations of tree series over term and general algebras.

In this paper, we consider systems of equations of stochastic tree series, i.e., systems of the form

$$\text{(E)} \qquad x_1 = p_1, \ldots, x_n = p_n,$$

T. Muntean, D. Poulakis, and R. Rolland (Eds.): CAI 2013, LNCS 8080, pp. 173–185, 2013.

where p_i, for every $1 \leq i \leq n$, is a tree series with finite support whose coefficients sum up to 1. We consider solutions of such systems using *[IO]*- and *OI*-substitutions of tree series. The existence of least *[IO]*- and *OI*-solutions is ensured by the fixpoint theorem of Tarski. Tree series obtained as components of least *[IO]*- and *OI*-solutions of our systems are called stochastically *[IO]*-equational (resp. *OI*-equational) and they are proved to be stochastic (whenever they differ from the constant series $\widetilde{0}$). We show that a Kleene characterization holds for stochastically *OI*-equational tree series. More precisely, the class of stochastically *OI*-equational tree series is the smallest convex set containing the constant tree series $\widetilde{0}$ and characteristic series of trees and being closed under *OI*-substitution and star operation. Furthermore, we state that the closure of the class of stochastically *OI*-equational tree series under nondeleting tree homomorphisms coincides with the class of stochastically *[IO]*-equational ones.

We consider also stochastic Σ-algebras (Σ a ranked alphabet), and we solve systems of equations of the above form in such algebras. Stochastic functions obtained as components of least *[IO]*-solutions (resp. *OI*-solutions) are called stochastically *[IO]*-regular (resp. *OI*-regular). We show the robustness of our theory, by proving a Mezei-Wright result relating stochastically *OI*-equational tree series to stochastically *OI*-regular functions, and stochastically *[IO]*-equational tree series to stochastically *[IO]*-regular functions.

2 Preliminaries

We denote by $\mathbb{R}_+$ the set of nonnegative reals. Multiplication in real numbers will be denoted simply by concatenation. We recall a notation from [5]. More precisely, let A be a set, $n \geq 1$, $1 \leq i_1 < \ldots < i_k \leq n$, and $a_1, \ldots, a_k \in A$. We let

$$A^n|_{(i_1,a_1)\ldots(i_k,a_k)} = \{(b_1, \ldots, b_n) \in A^n \mid b_{i_1} = a_1, \ldots, b_{i_k} = a_k\}$$

i.e., $A^n|_{(i_1,a_1)\ldots(i_k,a_k)}$ is the set of those elements of A^n, each of which has a_j as its i_jth component for $j = 1, \ldots, k$. For $n = 0$, we define $A^0 = \{(\,)\}$ (even if $A = \emptyset$), where $(\,)$ is the empty vector.

A *partially ordered set* (*poset* for short) is a pair $(A, \leq)$, where A is a set and $\leq$ is a *partial order*, i.e., a reflexive, antisymmetric, and transitive relation on A. We will write simply A for $(A, \leq)$ and, for every $A' \subseteq A$, we denote by $\sup A'$ the supremum of A' in A, if it exists. A poset A is called ω*-complete* if it has a least element $\perp$ and every ω-chain $a_0 \leq a_1 \leq \ldots$ in A has a supremum in A, denoted by $\sup_{i \geq 0} a_i$.

Let $f : A \to A$ be a mapping. A *fixpoint of* f is an element $a \in A$ such that $f(a) = a$. A fixpoint a of f is the *least fixpoint* if $a \leq a'$ for every fixpoint a' of f. Moreover, f is called ω*-continuous* if for every ω-chain $a_0 \leq a_1 \leq \ldots$ in A which has a supremum, the supremum of $\{f(a_i) \mid i \geq 0\}$ exists and $f(\sup_{i \geq 0} a_i) = \sup\{f(a_i) \mid i \geq 0\}$. It is obvious that if f is ω-continuous, then it is *monotonic*, meaning that $f(a) \leq f(a')$ whenever $a \leq a'$ for every $a, a' \in A$. The subsequent result is known as the *fixpoint theorem* (cf. e.g. [16, Sect. 1.5, Thm. 7]).

Proposition 1. *Let* $(A, \leq)$ *be an* ω*-complete poset and* $f : A \to A$ *an* ω*-continuous mapping. Then* f *has a least fixpoint* $\mathrm{fix}f$*, and* $\mathrm{fix}f = \sup\{f^{(i)}(\perp) \mid i \geq 0\}$*, where* $f^{(i)}$ *denotes the i-fold composition of* f*.*

Let $(x_i)_{i\in I}$ be a family of elements in $\mathbb{R}_+$ such that $\sup_{I'\subseteq_{fin} I}\sum_{i\in I'} x_i$ exists, where the notation $I' \subseteq_{fin} I$ means that I' is a finite subset of I. Then, we say that the sum $\sum_{i\in I} x_i$ exists and we set

$$\sum_{i\in I} x_i = \sup_{I'\subseteq_{fin} I} \sum_{i\in I'} x_i.$$

Furthermore, if the sequence $(x_{ij})_{j\geq 0}$ is increasing and $\sup_{j\geq 0} x_{ij}$ exists for every $i \in I$, and $\sum_{i\in I'} x_{ij} \leq 1$ for every $I' \subseteq_{fin} I, j \geq 0$, then

$$\sum_{i\in I} \sup_{j\geq 0} x_{ij} = \sup_{j\geq 0} \sum_{i\in I} x_{ij}.$$

Let $f : A \to [0, 1]$ be a function. The sum $\sum_{a\in A} f(a)$, whenever it exists, is called the *content of* f and is denoted by *cont*(f). If *cont*$(f) = 1$, then f is called a *stochastic function (over* A*)*. The class of all stochastic functions over A is denoted by $STOCH(A)$. The *support of* $f : A \to [0,1]$ is the set of all elements of A with non-vanishing image, i.e., *supp*$(f) = \{a \in A \mid f(a) > 0\}$. A stochastic function $f : A \to [0, 1]$ with finite support is a *stochastic polynomial.* We denote by *Stoch*(A) the family of stochastic polynomials over A. Let $(f_i)_{i\in I}$ be a family of functions $f_i : A \to \mathbb{R}_+$ such that for every $a \in A$ the sum $\sum_{i\in I} f_i(a)$ exists. Then the assignment

$$a \mapsto \sum_{i\in I} f_i(a)$$

defines a function from A to $\mathbb{R}_+$ denoted by $\sum_{i\in I} f_i$, i.e.,

$$\left(\sum_{i\in I} f_i\right)(a) = \sum_{i\in I} f_i(a)$$

for every $a \in A$.

A family $(\lambda_i)_{i\in I}$ of numbers in $[0, 1]$ is called *stochastic* if $\sum_{i\in I} \lambda_i = 1$. Let $(\lambda_i)_{i\in I}$ be a stochastic family and $f_i \in STOCH(A)$ for every $i \in I$. Then for every finite subset $I' \subseteq I$ and $a \in A$ we have $\sum_{i\in I'} \lambda_i f_i(a) \leq 1$, hence $\sup_{I'\subseteq_{fin} I} \sum_{i\in I'} \lambda_i f_i(a)$ exists, and

$$\sum_{i\in I} \lambda_i f_i(a) = \sup_{I'\subseteq_{fin} I} \sum_{i\in I'} \lambda_i f_i(a) \leq 1.$$

Therefore, the function $\sum_{i\in I} \lambda_i f_i$ is well-defined. Moreover, it is stochastic. Thus, we get the following result which is called the *Strong Convexity Lemma* (*SCL* for short).

Lemma 1 (Strong Convexity Lemma). *The set* $STOCH(A)$ *is a* strongly convex set, *i.e., if* $(\lambda_i)_{i\in I}$ *is a stochastic family and* $f_i \in STOCH(A)$ *for every* $i \in I$*, then the function* $\sum_{i\in I} \lambda_i f_i$ *is stochastic.*

The class of all functions $[0,1]^A$ is a poset where the order of functions is defined elementwise, i.e., $f \leq f'$ iff $f(a) \leq f'(a)$ for every $a \in A$. Moreover, it is ω-complete since if $(f_i)_{i \geq 0}$ is an ω-chain in $[0,1]^A$, then the function $\sup_{i\geq 0} f_i$ is well-defined by $\left(\sup_{i\geq 0} f_i\right)(a) = \sup_{i\geq 0} f_i(a)$ for every $a \in A$. If $(f_i)_{i\geq 0}$ is an ω-chain in $STOCH(A)$, then the function $\sup_{i\geq 0} f_i$ is also stochastic. Nevertheless, the class $STOCH(A)$ fails to be ω-complete since the least element of $[0,1]^A$, i.e., the constant function assigning the value 0 to every $a \in A$ is not stochastic.

Next let B be a further set and $h : A \to STOCH(B)$ a mapping. Then h can be extended to a mapping $\overline{h} : STOCH(A) \to STOCH(B)$ by setting

$$\overline{h}(f) = \sum\nolimits_{a\in A} f(a)h(a)$$

for every $f \in STOCH(A)$. Indeed, since $(f(a))_{a\in A}$ is a stochastic family of numbers and $h(a)$ is, by definition, stochastic for every $a \in A$, we get by the *SCL*, that $\overline{h}(f)$ is also stochastic. In particular, every mapping $h : A \to B$ is extended to a mapping $\overline{h} : STOCH(A) \to STOCH(B)$ using the same as above formula. Moreover, if $(\lambda_i)_{i\in I}$ is a stochastic family of numbers and $(f_i)_{i\in I}$ a family of elements in $STOCH(A)$, then we can show that

$$\overline{h}\left(\sum\nolimits_{i\in I} \lambda_i f_i\right) = \sum\nolimits_{i\in I} \lambda_i \overline{h}\left(f_i\right).$$

3 Stochastic Tree Series

A *ranked alphabet* is a pair (Σ, rk) (simply denoted by Σ) where Σ is a finite set and $rk : \Sigma \to \mathbb{N}$ is the rank function. As usual, we set $\Sigma_k = \{\sigma \in \Sigma \mid rk(\sigma) = k\}$ for every $k \geq 0$.

Let $\Sigma_0 \neq \emptyset$ and V be a finite set with $V \cap \Sigma = \emptyset$. The set $T_\Sigma(V)$ of finite trees over Σ and V is defined by induction to be the least set T such that (i) $V \subseteq T$ and (ii) if $k \geq 0$, $\sigma \in \Sigma_k$, and $t_1, \ldots, t_k \in T$, then $\sigma(t_1, \ldots, t_k) \in T$. If $\sigma \in \Sigma_0$, then we write just σ for $\sigma()$ and we write T_Σ for $T_\Sigma(\emptyset)$. Note that $T_\Sigma \neq \emptyset$ since $\Sigma_0 \neq \emptyset$. Every subset of $T_\Sigma(V)$ is called a *tree language*. Let $X = \{x_1, x_2, \ldots\}$ be a countably infinite set of *variables*, which is disjoint from every ranked alphabet considered in the paper. We set $X_n = \{x_1, \ldots, x_n\}$ for $n \geq 0$, hence $X_0 = \emptyset$. Let $t \in T_\Sigma(X_n)$ be a tree. The *set* $var(t) \subseteq X_n$ *of variables in* t is defined such that $var(t) = \{t\}$ if $t \in X_n$, and $var(t) = \bigcup_{i=1}^k var(t_i)$ if $t = \sigma(t_1, \ldots, t_k)$ for some $k \geq 0$, $\sigma \in \Sigma_k$, and $t_1, \ldots, t_k \in T_\Sigma(X_n)$. We denote by $|t|_{x_i}$ the number of occurrences of x_i in t for every $1 \leq i \leq n$. Then t is called (X_n-)*linear* (resp. *nondeleting*) if $|t|_{x_i} \leq 1$ (resp. $|t|_{x_i} \geq 1$) for every $1 \leq i \leq n$. A subset $L \subseteq T_\Sigma(X_n)$ is *linear* (resp. *nondeleting*), if every $t \in L$ is linear (resp. nondeleting).

Next we recall tree substitution. Let $t \in T_\Sigma(X_n)$ and $t_1, \ldots, t_n \in T_\Sigma(V)$. We denote by $t[t_1/x_1, \ldots, t_n/x_n]$ or simply by $t[t_1, \ldots, t_n]$ the tree which we obtain by substituting simultaneously t_i for every occurrence of x_i in t for every $1 \leq i \leq n$.

Remark 1. If $x_i \notin var(t)$, then $t[t_1,\ldots,t_i,\ldots,t_n] = t[t_1,\ldots,t',\ldots,t_n]$ for every $t' \in T_\Sigma(V)$.

Let now Δ be a further ranked alphabet and $\Xi = \{\xi_1, \xi_2, \ldots\}$ be another set of variables, which is disjoint from every ranked alphabet considered in the paper, and let $\Xi_n = \{\xi_1,\ldots,\xi_n\}$ for every $n \geq 0$. A *tree homomorphism from Σ to Δ* is a family of mappings $(h_k)_{k\geq 0}$ such that for every $k \geq 0$, $h_k : \Sigma_k \to T_\Delta(\Xi_k)$. Such a tree homomorphism is called *linear* (resp. *nondeleting)* if for every $k \geq 1$ and $\sigma \in \Sigma_k$ the tree $h_k(\sigma)$ is Ξ_k-linear (resp. nondeleting).

For every finite set V, the tree homomorphism $(h_k)_{k\geq 0}$ induces a mapping $h : T_\Sigma(V) \to T_\Delta(V)$ defined inductively in the following way. For every $t \in T_\Sigma(V)$ we let

- $h(t) = t$ if $t \in V$, and
- $h(t) = h_k(\sigma)[h(t_1)/\xi_1,\ldots,h(t_k)/\xi_k]$ if $t = \sigma(t_1,\ldots,t_k)$ with $k \geq 0, \sigma \in \Sigma_k$, and $t_1,\ldots,t_k \in T_\Sigma(V)$.

As usual, we also call the induced mapping h tree homomorphism. We shall denote by $\mathcal{H}^{nd}_{\Sigma,\Delta}$ (cf. [13]) the class of all nondeleting tree homomorphisms from T_Σ to T_Δ.

A function $s : T_\Sigma(V) \to [0,1]$ $(n \geq 0)$ is usually called a *tree series over Σ, V, and* $[0,1]$. For every $t \in T_\Sigma(V)$ we write (s,t) for $s(t)$ and refer to it as the *coefficient of t in s*. In case $V = X_n$, the tree series s is *linear* (resp. *nondeleting*) if its support is linear (resp. nondeleting). We denote by $\widetilde{0}$ the *constant series* defined by $(\widetilde{0},t) = 0$ for every $t \in T_\Sigma(V)$. A tree series $s : T_\Sigma(V) \to [0,1]$ is called *stochastic over Σ and V* if the sum of all its coefficients exists and equals to 1, i.e., $cont(s) = 1$. We shall denote by $STOCH(\Sigma,V)$ the set of all stochastic tree series over Σ and V. In case $V = \emptyset$, we shall write $STOCH(\Sigma)$. For every $L \subseteq T_\Sigma(V)$, the *characteristic series* 1_L *of* L is defined as usual by $(1_L,t) = 1$ if $t \in L$ and 0 otherwise. Obviously, the characteristic series $1_{\{t\}}$, denoted for simplicity by 1_t, is stochastic for every $t \in T_\Sigma(V)$.

Let $h : T_\Sigma(V) \to T_\Delta(V)$ be a tree homomorphism. For every element $s \in STOCH(\Sigma,V)$ we let

$$h(s) = \sum_{t\in T_\Sigma(V)} (s,t)h(t).$$

By *SCL* $h(s)$ is a stochastic tree series over Δ and V, hence we derive a mapping $h : STOCH(\Sigma,V) \to STOCH(\Delta,V)$.

Proposition 2. *The set $STOCH(\Sigma,V)$ is strongly convex, i.e., for every stochastic family $(\lambda_i)_{i\in I}$ of elements of $[0,1]$ and every family $(s_i)_{i\in I}$ of tree series in $STOCH(\Sigma,V)$ the tree series $\sum_{i\in I}\lambda_i s_i$ exists and belongs to $STOCH(\Sigma,V)$.*

Let $k \geq 1$ and $\sigma \in \Sigma_k$. The *σ-catenation $\sigma(s_1,\ldots,s_k)$ of the tree series* $s_1,\ldots,s_k : T_\Sigma(V) \to [0,1]$ is defined by

$$\sigma(s_1,\ldots,s_k) = \sum_{t_1,\ldots,t_k\in T_\Sigma(V)} (s_1,t_1)\ldots(s_k,t_k)\sigma(t_1,\ldots t_k)$$

and clearly has its coefficients in $[0,1]$.

Now we recall (cf. [3,6]) the *[IO]*-[1] and *OI*-substitutions of tree series. More precisely, let $t \in T_\Sigma(X_n)$ with $var(t) = \{x_{i_1}, \dots, x_{i_k}\}$ and $s_1, \dots, s_n : T_\Sigma(V) \to [0,1]$ be trees series. The *[IO]-substitution of* $s_1, \dots, s_n$ *in* t is the tree series

$$t[s_1, \dots, s_n]_{[IO]} = \sum_{t_1, \dots, t_k \in T_\Sigma(V)} (s_{i_1}, t_1) \dots (s_{i_k}, t_k) t[v_1, \dots, v_n]$$

where for every $t_1, \dots, t_k \in T_\Sigma(V)$, the sequence $v_1, \dots, v_n$ is an arbitrary element of $T_\Sigma(V)^n|_{(i_1,t_1)\dots(i_k,t_k)}$. The right-hand side of the above equality is, by Remark 1, independent of the choice of the sequences $v_1, \dots, v_n$, hence the *[IO]*-substitution is well-defined.

The *OI-substitution of* $s_1, \dots, s_n$ *in* t is the tree series $t[s_1, \dots, s_n]_{OI}$ which is defined inductively on the structure of t as follows.

(i) If $t = c \in \Sigma_0$, then $t[s_1, \dots, s_n]_{OI} = c$.
(ii) If $t = x_i$, then $t[s_1, \dots, s_n]_{OI} = s_i$.
(iii) If $t = \sigma(t_1, \dots, t_k)$, for some $k \geq 1$, $\sigma \in \Sigma_k$ and $t_1, \dots, t_k \in T_\Sigma(X_n)$, then $t[s_1, \dots, s_n]_{OI} = \sigma\left(t_1[s_1, \dots, s_n]_{OI}, \dots, t_k[s_1, \dots, s_n]_{OI}\right)$.

Let $s \in STOCH(\Sigma, X_n)$ and $s_1, \dots, s_n : T_\Sigma(V) \to [0,1]$. The *u-substitution of* $s_1, \dots, s_n$ *in* s is the tree series $s[s_1, \dots, s_n]_u : T_\Sigma(V) \to [0,1]$ defined by

$$s[s_1, \dots, s_n]_u = \sum_{t \in T_\Sigma(X_n)} (s,t) t[s_1, \dots, s_n]_u$$

for *u=[IO], OI*.

Proposition 3. *Let* $s \in STOCH(\Sigma, X_n)$ *and* $s_1, \dots, s_n \in STOCH(\Sigma, V)$. *Then the tree series* $s[s_1, \dots, s_n]_u$ *is stochastic for u=[IO],OI.*

Next, for every $s \in STOCH(\Sigma, X_n)$, *u=[IO],OI* we define the mapping

$$\Phi_{s,u} : \left([0,1]^{T_\Sigma(V)}\right)^n \to [0,1]^{T_\Sigma(V)}, \qquad (s_1, \dots, s_n) \mapsto s[s_1, \dots, s_n]_u$$

for every $(s_1, \dots, s_n) \in \left([0,1]^{T_\Sigma(V)}\right)^n$.

Lemma 2. *For every* $s \in STOCH(\Sigma, X_n)$ *and u=[IO],OI, the mapping* $\Phi_{s,u}$ *is* ω*-continuous.*

4 Systems of Equations of Stochastic Tree Series

In this section, we deal with systems of equations of stochastic tree series. More precisely, a *system of equations of stochastic tree series over* Σ *and* X_n is a system

$$\text{(E)} \qquad x_1 = p_1, \dots, x_n = p_n,$$

[1] We should note that the *[IO]*- differs from the *IO*-substitution mode. Due to space limitations, we refer the reader to [6,13] for details.

where $p_i \in Stoch(\Sigma, X_n)$, i.e., p_i is a stochastic polynomial over Σ and X_n for every $1 \le i \le n$. The system (E) is called *linear* (resp. *nondeleting*) if p_i is linear (resp. nondeleting) for every $1 \le i \le n$.

We associate with (E) the mapping

$$\Phi_{E,u} : \left([0,1]^{T_\Sigma}\right)^n \to \left([0,1]^{T_\Sigma}\right)^n$$

which is defined by $\Phi_{E,u}(s_1, \ldots, s_n) = (\Phi_{p_1,u}(s_1, \ldots, s_n), \ldots, \Phi_{p_n,u}(s_1, \ldots, s_n))$ for u=*[IO],OI* and $(s_1, \ldots, s_n) \in \left([0,1]^{T_\Sigma}\right)^n$. The mapping $\Phi_{E,u}$ is ω-continuous since by Lemma 2, the mapping $\Phi_{p_i,u}$ is ω-continuous for every $1 \le i \le n$. Therefore, by Proposition 1, the least fixpoint fix$\Phi_{E,u}$ exists for u=*[IO],OI.* More precisely, we have fix$\Phi_{E,u} = \sup_{k \ge 0}((s_{1,k,u}, \ldots, s_{n,k,u}))$ where for every $1 \le i \le n$

$$s_{i,0,u} = \widetilde{0} \quad \text{and} \quad s_{i,k+1,u} = p_i\left[s_{1,k,u}, \ldots, s_{n,k,u}\right]_u, \text{ for } k \ge 0.$$

In the sequel, we shall call a fixpoint of $\Phi_{E,u}$ a *u-solution of* (E) and fix$\Phi_{E,u}$ the *least u-solution of* (E).

A tree series $s : T_\Sigma \to [0,1]$ is called *stochastically u-equational* (*over* Σ) if it is a component of the least u-solution of a system of equations of stochastic tree series over Σ and X_n. For u=*[IO],OI*, we denote by $StochEq_u(\Sigma)$ the class of stochastically u-equational tree series over Σ. The constant tree series $\widetilde{0}$ is stochastically u-equational for every u=*[IO],OI.* For instance, if $supp(p_i) \cap T_\Sigma = \emptyset$ for every $1 \le i \le n$, then $\left(\widetilde{0}, \ldots, \widetilde{0}\right)$ is the least u-solution of (E) for u=*[IO],OI.*

Theorem 1. *Let*

$$\text{(E)} \qquad x_1 = p_1, \ldots, x_n = p_n,$$

be a system of equations of stochastic tree series. If either $supp(p_i) \cap T_\Sigma \neq \emptyset$ and $supp(p_i) \cap (T_\Sigma(X_n) \setminus T_\Sigma)$ is nondeleting for every $1 \le i \le n$ or $supp(p_i) \cap T_\Sigma = \emptyset$ for every $1 \le i \le n$, then the components of the least u-solution (u=[IO],OI) of (E) *either are stochastic tree series or equal to $\widetilde{0}$, respectively.*

Proof. (Sketch) Let fix$\Phi_{E,u} = (s_{1,u}, \ldots, s_{n,u})$ for u=*[IO],OI.* Observe that if $supp(p_i) \cap T_\Sigma = \emptyset$, for some $1 \le i \le n$, then $s_{i,u} = \widetilde{0}$ because of our assumption for (E). Therefore, let us assume that $supp(p_i) \cap T_\Sigma \neq \emptyset$ for every $1 \le i \le n$, and let $p_i^{(0)}$ be the restriction of p_i on T_Σ, i.e., $p_i^{(0)} = p_i|_{T_\Sigma}$ and $p_i^{(1)} = p_i - p_i^{(0)}$. Thus, we get $cont\,(p_i) = cont\left(p_i^{(0)}\right) + cont\left(p_i^{(1)}\right)$. If for some index $1 \le i \le n$ we have $p_i^{(1)} = \widetilde{0}$, then the ith component of the least u-solution of (E) equals to $p_i^{(0)} = p_i$, hence it is stochastic. Thus, without any loss, we may assume that (E) satisfies the condition $cont\left(p_i^{(1)}\right) > 0$ for every $1 \le i \le n$. Furthermore, let $cont\left(p_1^{(1)}\right) = \lambda$, hence by our assumption for fix$\Phi_{E,u}$, we have $0 < \lambda < 1$. For every $1 \le i \le n$, u=*[IO],OI* we let

$$s_{i,0,u} = \widetilde{0} \quad \text{and} \quad s_{i,k+1,u} = p_i\left[s_{1,k,u}, \ldots, s_{n,k,u}\right]_u, \text{ for } k \ge 0$$

and thus, $s_{i,u} = \sup_{k\geq 0} s_{i,k,u}$. By construction, the sequence $(s_{i,k,u})_{k\geq 0}$ is increasing and bounded, hence

$$\sup_{k\geq 0}(cont\,(s_{i,k,u})) = \lim_{k\to\infty} cont\,(s_{i,k,u}).$$

Now, for every $1 \leq i \leq n$, u=[IO],OI, we consider the sequence $(\overline{s_{i,k,u}})_{k\geq 1}$ of polynomials over Σ and X_n as follows

$$\overline{s_{i,1,u}} = p_i \quad \text{and} \quad \overline{s_{i,k+1,u}} = p_i\left[\overline{s_{1,k,u}}, \ldots, \overline{s_{n,k,u}}\right]_u, \quad \text{for } k \geq 0.$$

By Proposition 3, the polynomial $\overline{s_{i,k,u}}$ is stochastic for every $1 \leq i \leq n$, $k \geq 1$, and u=[IO],OI. Moreover, we have $\overline{s_{i,k,u}}^{(0)} = s_{i,k,u}$, and since $cont\,(\overline{s_{i,k,u}}) = cont\,(\overline{s_{i,k,u}}^{(0)}) + cont\,(\overline{s_{i,k,u}}^{(1)}) = 1$, we get

$$cont\,(s_{i,k,u}) + cont\left(\overline{s_{i,k,u}}^{(1)}\right) = 1. \tag{1}$$

We show that $\lim_{k\to\infty} cont\,(\overline{s_{i,k,u}}^{(1)}) = 0$, hence by (1) we get

$$\lim_{k\to\infty} cont\,(s_{i,k,u}) = 1$$

i.e., $\sup_{k\geq 0}(cont(s_{i,k,u})) = 1$. Using the last relation, we show that $cont\,(s_{i,u}) = 1$, for every $1 \leq i \leq n$, and we are done.

In the proof of theorem above, we did not use the finiteness of the supports of p_i's. Therefore, we can state the following result.

Theorem 2. *Let*

$$(\mathrm{E_g}) \qquad x_1 = p_1, \ldots, x_n = p_n,$$

be a generalized system of equations of stochastic tree series, i.e., $p_i \in STOCH(\Sigma, X_n)$ *for every* $1 \leq i \leq n$. *If either* $supp(p_i) \cap T_\Sigma \neq \emptyset$ *and* $supp(p_i) \cap (T_\Sigma(X_n) \setminus T_\Sigma)$ *is nondeleting for every* $1 \leq i \leq n$ *or* $supp(p_i) \cap T_\Sigma = \emptyset$ *for every* $1 \leq i \leq n$, *then the components of the least u-solution (u=[IO],OI) of* $(\mathrm{E_g})$ *either are stochastic tree series or equal to* $\widetilde{0}$, *respectively.*

Given $p \in STOCH(\Sigma, X_n)$ the kth *star of* p $(1 \leq k \leq n)$ is by definition the least *OI*-solution of the equation $x_k = p$ and it is denoted by $p^{*,k}$. We note that in this case the variables $x_1, \ldots, x_{k-1}, x_{k+1}, \ldots x_n$ are considered as letters of rank 0.

Theorem 3. *Let*

$$(\mathrm{E_g}) \qquad x_1 = p_1, \ldots, x_n = p_n$$

be a generalized system of equations of stochastic tree series and assume that $p_n^{*,n} \neq \widetilde{0}$. *Consider the system*

$$\begin{aligned}(\mathrm{E'_g})\ x_1 &= p_1\left[x_1, \ldots, x_{n-1}, p_n^{*,n}/x_n\right]_{OI}, \ldots,\\ x_{n-1} &= p_{n-1}\left[x_1, \ldots, x_{n-1}, p_n^{*,n}/x_n\right]_{OI}\end{aligned}$$

of $n-1$ equations of stochastic tree series. If $(s_1, \ldots, s_{n-1})$ is the least OI-solution of (E'_g)*, then* $(s_1, \ldots, s_{n-1}, p_n^{*,n}[s_1, \ldots, s_{n-1}])$ *is the least OI-solution of* (E_g)*.*

The last theorem allows us to solve equation by equation a generalized system of equations of stochastic tree series. Furthermore, it has nice consequences. First, we get that Theorem 2 remains valid, for the case u=*OI*, without "non-deleting" assumption on the supports of its right-hand side members.

Theorem 4. *Let*

$$(\mathrm{E}_\mathrm{g}) \qquad x_1 = p_1, \ldots, x_n = p_n,$$

be a generalized system of equations of stochastic tree series. If either $supp(p_i) \cap T_\Sigma \neq \emptyset$ *for every* $1 \leq i \leq n$ *or* $supp(p_i) \cap T_\Sigma = \emptyset$ *for every* $1 \leq i \leq n$*, then the components of the least OI-solution of* (E_g) *either are stochastic tree series or equal to* $\widetilde{0}$*, respectively.*

The second important consequence of Theorem 3 is a Kleene characterization for the class $StochEq_{OI}(\Sigma)$.

Theorem 5. *The class* $StochEq_{OI}(\Sigma, X)$ *is the smallest convex set containing the constant tree series* $\widetilde{0}$ *and characteristic series* 1_t *for every* $t \in T_\Sigma(X)$*, and being closed under OI-substitution and star operation.*

We conclude this section, by stating a relation among $StochEq_{OI}$ and $StochEq_{[IO]}$, i.e., the classes of all stochastically *[IO]*- and *OI*-equational tree series. We need the following notation. For a class of tree series $\mathcal{C}$, we let

$$\mathcal{H}^{nd}(\mathcal{C}) = \{h(s) \mid h \in \mathcal{H}^{nd}_{\Sigma,\Delta}, s : T_\Sigma \to [0,1], s \in \mathcal{C}\}.$$

Theorem 6. $\mathcal{H}^{nd}\,(StochEq_{OI}) = StochEq_{[IO]}$.

5 Stochastic Algebras

In this section, we introduce stochastic Σ-algebras and we state a Mezei-Wright type result for stochastically u-equational tree series. More precisely, a *stochastic Σ-algebra* is a pair $\mathcal{A} = (A, \Sigma_\mathcal{A})$ where the nonempty set A is the domain set of $\mathcal{A}$, and $\Sigma_\mathcal{A}$ is a family $(\sigma_\mathcal{A} \mid \sigma \in \Sigma)$ of operations on A such that for every $k \geq 0$ and $\sigma \in \Sigma_k$, we have $\sigma_\mathcal{A} : A^k \to STOCH(A)$. If $\sigma \in \Sigma_0$, then $\sigma_\mathcal{A} \in STOCH(A)$. If no confusion arises, then we drop $\mathcal{A}$ from $\Sigma_\mathcal{A}$. The mapping $\sigma_\mathcal{A}$ can be extended into a function $\overline{\sigma_\mathcal{A}} : STOCH(A)^k \to STOCH(A)$ by setting

$$\overline{\sigma_\mathcal{A}}(f_1, \ldots, f_k) = \sum_{a_1,\ldots,a_k \in A} f_1(a_1) \ldots f_k(a_k) \sigma_\mathcal{A}(a_1, \ldots, a_k)$$

for every $f_1, \ldots, f_k \in STOCH(A)$. Clearly, $\overline{\sigma_\mathcal{A}}(f_1, \ldots, f_k)$ is stochastic by the *SCL*.

Proposition 4. *The mapping $\overline{\sigma_{\mathcal{A}}}$ preserves strong convex combinations at every argument, i.e., it holds*

$$\overline{\sigma_{\mathcal{A}}}\left(\sum_{i_1\in I_1}\lambda_{1,i_1}f_{1,i_1},\ldots,\sum_{i_k\in I_k}\lambda_{k,i_k}f_{k,i_k}\right)$$
$$=\sum_{i_1\in I_1,\ldots,i_k\in I_k}\lambda_{1,i_1}\ldots\lambda_{k,i_k}\overline{\sigma_{\mathcal{A}}}(f_{1,i_1},\ldots,f_{k,i_k})$$

where $(\lambda_{1,i_1})_{i_1\in I_1},\ldots,(\lambda_{k,i_k})_{i_k\in I_k}$ are stochastic families of real numbers and $f_{1,i_1},\ldots,f_{k,i_k}\in STOCH(A)$ for every $i_1\in I_1,\ldots,i_k\in I_k$.

Let $\mathcal{B}=(B,\Sigma_{\mathcal{B}})$ be a further stochastic Σ-algebra. A mapping $H:A\to STOCH(B)$ can be extended to a mapping $\overline{H}:STOCH(A)\to STOCH(B)$ by letting for every $f\in STOCH(A)$

$$\overline{H}(f)=\sum_{a\in A}f(a)H(a).$$

Then H is called a *morphism of stochastic Σ-algebras* if

$$\overline{H}\left(\sigma_{\mathcal{A}}(a_1,\ldots,a_k)\right)=\overline{\sigma_{\mathcal{B}}}\left(H(a_1),\ldots,H(a_k)\right)$$

for every $k\geq 0$, $\sigma\in\Sigma_k$, and $a_1,\ldots,a_k\in A$. This implies that $\overline{H}\left(\sigma_{\mathcal{A}}\right)=\sigma_{\mathcal{B}}$ for every $\sigma\in\Sigma_0$.

It should be clear that every Σ-algebra $\mathcal{A}$ can be considered as a stochastic Σ-algebra, in particular the *term algebra* $\mathcal{T}_\Sigma(V)=(T_\Sigma(V),\Sigma)$ of all trees over Σ and V, where $\sigma_{\mathcal{T}_\Sigma(V)}(t_1,\ldots,t_k)=\sigma(t_1,\ldots,t_k)$ for every $k\geq 0$, $\sigma\in\Sigma_k$, and $t_1,\ldots,t_k\in T_\Sigma(V)$. In fact, it is the *free stochastic Σ-algebra generated by V* in the class of all Σ-algebras, i.e., for every Σ-algebra $\mathcal{A}$, every mapping $H:V\to STOCH(A)$ extends uniquely to a Σ-algebra morphism $H:\mathcal{T}_\Sigma(V)\to STOCH(A)$. If $V=\emptyset$, then we denote the unique morphism from $\mathcal{T}_\Sigma$ to $\mathcal{A}$ by $H_{\mathcal{A}}$.

Let $\mathcal{A}=(A,\Sigma)$ be a Σ-algebra, $t\in T_\Sigma(X_n)$, $a_1,\ldots,a_n\in A$, and $H:X_n\to A$ be a mapping with $H(x_i)=a_i$ $(1\leq i\leq n)$. For every $t\in T_\Sigma(X_n)$ we denote $H(t)$ by $t_{\mathcal{A}}[a_1,\ldots,a_n]$ and call it the *substitution of $a_1,\ldots,a_n$ in t*. In case $\mathcal{A}=\mathcal{T}_\Sigma(X)$, then $t_{\mathcal{T}_\Sigma(X)}[a_1,\ldots,a_n]$ coincides with the substitution of a_i at x_i $(1\leq i\leq n)$ defined in Section 3. Next, we define the *[IO]*- and *OI*-substitutions of stochastic functions over A in tree series over Σ and X_n. More precisely, let $t\in T_\Sigma(X_n)$ with $var(t)=\{x_{i_1},\ldots,x_{i_k}\}$ and $f_1,\ldots,f_n\in STOCH(A)$. The *[IO]-substitution of $f_1,\ldots,f_n$ in t*, is defined as follows:

$$t_{\mathcal{A}}[f_1,\ldots,f_n]_{[IO]}=\sum_{a_1,\ldots,a_k\in A}f_{i_1}(a_1)\ldots f_{i_k}(a_k)t_{\mathcal{A}}[b_1,\ldots,b_n]$$

where for every $a_1,\ldots,a_k\in A$, the sequence $b_1,\ldots,b_n$ is an arbitrary element of $A^n|_{(i_1,a_1)\ldots(i_k,a_k)}$. The *OI-substitution of $f_1,\ldots,f_n$ in t* is defined inductively in the following way.

(i) If $t = x_i$, then $t_{\mathcal{A}}[f_1, \ldots, f_n]_{OI} = f_i$.
(ii) If $t = \sigma(t_1, \ldots, t_k)$ for some $k \geq 0$, $\sigma \in \Sigma_k$ and $t_1, \ldots, t_k \in T_\Sigma(X_n)$, then $t_{\mathcal{A}}[f_1, \ldots, f_n]_{OI} = \overline{\sigma_{\mathcal{A}}}(t_{1,\mathcal{A}}[f_1, \ldots, f_n]_{OI}, \ldots, t_{k,\mathcal{A}}[f_1, \ldots, f_n]_{OI})$.

Let $s \in STOCH(\Sigma, X_n)$, $f_1, \ldots, f_n \in STOCH(A)$. The *u-substitution of* $f_1, \ldots, f_n$ in s is the stochastic function $s_{\mathcal{A}}[f_1, \ldots, f_n]_u \in STOCH(A)$ defined by

$$s_{\mathcal{A}}[f_1, \ldots, f_n]_u = \sum_{t \in T_\Sigma(X_n)} (s,t) t_{\mathcal{A}}[f_1, \ldots, f_n]_u.$$

Next, for every $s \in STOCH(\Sigma, X_n)$, *u=[IO],OI* we define the mapping

$$\Phi^{\mathcal{A}}_{s,u} : \left([0,1]^A\right)^n \to [0,1]^A, \qquad (f_1, \ldots, f_n) \mapsto s_{\mathcal{A}}[f_1, \ldots, f_n]_u$$

for every $(f_1, \ldots, f_n) \in \left([0,1]^A\right)^n$.

Lemma 3. *For every $s \in STOCH(\Sigma, X_n)$ and u=[IO],OI the mapping $\Phi^{\mathcal{A}}_{s,u}$ is ω-continuous.*

In the sequel, we consider least u-solutions of generalized systems of equations of stochastic tree series

$$(\mathrm{E_g}) \qquad x_1 = p_1, \ldots, x_n = p_n,$$

in the Σ-algebra $\mathcal{A}$. More precisely, we associate with $(\mathrm{E_g})$ the mapping

$$\Phi^{\mathcal{A}}_{E,u} : \left([0,1]^A\right)^n \to \left([0,1]^A\right)^n$$

which is defined by $\Phi^{\mathcal{A}}_{E,u}(f_1, \ldots, f_n) = \left(\Phi^{\mathcal{A}}_{p_1,u}(f_1, \ldots, f_n), \ldots, \Phi^{\mathcal{A}}_{p_n,u}(f_1, \ldots, f_n)\right)$ for every *u=[IO],OI* and $(f_1, \ldots, f_n) \in \left([0,1]^A\right)^n$. The mapping $\Phi^{\mathcal{A}}_{E,u}$ is ω-continuous since by Lemma 3, the mapping $\Phi^{\mathcal{A}}_{p_i,u}$ is ω-continuous for every $1 \leq i \leq n$. Therefore, by Proposition 1, the least fixpoint fix$\Phi^{\mathcal{A}}_{E,u}$ exists for *u=[IO],OI.* We call a fixpoint of $\Phi^{\mathcal{A}}_{E,u}$ a *u-solution of* $(\mathrm{E_g})$ *in* $\mathcal{A}$ and fix$\Phi^{\mathcal{A}}_{E,u}$ the *least u-solution of* $(\mathrm{E_g})$ *in* $\mathcal{A}$. A function $f : A \to [0,1]$ is called *stochastically u-regular* if it is a component of the least u-solution of a system of equations of stochastic tree series over Σ and X_n whose right-hand sides are polynomials. We denote by $StochReg_u(A)$ the class of all stochastically u-regular functions over A.

Theorem 7. *The class $StochReg_{OI}(\mathcal{A})$ is a convex set closed under stochastically regular substitution, i.e., if s is in $STOCH(\Sigma, X_n)$ and if $f_1, \ldots, f_n$ are in $StochReg_{OI}(A)$, then $s_{\mathcal{A}}[f_1, \ldots, f_n]_{OI} \in StochReg_{OI}(A)$ as well. Moreover, if the right-hand sides of a system $(\mathrm{E_g})$ consist of stochastically OI-equational tree series, then every component of the least OI-solution of $(\mathrm{E_g})$ is stochastically OI-regular.in*

Finally, we present the announced Mezei-Wright type result.

Theorem 8. *Let $\mathcal{A} = (A, \Sigma)$ be a stochastic Σ-algebra. A function $f \in [0,1]^A$ is stochastically u-regular iff there exists a stochastically u-equational tree series s over Σ such that $H_{\mathcal{A}}(s) = f$, for u=[IO],OI.*

6 Conclusion

We introduced systems of equations of stochastic tree series and we proved the existence of their least *[IO]*- and *OI*-solutions whose non-zero components are stochastic tree series. We gave a Kleene characterization for the class of stochastically *OI*-equational tree series. Furthermore, we proved that the class of stochastically *[IO]*-equational tree series is the closure of the class of stochastically *OI*-equational tree series under nondeleting tree homomorphisms. We considered also stochastic algebras and established a Mezei-Wright result showing the robustness of our theory. We note that systems of equations of stochastic polynomials over non-commuting variables are a special case of our systems over stochastic Σ-algebras.

Several open problems arise for our systems of equations of stochastic tree series. The behaviors of existing models of probabilistic tree automata fail to have an equational characterization as well a Kleene type one. It is our next task to introduce a reasonable model of stochastic tree automata having these properties. On the other hand, the fixpoint theory ensures the existence of the least *[IO]*- and *OI*-solutions which can be determined by the suprema of the corresponding approximating sequences. Nevertheless, it is shown in other setups of stochastic systems of equations that the determination of these suprema is exponentially [9]. Therefore, the complexity of the computation of our stochastically equational tree series is an interesting open problem for further investigation.

References

1. Berstel, J., Reutenauer, C.: Recognizable formal power series on trees. Theoret. Comput. Sci. 18, 115–148 (1982)
2. Bloom, S.L., Ésik, Z.: An extension theorem with an application to formal tree series. J. Autom. Lang. Comb. 8, 145–185 (2003)
3. Bozapalidis, S.: Equational elements in additive algebras. Theory of Comput. Syst. 32, 1–33 (1999)
4. Bozapalidis, S., Fülöp, Z., Rahonis, G.: Equational tree transformations. Theoret. Comput. Sci. 412, 3676–3692 (2011)
5. Bozapalidis, S., Fülöp, Z., Rahonis, G.: Equational weighted tree transformations. Acta Inform. 49, 29–52 (2012)
6. Bozapalidis, S., Rahonis, G.: On the closure of recognizable tree series under tree homomorphisms. J. Autom. Lang. Comb. 10, 185–202 (2005)
7. Ésik, Z., Kuich, W.: Formal tree series. J. Autom. Lang. Comb. 8, 219–285 (2003)
8. Esparza, J., Gaizer, A., Kiefer, S.: Computing least fixed points of probabilistic systems of polynomilas. In: Proceedings of STACS 2010. LIPIcs, vol. 5, pp. 359–370. Schloss Dagstuhl–Leibniz-Zentrum fuer Informatik (2010)
9. Etessami, K., Yannakakis, M.: Recursice Markov chains, stochastic grammars, and monotone systems of non-linear equations. J. ACM 56, 1–66 (2009)
10. Etessami, K., Stewart, A., Yannakakis, M.: Polynomial time algorithms for multi-type branching processes and stochastic context-free grammars. In: Proceedings of the 44th Symposium on Theory of Computing, STOC 2012, pp. 579–588. ACM, New York (2012)

11. Etessami, K., Yannakakis, M.: Model checking of recursive probabilistic systems. ACM Trans. Comput. Logic 13(2), 12:1–12:40 (2012)
12. Fülöp, Z., Rahonis, G.: Equational weighted tree transformations with discounting. In: Kuich, W., Rahonis, G. (eds.) Algebraic Foundations in Computer Science. LNCS, vol. 7020, pp. 112–145. Springer, Heidelberg (2011)
13. Fülöp, Z., Vogler, H.: Weighted tree automata and tree transducers. In: Handbook of Weighted Automata. Monographs in Theoretical Computer Science, An EATCS Series, pp. 313–404. Springer (2009)
14. Kuich, W.: Formal series over algebras. In: Nielsen, M., Rovan, B. (eds.) MFCS 2000. LNCS, vol. 1893, pp. 488–496. Springer, Heidelberg (2000)
15. Mezei, J., Wright, J.B.: Algebraic automata and context-free sets. Inform. Control 11, 3–29 (1967)
16. Wechler, W.: Universal Algebra for Computer Scientists. EATCS Monographs on Theoretical Computer Science, vol. 25. Springer (1992)

On Gröbner Bases in the Context of Satisfiability-Modulo-Theories Solving over the Real Numbers

Sebastian Junges, Ulrich Loup, Florian Corzilius, and Erika Ábrahám*

RWTH Aachen University, Germany

Abstract. We address satisfiability checking for the first-order theory of the real-closed field (RCF) using *satisfiability-modulo-theories (SMT)* solving. SMT solvers combine a *SAT solver* to resolve the Boolean structure of a given formula with *theory solvers* to verify the consistency of sets of theory constraints.

In this paper, we report on an integration of *Gröbner bases* as a theory solver so that it conforms with the requirements for efficient SMT solving: (1) it allows the incremental adding and removing of polynomials from the input set and (2) it can compute an inconsistent subset of the input constraints if the Gröbner basis contains 1.

We modify Buchberger's algorithm by implementing a new update operator to optimize the Gröbner basis and provide two methods to handle inequalities. Our implementation uses special data structures tuned to be efficient for huge sets of sparse polynomials. Besides solving, the resulting module can be used to simplify constraints before being passed to other RCF theory solvers based on, e.g., the cylindrical algebraic decomposition.

1 Introduction

Formulas of first-order logic over the theory of the *real-closed field* (RCF) are Boolean combinations of polynomial constraints with real-valued variables. Be it the analysis of real-time systems, the optimization of railway schedules or the computation of dense sphere packings in Euclidean space, many practical and theoretical problems can be expressed in this logic. Sophisticated decision procedures and increased computational power have led to efficient tools to analyze such formulas.

Boolean formulas are well-suited for the description of discrete systems, e.g., digital controllers. State-of-the-art *SAT solvers*, dedicated programs to determine the satisfiability of Boolean formulas, are highly tuned for efficiency. They can handle formulas with millions of literals and are frequently used not only in academic research but also in industry.

* This work has been partially supported by the German Research Council (DFG) as part of the Research Training Group "AlgoSyn" (GRK 1298, `http://www.algosyn.rwth-aachen.de/`).

T. Muntean, D. Poulakis, and R. Rolland (Eds.): CAI 2013, LNCS 8080, pp. 186–198, 2013.

The success of SAT solvers has led to an approach called *satisfiability-modulo-theories* (*SMT*) solving for handling first-order logic over certain theories. This approach combines the high efficiency of SAT solvers to handle the Boolean structure with dedicated *theory solvers* to check sets of constraints from the given theory for consistency. For the optimal combination of these modules, theory solvers should be *SMT compliant*: they should support the extension of the constraint set (*incrementality*), the removal of constraints (*backtracking*) and the generation of small *infeasible subsets* in case of inconsistency [2][Ch. 26].

In this paper, we consider the existential fragment of the first-order logic over the theory of the RCF. Immense advances have been made in this area in the last decades. Besides complete decision procedures as the *cylindrical algebraic decomposition* (CAD) method [4], e.g. implemented in the tool `QEPCAD`, also incomplete methods such as the *virtual substitution* (VS) method [15], e.g. available in the package `Redlog` of the computer algebra system `Reduce`, *simplex* [8] or *interval constraint propagation* [9], e.g. implemented in `iSAT`, are available. In addition to such explicit methods working on the solution space, some symbolic approaches find application in SMT solving for preprocessing by using simple rules and basic Gröbner basis computations, or outside of SMT solvers in standalone tools, often based on some application of the Positivstellensatz [13] such as in the tool `KeYmaera`.

We aim to improve the integration of the *Gröbner bases* methodology in SMT solving, thereby enhancing speed and effectiveness. To reach this goal, we have to overcome several challenges. (1) The methodology has to be adapted to be *SMT compliant* and (2) to cope with typical *SMT-problem structures*, which often significantly differ from algebraically hard problems. (3) As we are solving over the RCF, we are more interested in the *real radical* than the ideal of our input polynomials. (4) Finally, we need to handle *inequalities* as well.

Gröbner basis computations are used for preprocessing in [7] and [11]. [13] proposes a combination of Gröbner basis computations with the Fourier-Motzkin method. However, this work is not directly related to SMT. Direct relation to SMT can be found in [6] for finding minimal infeasible subsets, and in [12] for coping with the special structure. Saturation to approximate the real radical is used in [11] and in [13].

We implement our approaches as a module in the SMT-solving framework `SMT-RAT`, which is a `C++` toolbox allowing the combination of different theory solvers in a user-defined strategy. Our *Gröbner bases module* can be applied both as a preprocessing and as a solving technique.

Regarding (1), our Gröbner bases module supports the adding and removal of constraints as well as the computation of small infeasible subsets. The basic features of this module are the simplification of equations and the check whether there are common zeros of the input equations. To tackle (2), we utilize some ideas from [12] and [14] to develop data structures that can handle a large number of variables and huge sets of sparse input polynomials, not necessarily of low degree, as they frequently occur in our setting. For (3), we further adapt Buchberger's algorithm in that we prune polynomials without real zeros in the

Gröbner basis. We implemented two different strategies to realize (4): firstly, we can encode all inequalities as equations and compute a Gröbner basis of the extended set of polynomials, or secondly, we reduce the polynomials belonging to the inequalities modulo the Gröbner basis for the equations.

The rest of the paper is structured as follows: In Section 2 we recall some basics for Gröbner bases. In Section 3 we describe our SMT framework before explaining our methods and their integration in Section 4. After giving some experimental results in Section 5, we conclude the paper in Section 6.

2 Preliminaries

We denote the set of *real*, *rational* and *natural* numbers by $\mathbb{R}$, $\mathbb{Q}$ and $\mathbb{N}$ ($0 \in \mathbb{N}$) respectively. We use $\mathbb{R}$ and $\mathbb{Q}$ also for the corresponding (*ordered*) *fields* over the arithmetic operations $+$, $\cdot$ and the ordering relation $<$. W.l.o.g., we refer to $\mathbb{R}$ as the *real-closed field* (*RCF*). We omit the symbol $\cdot$ when the context is clear. We abbreviate sequences of variables $x_1, \ldots, x_n$, $n \geq 1$, by $\overline{x}$.

Let K be a field. $K[\overline{x}]$ denotes the *polynomial ring* over K in the variables $\overline{x}$. We call a product $m = \prod_{1 \leq i \leq n} x_i^{d_i}$ with $d_i \in \mathbb{N}$ a *monomial* having the *degree* $\deg(m) := \sum_{1 \leq i \leq n} d_i$. With $M_{\overline{x}}$ we denote the set of all monomials in $\overline{x}$. A product $a \cdot m$ with $a \in K$ and $m \in M_{\overline{x}}$ is called a *term* and a the *coefficient of* m. Hence, a polynomial $p \in K[\overline{x}]$ is a sum of terms. We say that $x_i \in p$ if x_i occurs in the polynomial $p \in K[\overline{x}]$. We define the *total degree* of p as $\text{tdeg}(p) := \max\{\deg(m) \mid m \text{ monomial in } p\}$. A *monomial ordering* is a linear well-ordering on monomials respecting multiplication of monomials, i.e., a linear ordering $\prec$ with a minimal element such that $m_1 \prec m_2$ entails $m_1 m_3 \prec m_2 m_3$ for all $m_1, m_2, m_3 \in M_{\overline{x}}$. By $\text{lm}(p)$ we denote the *leading monomial of* p, i.e., the maximal monomial w.r.t. the current ordering. Analogously, we define $\text{lt}(p)$ to be the *leading term of* p. The coefficient of $\text{lt}(p)$ is called *leading coefficient*, denoted by $\text{lc}(p)$. It holds that $\text{lt}(p) = \text{lc}(p)\text{lm}(p)$ for all polynomials $p \in K[\overline{x}]$.

Let $p \in \mathbb{Q}[\overline{x}]$. We call $p \sim 0$ a (*polynomial*) *constraint over* p if and only if $\sim \; \in \{=, >, \geq, \neq\}$. For $P \subseteq K[\overline{x}]$ and C a set of constraints over P we define $\text{pol}(C) := P$. Our input formulas are quantifier-free first-order formulas over polynomial constraints, i.e., Boolean combinations connected by $\wedge$, $\vee$, $\neg$ of constraints. We refer to such formulas as *RCF formulas.* Note that we only consider the existential fragment of the first-order theory of the RCF here.

2.1 Gröbner Bases

We briefly introduce *Gröbner bases* and an application to solve real-algebraic constraint systems. More information can be found in [1].

Let $R = \mathbb{Q}[\overline{x}]$ with a fixed monomial ordering. Given a finite set of polynomials $P \subseteq R$, we define the *ideal generated by* P as the set $\langle P \rangle := \{\sum_{p \in P} r_p p \mid r_p \in R \text{ for each } p \in P\}$. Note that the more general notion of an ideal is also covered by our definition because, due to Hilbert's basis theorem, every ideal in R has a finite set of generators. By $\mathcal{V}_K(\langle P \rangle) := \{a \in K \mid p(a) = 0 \text{ for all } p \in P\}$ we denote the *K variety of* P, i.e., the set of common zeros of P in K.

Reduction. Let $p, p', f \in R$ with $p, f \neq 0$, $p = \sum_{i=0}^{k} a_i m_i$, $k \in \mathbb{N}$ and let $F \subseteq R$. If $p' = p - sf$ for some $s \in R$ such that $s \cdot \mathrm{lt}(f) = a_i m_i$ for some $i \in \{1, \ldots, k\}$ then p *reduces to* p' *modulo* f, written $p \xrightarrow{f} p'$. We call f the *reductor of* p. We say that p *reduces to* p' *modulo* F, written $p \xrightarrow{F} p'$, if $p \xrightarrow{f} p'$ for some $f \in F$. If no $f \in F$ with $p \xrightarrow{f} p'$ exists, p is *in normal form modulo* F. If $p \xrightarrow{F} \ldots \xrightarrow{F} p'$ and p' is in normal form modulo F then we call p' the *normal form of* P *modulo* F, denoted by $\mathrm{red}_F(p)$.

Definition 1 (Gröbner basis). *Let* $P \subseteq R$. *A finite set* $G \subseteq \langle P \rangle$ *is called a* Gröbner basis (GB) *of* $\langle P \rangle$ *if* $\langle \{\mathrm{lt}(g) \mid g \in G\} \rangle = \langle \{\mathrm{lt}(p) \mid p \in P\} \rangle$. *Let* $\mathrm{lc}(p) = 1$ *for all* $p \in G$. *We call* G minimal *if* $\mathrm{lt}(g) \notin \langle \mathrm{lt}(\tilde{g}) \mid \tilde{g} \in G \setminus \{g\} \rangle$ *for all* $g \in G$, *and* reduced *if* $m \notin \langle \mathrm{lt}(\tilde{g}) \mid \tilde{g} \in G \setminus \{g\} \rangle$ *for all monomials* m *of* g.

We always regard a reduced GB, which is unique for a given monomial ordering. If the reduced Gröbner basis of $\langle P \rangle$ is $\{1\}$ then $\mathcal{V}_{\mathbb{R}}(\langle P \rangle) = \emptyset$, i.e., P has no common zeros.

Buchberger's Algorithm. In his PhD thesis, Bruno Buchberger suggested a simple fixed-point iteration algorithm for computing a Gröbner basis [3] (see Listing (1) of Table 1). The most important tool in Buchberger's algorithm is the S-polynomial: Let $p, q \in R$ with $\mathrm{lm}(p) = \prod_{i=1}^{n} x_i^{d_i}$ and $\mathrm{lm}(q) = \prod_{i=1}^{n} x_i^{\tilde{d}_i}$, then the least common multiple of $\mathrm{lm}(p)$ and $\mathrm{lm}(q)$ is $\mathrm{lcm}(\mathrm{lm}(p), \mathrm{lm}(q)) = \prod_{i=1}^{n} x_i^{\max(d_i, \tilde{d}_i)} =: l$. We define $S(p, q) := \frac{l}{\mathrm{lt}(p)} \cdot p - \frac{l}{\mathrm{lt}(q)} \cdot q$ to be the *S-polynomial of* p *and* q. All possible S-polynomials are computed during Buchberger's algorithm. We refer to a pair (p, q) whose S-polynomial is not yet computed as *S-pair.*

We call a mapping $U : 2^R \times R \to 2^R$ an *update operator*, where 2^R denotes the power set of R. Buchberger's algorithm uses the *standard update operator* $\mathrm{U}_{\mathrm{std}}(G, s) = G \cup \{s\}$.

A reduced Gröbner basis can be obtained by iteratively removing each polynomial whose leading term is a multiple of another leading term, and applying reduction modulo $G \setminus \{p\}$ for the remaining $p \in G$, see [1][Table 5.5].

3 SMT-RAT

In this section, we give a short overview of our toolbox `SMT-RAT` [5], in which we embed our Gröbner bases implementation. The core procedure of Buchberger's algorithm and it's underlying data structures are implemented in the extension `GiNaCRA` of the `GiNaC` library.

Framework. `SMT-RAT` is a `C++` library consisting of (1) a collection of SMT-compliant theory solver modules which can be used to extend an existing SMT solver to RCF and (2) an SMT solver in which these modules can be (and most of them are) integrated to tackle RCF. The latter is intended to be a testing

Table 1. Buchberger's algorithm and GB module consistency check

Listing (1) Buchberger's algorithm.	**Listing (2)** GB module consistency check.
Input: Set of polynomials P	Input: C_{rcv}, state (A, G)
Output: Gröbner basis G for $\langle P \rangle$	Output: (ans, C_{inf}),
	with $C_{inf} \subseteq C_{rcv}$
$G := P$	and $ans \in \{$sat, unsat, unknown$\}$
while true:	
$G' := G$	if $A \neq \emptyset$:
for each $\{p, q\} \subseteq G', p \neq q$	G := Groebner($G \cup A$)
$s := \text{red}_G(S(p, q))$	A := $\emptyset$
if $s \neq 0$:	if G = $\{1\}$:
$G := \text{U}_{\text{std}}(G, s)$	return $(unsat, C_{\text{rsn}}(1 = 0))$
if $G = G'$:	C_{pas} := $(C_{\text{rcv}} \setminus C_{\text{rcv}}[=])$
break	$\cup \; \{p = 0 \mid p \in G\}$
return G	(r, C'_{inf}) := runBackends(C_{pas})
	determine C_{inf} from C'_{inf}
	return (r, C_{inf})

environment for the development of SMT-compliant theory solvers, as the one presented in this paper. SMT-RAT defines three types of components (see [10, Appendix B]): *manager*, *strategy* and *module*. In the following we first describe the functionality of a module and show how the manager composes different modules according to a strategy to a solver.

Modules. The main procedure of a module is check(C_{rcv}). For a given set C_{rcv} of RCF formulas, called *the set of received formulas*, the procedure either decides whether C_{rcv} is satisfiable or not returning *sat* or *unsat*, respectively, or returns *unknown*. Note, that a set of formulas is semantically defined by their conjunction. We can manipulate the set of received formulas by adding (removing) formulas φ to (from) it with add(φ) (remove(φ)). Since in the SMT embedding C_{rcv} is usually changed between two consecutive check(C_{rcv}) calls only by adding/removing constraints, the solver's performance can be significantly improved if the modules can make use of the results of previous checks (*incrementality* and *backtracking*). In case that the module determines the unsatisfiability of C_{rcv}, it is expected to compute at least one preferably small *infeasible subset* $C_{\text{inf}} \subseteq C_{\text{rcv}}$. Moreover, a module has the possibility to name lemmas, which are RCF tautologies. These lemmas should encapsulate information which can be extracted from a module's internal state and propagated among other SMT-RAT modules. Furthermore, SMT-RAT provides the feature that a module itself can ask other modules for the satisfiability of a set C_{pas} of RCF formulas, called *the set of passed formulas*, using the procedure runBackends(C_{pas}) which is controlled by the manager.

This paper presents the implementation of a new SMT-RAT module called $\mathtt{M}_{\mathtt{GB}}$ based on Gröbner bases; the next section gives details on its implementation.

SMT-RAT already contains various modules implementing, among others a conjunctive normal form transformer $\mathtt{M_{CNF}}$, a SAT solver $\mathtt{M_{SAT}}$ and the modules $\mathtt{M_{LRA}}$ for simplex, $\mathtt{M_{VS}}$ for VS and $\mathtt{M_{CAD}}$ for CAD. Note that most of these procedures are not complete. If a module cannot solve a problem then it either returns *unknown* or consults another module as explained below.

Manager and Strategy. A *strategy* is a directed tree $T := (V, E)$ with a set V of module instances as nodes and $E \subseteq V \times \Omega \times V$, where Ω is a set of conditions. Initially, the *manager* calls the method check(C_{rcv}) of the module instance given by the root of the strategy, where C_{rcv} is a set of RCF formulas. Whenever a module instance $m \in V$ calls runBackends(C_{pas}), the manager calls check(C_{pas}) of each module m', for which an edge $(m, \omega, m') \in E$ exists such that ω holds for C_{pas}, and passes the results back to m. Furthermore, it also passes back the infeasible subsets and lemmas provided by the invoked modules. The module m can now benefit in its solving and reasoning process from this shared information. In the following we write short (m, m') for (m, ω, m) if $\omega = \text{True}$.

Usually, the root module $\mathtt{M_{CNF}}$ transforms its set of received formulas C_{rcv} to an equisatisfiable set of clauses C_{pas} and calls runBackends(C_{pas}). The backend is a SAT-solver module $\mathtt{M_{SAT}}$, which runs DPLL-style SAT-solving on the Boolean abstraction of the set of received clauses C_{rcv}. $\mathtt{M_{SAT}}$ might call runBackends(C_{pas}) for partial Boolean assignments on the corresponding set of formulas C_{pas}; we refer to such a backend call as *theory call*. The Boolean abstraction of the obtained infeasible subsets and lemmas are stored as additional clauses. Infeasible subsets and lemmas, which contain only formulas from C_{rcv}, prune the Boolean search space and hence the number of theory calls. Smaller infeasible subsets are usually more advantageous, because they make larger cuts in the search space. Other types of lemmas contain new formulas, so-called *inventive lemmas* (*non-inventive* otherwise) and might enlarge the Boolean search space, but they can reduce the complexity of later theory calls. This way we can compose SMT solvers for RCF, e.g., using the simple strategy defined by the nodes $I_{\mathtt{M_{CNF}}}$, $I_{\mathtt{M_{SAT}}}$ and $I_{\mathtt{M_{CAD}}}$ and the edges $(I_{\mathtt{M_{CNF}}}, I_{\mathtt{M_{SAT}}})$ and $(I_{\mathtt{M_{SAT}}}, I_{\mathtt{M_{CAD}}})$.

4 Applying Gröbner Bases

In this section we describe our SMT-RAT module $\mathtt{M_{GB}}$ applying Gröbner bases (GB) computations. In Section 4.1 we discuss how its design wraps a GB procedure such as Buchberger's algorithm, while leaving the GB procedure itself untouched. In turn, Section 4.2 comprises how Buchberger's algorithm can be adapted to work inside an SMT-RAT module. Moreover, we show how to treat inequalities in Section 4.3, how to realize a tighter SMT integration by giving lemmas in Section 4.4, and an extension to the GB module $\mathtt{M_{GB}}$ which makes it more suitable for preprocessing in Section 4.5.

In this section we assume C_{rcv} to be a set of constraints. Given a constraint c, a set of constraints C and a set of polynomials P, we use $C[\sim] = \{p \sim 0 \mid p \sim 0 \in$

$C\}$ to select constraints and $C_{\sim}(P) := \{p \sim 0 \mid p \in P\}$ to construct constraints from polynomials. We call $C_{\text{rsn}}(c) \subseteq C_{\text{rcv}}$ a *reason set of* c if $\bigwedge_{r \in C_{\text{rsn}}(c)} r \implies c$ and $C_{\text{rsn}}(C) \doteq \bigcup_{c \in C} C_{\text{rsn}}(c)$ a reason set of C.

4.1 SMT-Compliant Consistency Checking

In this section we show how consistency checking in an SMT-RAT module based on a Gröbner bases core procedure can be accomplished. We do not further specify this core procedure here. It is thus possible to plug in an off-the-shelf GB procedure implementation such as the one in Singular.

The input consists of a set C_{rcv} of received constraints and the set of constraints arrived since the last consistency check. We call a tuple $(A, G) \subseteq \mathbb{Q}[\overline{x}] \times \mathbb{Q}[\overline{x}]$ a (*GB module*) *state* if $A \subseteq \text{pol}(C_{\text{rcv}})$ is the set of polynomials added since the last consistency check and G is a Gröbner basis for $\langle \text{pol}(C_{\text{rcv}}[=]) \setminus A \rangle$.

The *incremental* consistency check procedure is given in Listing (2) of Table 1. It operates on C_{rcv} and the state (A, G). The procedure possibly updates the state (A, G) and outputs, first, an answer as to whether C_{rcv} is *sat*, *unsat* or its consistency is *unknown*, and second, a subset of C_{rcv} building an infeasible subset C_{inf} in case of the answer *unsat*. The first step in the procedure is the computation of a Gröbner basis of all polynomials appearing on the left-hand-side in $C_{\text{rcv}}[=]$ (line 6 in Listing (2) of Table 1). Thereby we recompute the GB only if $A \neq \emptyset$. Then, we reuse G for the computation of the GB of $\text{pol}(C_{\text{rcv}}[=])$, what is possible because $\langle \text{pol}(C_{\text{rcv}}[=]) \rangle = \langle G \cup A \rangle$. If the Gröbner basis is $\{1\}$, the polynomials have no common real zeros; hence, we determine the infeasible subset C_{inf} as reason set of $1 = 0$ (details below) and return *unsat*. Otherwise, we call a module with the same inequations, and instead of the original equations, we pass equations formed by the Gröbner basis. In the following, we describe the extensions around the algorithm in Listing (2) of Table 1 to provide the SMT compliance.

Backtracking. As in SMT solving constraints can be removed from theory solvers, we make bookkeeping of the GB module states. Because SAT solvers mostly use *chronological backtracking* we use a stack of states $((A_0, G_0), \ldots, (A_k, G_k))$, $k \in \mathbb{N}$, illustrated in Figure 1: We start with an empty stack. Whenever an equality is added, we add a state to the stack (a). After each consistency check, we update the topmost state from the stack (b). If an equality is removed, we remove all states from the stack which were added afterwards (c). Then, we add the polynomials which were added after the just removed equality iteratively, like a new equality (d).

Infeasible Subsets. As argued before, the module is expected to return a subset $C_{\text{inf}} \subset C_{\text{rcv}}$ in case the set of received constraints C_{rcv} is inconsistent.

To determine such a subset, in [6] *certificates* for inconsistency were introduced. It was also shown that minimality of these certificates is a problem which is as hard as calculating the Gröbner basis. These certificates are basically tuples of polynomials $(h_1, \ldots, h_n)$ such that for an ideal $I = \langle f_1, \ldots, f_n \rangle$ and a

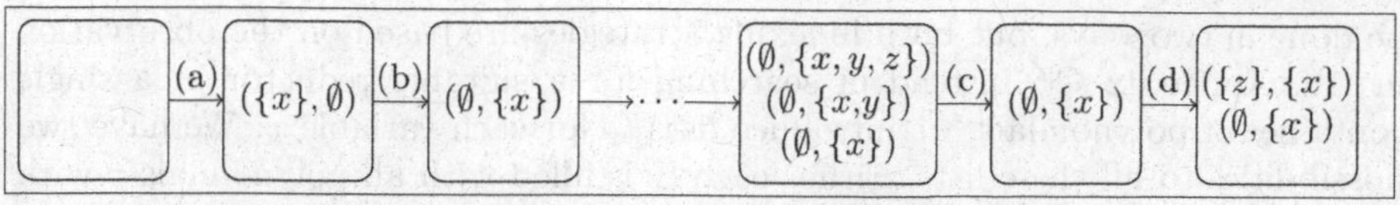

Fig. 1. The state stack in the GB module

polynomial $p \in I$ we have $\sum_{i=1}^{n} h_i f_i = p$ for suitable $h_i \in K[\bar{x}]$. In the case of inconsistency, we have $p = 1$. Calculating certificates requires the reductions within the Gröbner basis calculation to be extended to ordinary divisions, which is certainly less efficient. As we do this calculation for all reason sets, we implemented a more naive way. The realization of smaller reason and infeasible sets is obvious under the assumption that our GB procedure returns reason sets for each $p \in G$, with G a GB.

4.2 Our Gröbner Bases Procedure

We describe the adaptions to Buchberger's algorithm according to our setting of being called in an SMT-compliant way. The implementation is based on the description in [1].

Incrementality. As we call the GB procedure incrementally (cf. line 7 in Listing (2) of Table 1), we usually have to calculate Gröbner bases of $G \cup A$ for some set of polynomials A where G is a GB already. Instead of using Buchberger's algorithm from scratch, we skip all S-pairs (g_1, g_2), $g_1, g_2 \in G$ as they reduce to zero.

Reason Set Calculation. We calculate the *origin set* $C_{\text{org}}(p)$ *of a polynomial* p as follows: If p is added to our module, $C_{\text{org}}(p) = \{p\}$. Furthermore, for $p = S(p_1, p_2)$ and $p_1 \xrightarrow{p_2} p$, we set $C_{\text{org}}(p) = C_{\text{org}}(p_1) \cup C_{\text{org}}(p_2)$. Then $C_{\text{rsn}}(p = 0) = C_{\text{org}}(p)$. The set representations are realized by bit vectors and therefore taking the union costs at most a couple of machine operations.

Data Structures. We base our implementation of data structures on [14], e.g., we use a *compressed heap* during the reduction and for storing S-pairs. However, the term and ideal representations are adapted based on the following observations: The number of variables in the system is usually high and, due to incrementality, we do not have a fixed bound on the number of variables at initialization. However, most polynomials appearing are sparse, i.e., they consist of only few terms, each having small number of variables.

A term $a \cdot \prod_{i=1}^{n} x_i^{d_i}$ is represented as $(a, [(x_{i_1}, d_{i_1}), \ldots, (x_{i_k}, d_{i_k})], \sum_{i=1}^{n} d_i)$ with $d_{i_j} \neq 0$ for all $1 \leq j \leq k$ and $i_j < i_{j+1}$ for all $1 \leq j < k$. In our context, this representation seems more suitable than those from [14]. The degree is saved for fast access. For the ideal representation, we propose the adaption of the index structure from [12], which reduces the number of potential reductors. This can be done in two ways, but both indexing strategies are based on the observation in [10, Appendix C]. Instead of searching for a suitable reductor in a single container of polynomials, we introduce lists l_x for each variable x. We have two possibilities to fill these lists. Either each l_x is filled with all polynomials p with $x \in \mathrm{lm}(p)$ and during reduction of p we only search in an arbitrary l_x wheres $x \in \mathrm{lm}(p)$, or for each polynomial p we fill one arbitrary l_x with $x \in \mathrm{lm}(p)$ with p and during reduction of p, we search in all l_x where $x \in \mathrm{lm}(p)$. To reduce the number of terms which appear during the reduction, we order the polynomials in the index explained above according to the number of terms.

Real Radical. Among others, [11] discusses the problem that calculating the real radical is hard. They both propose the iterative application of simple rules to the ideal and thereby approximating the real radical. We propose to take this one step further. Instead of alternately calculating the GB and applying such rules, we integrate the rules within the calculation of the GB. For a given set of polynomials P, such a procedure thus no longer yields a GB for the ideal. However, we neither require the procedure the calculate the real radical of P. We only require that it *preserves* the common real zeroes.

Definition 2 (Real-radical preserving GB procedure). *A procedure $\mathcal{G}$ is called a* real-radical preserving GB procedure *if $\mathcal{V}_{\mathbb{R}}(P) = \mathcal{V}_{\mathbb{R}}(\mathcal{G}(P))$ and $\mathcal{G}(P)$ is a GB.*

To achieve such a procedure, we modify the update operator in Buchberger's algorithm (line 9, in Listing (1) of Table 1).

Definition 3 (Real-radical preserving update operator). *Let U be an update operator, $\prec$ be a monomial ordering. U is said to be* real-radical preserving *if for $P \subset \mathbb{Q}[\overline{x}]$ and $s \in \mathbb{Q}[\overline{x}]$ we have that $U(P, s) = P \cup Q$, where $Q \subset \mathbb{Q}[\overline{x}]$ such that $\mathcal{V}_{\mathbb{R}}(Q) = \mathcal{V}_{\mathbb{R}}(\langle s \rangle)$ and q is normal form modulo P for all $q \in Q$.*

The following theorem formalizes the relation between the used update operator and the GB procedure.

Theorem 1. *If the update operator in the Buchberger algorithm is modified into a real-radical preserving update operator, then the modified Buchberger algorithm is a real-radical preserving GB procedure.*

The proof is included in [10, Appendix D]. In [10, Appendix E] we give some computationally cheap rules implemented.

4.3 The Handling of Inequalities

Our implementation offers two different approaches to deal with a received inequality $p \sim 0$. The first approach *equalizes* the inequation by introducing a new variable y according to the following valid equivalences [13]:

$$p \geq 0 \Leftrightarrow \exists y. p - y^2 = 0, \quad p > 0 \Leftrightarrow \exists y. py^2 - 1 = 0, \quad p \neq 0 \Leftrightarrow \exists y. py - 1 = 0$$

The resulting equation can then be handled as before.

In the second approach we *reduce* p to $q := \text{red}_P(p)$ w.r.t. some subset P of a GB G. If $q \in \mathbb{Q}$, then either $q \sim 0$ and we do not have to pass it to our backends, or $q \not\sim 0$ and we obtain $C_{\text{rsn}}(C_=(P)) \cup \{p \sim 0\}$ as infeasible subset and return *unsat*. In order to allow the correct interaction of the reduction of $p \sim 0$ with the GB module stack, we store the most relevant reductions in a *reduction chain* $\text{RC}(p \sim 0) \subseteq \mathbb{Q}[\overline{x}] \times \mathbb{N}$: Assuming our stack is $((A_0, G_0), \ldots, (A_k, G_k))$, then $\text{RC}(p \sim 0) = \{(p, 0)\} \cup \{(\text{red}_{G_k}(p), k) \mid \text{red}_{G_k}(p) \neq p\}$. If a new state (A_j, G_j) is added to the stack, we set $\text{RC}(p \sim 0) = \text{RC}(p \sim 0) \cup \{(\text{red}_{G_j}(p_m), j) \mid \text{red}_{G_j}(p_m) \neq p_m\}$ where $m = \max\{i \in \mathbb{N} \mid (p, i) \in \text{RC}(p \sim 0)\}$. If an equality is removed such that the new stack size is k', then we remove all (p, i), $i > k'$ from $\text{RC}(p \sim 0)$. If $p \sim 0$ is removed, we simply delete $\text{RC}(p \sim 0)$.

4.4 Learning

In the following we consider that a constraint $p \sim 0$ is deduced from C_{rcv} by the module. If we achieve a constant value, i.e. $q \in \mathbb{Q}$ and $q \sim 0$ holds, we obtain the non-inventive lemma $C_{\text{rsn}}(C_=(P)) \rightarrow (p \sim 0)$. If q is a linear polynomial and P contains at least one nonlinear constraint, we share the inventive lemma $C_{\text{rsn}}(C_=(P)) \rightarrow (q = 0)$. Successive theory calls might then be solved by a more efficient linear solver. Note, that linear solvers are usually capable of detecting such deductions where P consists only of linear constraints. Finally, if $q := \sum t_i x$ and $\text{tdeg}(q)$ is sufficiently small, for instance less then the maximum degree occurring in C_{rcv}, we learn the inventive lemma $C_{\text{rsn}}(C_=(P)) \rightarrow (x = 0 \vee \sum t_i = 0)$. It forms a case splitting and at least one case reduces the complexity of the subsequent theory call significantly.

4.5 Iterative Variable Elimination

In Section 4.2 we have discussed the embedding of saturation rules for the real radical into the GB procedure. However, some saturation rules from [11][13] are not (yet) suitable for this kind of integration, e.g., rules involving a case splitting, which is optimally resolved by learning as discussed in the previous subsection.

Another example is the *iterative variable elimination* (IVE) as introduced in [13]. In practice, a GB G contains a lot of identities of the form $t - x$, where t is a term not containing x. IVE removes the respective identity and substitutes x by t in G, in symbols $G[t \backslash x]$, yielding $G' = (G \setminus \{t - x\})[t \backslash x]$, which is in general not a GB. Then, it applies the GB procedure to obtain a GB and repeats these two steps until we reach a fixpoint. The strict embedding of this saturation rule into the GB procedure is not straightforward, as potentially all GB elements are affected. Furthermore, we apply the encountered substitutions to the GB module's received inequalities.

When applying IVE, we have to preserve the module's SMT compliance, which turns out to be rather straightforward for the provided mechanisms. The incrementality can be guaranteed as all substitutions can be applied to

Table 2. # instances more than δ ms faster/slower than SMT-RAT with S_{ref}

Set (# instances)	δ	GB_{np}	$\text{GB}_{\text{np}}^{\text{IVE}}$	GB_{p}	$\text{GB}_{\text{p}}^{\text{IVE}}$	GB_{t}	$\text{GB}_{\text{t}}^{\text{IVE}}$	Any
KEY(421)	5	102/36	120/44	110/46	119/51	183/45	178/56	252/4
	500	29/0	29/1	28/5	27/6	31/2	35/0	36/0
MET(8276)	25	267/231	175/416	352/434	254/613	167/1410	239/1401	698/77
BOUNCE(180)	500	0/0	0/1	10/11	77/7	0/0	1/0	78/0

the polynomials of added constraints belatedly. In order to provide backtrackability, we add the substitutions to the stored module state. We define the reason set of a constraint $c' := c[t\backslash x]$ we obtained by applying a substitution to be $C_{\text{rsn}}(c') := C_{\text{rsn}}(c) \cup C_{\text{rsn}}(t - x = 0)$ and identify infeasible subsets as before.

With IVE we are able to detect the infeasibitity of a set of constraints more often. Moreover, the constraints we pass to our backends contain less variables by the cost of an in general higher complexity in the remaining variables. A drawback of IVE is that it blows up the reason sets of the constraints and therefore leads to greater infeasible subsets.

5 Experimental Results

The symbolic computations we present in this paper can significantly improve the performance of an SMT-RAT solver instance. We tested six different M_{GB} settings with the SMT-RAT strategy $S := (V, E)$ where $V := \{I_{\text{M}_{\text{CNF}}}, I_{\text{M}_{\text{SAT}}}, I_{\text{M}_{\text{LRA}}}, I_{\text{M}_{\text{GB}}}, I_{\text{M}_{\text{VS}}}, I_{\text{M}_{\text{CAD}}}\}$ with I_M an instance of module M and $E := \{(I_{\text{M}_{\text{CNF}}}, I_{\text{M}_{\text{SAT}}}), (I_{\text{M}_{\text{SAT}}}, I_{\text{M}_{\text{LRA}}}), (I_{\text{M}_{\text{LRA}}}, I_{\text{M}_{\text{GB}}}), (I_{\text{M}_{\text{GB}}}, I_{\text{M}_{\text{VS}}}), (I_{\text{M}_{\text{VS}}}, I_{\text{M}_{\text{CAD}}})\}$. Since M_{LRA} performs significantly faster on many instances containing linear constraints, it is positioned before M_{GB}. All M_{GB} settings implement the approaches explained in the Sections 4.1 and 4.2. The settings GB_{np} and GB_{p} reduce inequalities, GB_{t} transforms them. GB_{np} and GB_{t}, however, set $C_{\text{pas}} = C_{\text{rcv}}$, while GB_{p} passes constraints as described in Section 4.1. $\text{GB}_{\text{p}}^{\text{IVE}}$, $\text{GB}_{\text{np}}^{\text{IVE}}$, $\text{GB}_{\text{t}}^{\text{IVE}}$ are the extensions of the aforementioned settings by IVE. The computational effort and thus the room for optimization stepwise increases with enabling transformation and IVE. Passing the constraints has a major influence on the backends. We compared all settings with the reference strategy $S_{\text{ref}} := (V_{\text{ref}}, E_{\text{ref}})$ where $V_{\text{ref}} := V \setminus \{I_{\text{M}_{\text{GB}}}\}$ and $E_{\text{ref}} := (E \setminus \{(I_{\text{M}_{\text{LRA}}}, I_{\text{M}_{\text{GB}}}), (I_{\text{M}_{\text{GB}}}, I_{\text{M}_{\text{VS}}})\}) \cup \{(I_{\text{M}_{\text{LRA}}}, I_{\text{M}_{\text{VS}}})\}$. We regard three example sets: BOUNCE is an extension of examples introduced in [5]. KEY and MET originate from the tools KeYmaera and MetiTarski. Details of our benchmarks can be found in [10, Appendix A], here we give a summary.

Table 2 shows for each setting how many instances ran more than δ milliseconds faster/slower than the reference solver. In the last column, we give results for a hypothetical optimal solver, which always takes the setting yielding the best running time. Although many instances are not significantly influenced by M_{GB} in terms of running time, we observe a critical speed-up on specific instances. For KEY, improvements are gained by detecting unsatisfiability, which in most cases occurs during the reduction of inequalities. Here the received constraints are more suitable for passing. For BOUNCE, $\mathtt{M_{GB}}$ has only effect if the resolved identities are passed by $\mathtt{GB}_{\mathtt{P}}^{\mathtt{IVE}}$. A heuristic choosing the right setting increases the overall performance, and is essential for MET.

6 Conclusion and Future Work

In this work, we made use of the strength of traditional computer algebra procedures to resolve weaknesses of SMT solving for RCF. In particular, we integrated Gröbner bases computations in a module of an SMT solver. Moreover, we adapted the implementation of the Buchberger algorithm and its data structures to reflect differences in treated problems. To meet our requirement of real solutions, we embedded saturation rules for the real radical within the Buchberger algorithm, which makes the module more powerful. Experimental results show that selected instances are solved a lot faster.

As a next step we want to optimize the heuristics used in our Gröbner bases module and do other improvements, e.g., by developing new saturation rules or by algorithmic improvements tailored towards special input problem structures. We are also interested in integrating further methods based on (lexicographic) Gröbner bases, and especially in realizing applications of the Positivstellensatz. Another open point is the choice of the `SMT-RAT` strategy. For instance, the interplay between the GB and the CAD module could be much more dynamic as compared to one fixed strategy with fixed CAD settings.

References

1. Becker, T., Weispfenning, V., Kredel, H.: Gröbner bases: a computational approach to commutative algebra. Graduate texts in mathematics. Springer (1993)
2. Biere, A., Heule, M.J.H., van Maaren, H., Walsh, T. (eds.): Handbook of Satisfiability. Frontiers in Artificial Intelligence and Applications, vol. 185. IOS Press (2009)
3. Buchberger, B.: Ein Algorithmus zum Auffinden der Basiselemente des Restklassenringes nach einem nulldimensionalen Polynomideal. PhD thesis, University of Innsbruck (1965)
4. Collins, G.E.: Quantifier elimination for real closed fields by cylindrical algebraic decomposition. In: Brakhage, H. (ed.) GI-Fachtagung 1975. LNCS, vol. 33, pp. 134–183. Springer, Heidelberg (1975)
5. Corzilius, F., Loup, U., Junges, S., Ábrahám, E.: SMT-RAT: An SMT-compliant nonlinear real arithmetic toolbox. In: Cimatti, A., Sebastiani, R. (eds.) SAT 2012. LNCS, vol. 7317, pp. 442–448. Springer, Heidelberg (2012)
6. de Moura, L., Passmore, G.O.: On locally minimal Nullstellensatz proofs. In: Proc. of SMT 2009, pp. 35–42 (2009)
7. Dolzmann, A., Sturm, T.: Simplification of quantifier-free formulas over ordered fields. Journal of Symbolic Computation 24, 209–231 (1997)
8. Dutertre, B., de Moura, L.: A fast linear-arithmetic solver for DPLL(T). In: Ball, T., Jones, R.B. (eds.) CAV 2006. LNCS, vol. 4144, pp. 81–94. Springer, Heidelberg (2006)

9. Gao, S., Ganai, M.K., Ivancic, F., Gupta, A., Sankaranarayanan, S., Clarke, E.M.: Integrating ICP and LRA solvers for deciding nonlinear real arithmetic problems. In: Proc. of FMCAD 2010, pp. 81–89. IEEE (2010)
10. Junges, S., Loup, U., Corzilius, F., Ábrahám, E.: On Gröbner bases in the context of satisfiability-modulo-theories solving over the real numbers. Technical Report AIB-2013-08, RWTH Aachen University (May 2013)
11. Passmore, G.O.: Combined Decision Procedures for Nonlinear Arithmetics, Real and Complex. PhD thesis, University of Edinburgh (2011)
12. Passmore, G.O., de Moura, L., Jackson, P.B.: Gröbner basis construction algorithms based on theorem proving saturation loops. In: Decision Procedures in Software, Hardware and Bioware. Dagstuhl Seminar Proc., vol. 10161 (2010)
13. Platzer, A., Quesel, J.D., Rümmer, P.: Real world verification. In: Schmidt, R.A. (ed.) CADE-22. LNCS, vol. 5663, pp. 485–501. Springer, Heidelberg (2009)
14. Roune, B.H., Stillman, M.: Practical Gröbner basis computation. In: Proc. of ISSAC 2012, pp. 203–210. ACM (2012)
15. Weispfenning, V.: Quantifier elimination for real algebra – the quadratic case and beyond. AAECC 8(2), 85–101 (1997)

Approximation of Large Probabilistic Networks by Structured Population Protocols*

Michel de Rougemont[1] and Mathieu Tracol[2]

[1] University Paris II & LIAFA-CNRS
mdr@liafa.fr

[2] University Paris South
tracol@lri.fr

Abstract. We consider networks of Markov Decision Processes (MDPs) where identical MDPs are placed on N nodes of a graph G. The transition probabilities of an MDP depend on the states of its direct neighbors in the graph, and runs operate by selecting a random node and following a random transition in the chosen device MDP. As the state space of all the configurations of the network is exponential in N, classical analysis are unpractical. We study how a polynomial size statistical representation of the system, which gives the densities of the subgraphs of width k, can be used to analyze its behaviors, generalizing the approximate Model Checking of an MDP. We propose a *Structured Population Protocol* as a new Population MDP where states are statistical representations of the network, and transitions are inferred from the statistical s tructure. Our main results show that for some large networks, the distributions of probability of the statistics vectors of the population MDP approximate the distributions of probability of the statistics vectors of the real process. Moreover, when the network has some regularity, both real and approximation processes converge to the same distributions.

1 Introduction

We consider large networks of probabilistic systems, where each system (or device) is a Markov Decision Process, i.e. a transition system with both non deterministic and probabilistic transitions. The device MDPs are placed at nodes of the graph of the network with N nodes. A policy σ determines the decisions for all device MDPs, and the network itself can be considered as an MDP whose state space is the set of configurations of the network, of size exponential in N. Given a policy and an initial distribution, we define a stochastic process on the set of configurations by selecting a random node and by following a transition in the chosen MDP, which may be deterministic or randomized. Sensors networks are typical applications where sensors are nodes of a graph connected to some neighbors, and other applications include system biology and statistical physics. The classical Ising model is a special case where the network is a grid and the

* Work supported by ANR-07-SESU-013 program of the French Research Computer Security program.

T. Muntean, D. Poulakis, and R. Rolland (Eds.): CAI 2013, LNCS 8080, pp. 199–210, 2013.

device MDP is a Markov chain with 2 states. We consider *Evaluation problems* which predict the global behavior when a policy is fixed, and *Reachability problems* which look for possible policies to ensure predictable behaviors with high probabilities.

In [7], we presented some techniques to approximately decide both questions on a given MDP by associating *frequency vectors* to runs. Given an MDP with n states, we built its *Polytope of frequency vectors* H which represents the k-frequencies of the different states in runs, in polynomial time. We can then decide if there is a run which *approximately* verifies some Property with high probability with simple geometrical procedures.

Given a network of N device MDPs, the polytope-based method remains exponential in N. In this paper, we introduce a new approximate method based on the statistics on graph neighborhoods of depth k of the network. The crucial point is that the set of k-statistics has size polynomially bounded in N. A *Structured Population Protocol with Decisions* (SPPD) will define a new *Population-MDP* whose states are statistics vectors and where transitions are determined by the graph. If we fix a precision for the values of the statistics densities, say 1%, the number of possible vectors becomes independent of N. The construction of the population-MDP becomes feasible and we can then apply the initial polytope-based method. In this context, the classical problems are:

- *Evaluation problems.* Given a fixed policy σ for all the device MDPs and an initial distribution C, can we reach configuration C' with probability greater than λ ? For a property $\mathcal{P}$ on the runs, decide if $\mathbb{P}^{\sigma,C}[\text{a run satisfies } \mathcal{P}] \geq \lambda$ where $\lambda \leq 1$ is a threshold value.
- *Reachability problem.* Is there a policy σ, such that we can we reach configuration C' from configuration C with probability greater than λ ? If the device MDPs have two states for example, *dead* and *alive*, we may ask if $\mathbb{P}^{\sigma,C}[\text{more than 80\% of the states are alive in a run }] \geq \frac{1}{2}$?

We map configurations to their statistics, and approach these problems by considering their *approximate* versions on the population MDP. The *approximate evaluation* is: given the statistics of the configuration C, can we reach the statistics of configuration C' with probability greater than λ? The other problems can be formulated in a similar way. The main results of the paper are:

- The k-SPPD associated to a network of MDPs is itself an MDP.
- Bounds on the approximation of the network of MDPs by the k-SPPD (proposition 3 and theorem 1).
- Sufficient conditions for the convergence of the approximate process induced by the k-SPPD towards the limit of the real process.
- The polytope associated to the k-SPPD approximates the polytope of the class of *statistics policies* on the network of MDPs (theorem 2)

In section 2 we review the approximation of Markov Decision Processes (MDPs) [7] and define the k-statistics on graphs. In section 3 we define our model of network of MDPs. In section 4 we introduce the general model of *k-Structured*

Population Protocols with Decisions on a graph (k-SPPD), and we present how to associate a k-SPPD to a network of MDPs. In section 5 we present sufficient conditions for good approximations of networks of MDPs by our k-SPPDs. The conditions rely on a notion of *mixed configurations.* We also study the convergence of the approximate process induced by a k-SPPD, and present the polytope associated to the set of *statistics policies* on a network of MDPs.

1.1 Comparison with Related Models

Various theoretical models of networks have been considered in a context of distributed computing and statistical physics. Models for distributed computing [3] also include Petri nets [10], computer networks models [13] and cellular automata [18] which can be seen as a deterministic and synchronous restriction of our model. In statistical physics, spatial models [9] have similar probabilistic transitions associated with physical neighborhoods, in particular the Ising model describing models of spins. These statistical models do not integrate the possibility to take decisions, and the associated processes induce Markov chains on the sets of configurations.

If we restrict to MDPs with no decisions, i.e. to Markov chains, our model lies between the totally non ordered model of *population protocols*, introduced by Angluin et al in [3], and the totally ordered model of *cellular automata.* We differentiate from the population protocol model of [3], as structured graphs neighborhoods are chosen according to some statistics, as opposed to pairs of devices. Our work is closer to [2] where the authors consider devices distributed on the vertex of a graph with non randomized interactions between couples of devices. Cellular automata and dynamical systems consider regular geometries such as linear or square grid graphs (see [18,16]), and update all devices synchronously. In [1], the model is close to our model of SPP since the update function is asynchronous and uniformly random among the devices, with the restriction that the transition functions are deterministic.

2 Preliminaries

We first review the approximation of MDPs and graphs. They allow for efficient approximate solutions to reachability problems [7], in the spirit of Property testing [11] . We want to extend them to networks of MDPs.

2.1 Markov Decision Processes

Let $\mathcal{D}(S)$ be the set of distributions on a set S. A *Markov Decision Process (MDP)* is a triple $\mathcal{S} = (S, \Sigma, P)$ where S is a finite set of *states*, Σ is a set of *actions*, and $P : S \times \Sigma \times S \to [0,1]$ is the *transition function*: $P(s,a,t)$, also written $P(t|s,a)$, is the probability to arrive in t in one step when the current state is s and action $a \in \Sigma$ is chosen for the transition. If action a is not allowed from state s, $P(t|s,a) = 0$ for all $t \in S$. A *run* on $\mathcal{S}$ is a finite or infinite sequence of states. Given a run r and $n \in \mathbb{N}$, we write $r_{|n}$ for the sequence of the first $n-1$ states in r. A *policy* on $\mathcal{S}$, see [17], is a function $\sigma : S \rightarrow \mathcal{D}(\Sigma)$ which resolves the non determinism of the system by choosing a distribution on the set of available actions for each state of the MDP (we restrict our model to stationary and possibly randomized policies). A policy σ and an initial distribution $\alpha \in \mathcal{D}(S)$ induce a probability distribution $\mathbb{P}^{\sigma,\alpha}$ on the σ-field $\mathcal{F}$ of the set of runs, generated by the cones $C_\rho = \{r \mid r_{|\rho|} = \rho\}$, (see [6,17]). When there is no decision for the MDP, i.e. when $|\Sigma| = 1$, the MDP is in fact a Markov chain.

The *frequency vector* $\text{freq}_T(r)$ of the prefix of length T of a run r on $\mathcal{S}$ is the density vector of dimension $|S|$ which measures the proportions of time spent on the different states of the MDP until time T. That is, given $s \in S$,

$$\text{freq}_T(r)[s] = \frac{\text{number of occurrences of s in } r_{|T}}{T}$$

Let σ be a policy on $\mathcal{S}$ and $T \geq 0$, and let $\hat{x}^T$ be the random variable on the set of runs which associates to all r its frequency vector of length T: $\hat{x}^T = freq_T(r)$. Given an initial distribution α, the *Expected frequency vector* $x^T_{\sigma,\alpha}$ is $\mathbb{E}_{\sigma,\alpha}[\hat{x}^T]$, the expectation of $\hat{x}^T$. Let $x^\infty_{\sigma,\alpha}$ be the empty set if $x^T_{\sigma,\alpha}$ does not converge as $T \rightarrow +\infty$, and the limit point if $x^T_{\sigma,\alpha}$ converges. We define:

$$\mathcal{H}(\alpha) = \bigcup_{\sigma \text{ policy}} x^\infty_{\sigma,\alpha}$$

If $\mathcal{S}$ is an irreducible Markov chain, then $\mathcal{H}(\alpha)$ is the stationary distribution on the states of the chain. For a general MDP, $\mathcal{H}(\alpha)$ is a convex combination of the set of stationary distributions which can be reached on the Markov chains induced by stationary policies on $\mathcal{S}$. Generalizing the classical linear characterization of the stationary distribution of an irreducible Markov chain, the authors of [8,15] give linear characterizations of $\mathcal{H}(\alpha)$ [15]. As a consequence, the set $\mathcal{H}(\alpha)$ is a polytope, characterized by a number of linear equation polynomial in the size of the system. This makes possible the evaluation of properties such that: *with high probability, is state s in a run followed by state t?* [7]. Moreover, H is also the convex hull of the limit frequency vectors associated to non randomized policies.

2.2 Graph Neighborhoods and Statistics

Let $\mathcal{G} = (V, E)$ be a graph with vertex set V and edge set E, and S be a finite set of labels. Let $N = |V|$. An *S-labeled graph* on $\mathcal{G}$ is a triple $(\mathcal{G}, C, S)$ where $C : V \rightarrow S$ is a labeling function which associates a label in S to each state in V. We will often write C for the labeled graph $(\mathcal{G}, C, S)$. We write $\mathcal{C}$ for the set of S-labeled graphs on $\mathcal{G}$. The density vectors of neighborhoods at distance k for graphs $\mathcal{G}$ and labeled graphs C, resp. $ustat_k(\mathcal{G})$ and $ustat_k(C)$ (uniform statistics) give the probabilities for a random $v \in V$ that its neighborhood at distance k appear in $\mathcal{G}$ (resp. C). If we restrict to classes of graphs with uniformly bounded degrees, then $|\{ustat_k(C) \mid C \in \mathcal{C}\}|$ is polynomial in the number of nodes of the graphs.

3 Networks of MDPs

Our network of MDPs is a labeled graph where the set of labels is the set of states of an MDP $\mathcal{S} = (S, \Sigma, P)$. We need to generalize the notion of MDP to make the transitions depend on the environment of a node v. An environment of a node v is a pointed S-labeled graph $((\mathcal{H}, C, S), v)$ where $\mathcal{H} = (V, E)$, $v \in V$, and each vertex in $\mathcal{H}$ is at distance at most 1 from v. Let $\mathcal{N}_1$ be the set of such environments. Notice that in particular, given any S-labeled graph $\mathcal{F}$ on a structure $\mathcal{G}$ and v a vertex in $\mathcal{F}$, the neighborhood $\mathcal{N}_1(\mathcal{F}, v, 1)$ is in $\mathcal{N}_1$.

A *device MDP* is a triple $\mathcal{S} = (S, \Sigma, P_D)$ where S is a finite state space, Σ is a finite set of actions, and P_D is the transition function: $P_D : \mathcal{N}_1 \times \Sigma \to \mathcal{D}(S)$. Given $H \in \mathcal{N}_1$, $s \in S$ and $a \in \Sigma$, $P_D(H, a)(s)$, also written $P_D(s|H, a)$, is the probability that the state of the device MDP is s after the transition, given its environment is H and action a is chosen. The classical definition of an MDP can be retrieved by restricting the transition function P_D so that its values depend only on a and on the label of the pointed node of H.

Definition 1 (Network of MDPs). *A* network of MDPs *is a couple* $\mathcal{M} = (\mathcal{G}, \mathcal{S})$, *where* $\mathcal{G} = (V, E)$ *is a graph and* $\mathcal{S} = (S, \Sigma, P_D)$ *is a device MDP.*

A *configuration* on $\mathcal{M}$ is a function $C : V \to S$ which assigns to each vertex of $\mathcal{G}$ a state of the associated device MDP. We write $\mathcal{C}$ for the set of configurations on $\mathcal{M}$. A configuration can be seen as an S-labeled graph on $\mathcal{G}$. We may write C indifferently for the configuration or for the associated S-labeled graph.

3.1 Transitions on a Network of MDPs

We define transitions on $\mathcal{M}$ and obtain a new MDP, $\mathcal{S}(\mathcal{M})$. The state space of $\mathcal{S}(\mathcal{M})$ is $\mathcal{C}$, the set of configurations. The set of actions is Σ, used by each device MDP. For a transition, a random device MDP is chosen and its state is updated according to the transition function P_D. We sample *uniformly at random* a node v (device MDP) to update, as for the *random independent scheme*, a classical model for asynchronous Cellular Automata and other models of computation.

Let $C \in \mathcal{C}$, $v \in V$ and $s \in S$. We define $C_{v \to s}$ as the function from V to S which coincide with C on every $w \in V - \{v\}$, and such that $C_{v \to s}(v) = s$.

Given v and its 1-neighborhood $H = \mathcal{N}(C, v, 1)$ in the current configuration C, let s be sampled randomly according to distribution $P_D(-|H, a)$, where $a \in \Sigma$ is the chosen action. The configuration is changed to $C' = C_{v \to s}$. This process defines a transition function P on the MDP $\mathcal{S}(\mathcal{M}) = (\mathcal{C}, \Sigma, P)$ as follows: let $a \in \Sigma$, and let $C, C' \in \mathcal{C}$. Recall that $N = |V|$.

1. If $C \neq C'$ and there exists $v \in V$ and $s \in S$ such that $C' = C_{v \to s}$, then we define $P(C'|C, a) = \dfrac{P_D(s|\mathcal{N}(C, v, 1), a)}{N}$

2. If $C = C'$, then we define $P(C'|C,a) = \dfrac{\sum_{v \in V} P_D(C(v)|\mathcal{N}(C,v,1),a)}{N}$
3. In the other cases, we define $P(C'|C,a) = 0$.

For all $a \in \Sigma$ and $C \in \mathcal{C}$, $P(-|C,a)$ is indeed a probability distribution on $\mathcal{C}$, and $\mathcal{S}(\mathcal{M}) = (\mathcal{C}, \Sigma, P)$ is an MDP. Notice that the *policies on* $\mathcal{S}(\mathcal{M})$ *are global*, i.e. they consider the configurations, not the particular devices.

We will use *shift vectors* to quantify the change in the statistics of the configurations induced by the update of the state of one device MDP in the network. Given $C \in \mathcal{C}$, $v \in V$ and $a \in \Sigma$, let:

$$\Delta_k(C, v \to s) = N \cdot [ustat_k(C_{v \to s}) - ustat_k(C)]$$

By extension, $\Delta_k(\mathcal{N}(C,v,\phi(k)), v \to s)$ is the similar vector when we restrict C to the neighborhood of v at distance $\phi(k)$. Given $k \in \mathbb{N}$, a *k-statistics shift vector* on $\mathcal{G}$ is a vector $\Delta \in [-N,N]^{\mathcal{N}_k(\mathcal{C})}$ whose components in $\mathbb{Z}$ sum to zero. Clearly, vectors of the type $\Delta_k(C, v \to s)$ are k-shift vectors. The following proposition shows that when the state of one of the vertices of a labeled graph is changed, the variation on the k-statistics depends only on bounded neighborhoods around the changed vertex.

Proposition 1 **($(k, \phi(k))$-locality).** *Let $k \in \mathbb{N}$. Let $\phi : \mathbb{N} \to \mathbb{N}$ be such that $\phi(0) = 1$ and $\phi(k) = 2 \cdot k$ if $k \geq 1$. Then for all $C \in \mathcal{C}$, $v \in V$ and $s \in S$ we have $\Delta_k(C, v \to s) = \Delta_k(\mathcal{N}(C,v,\phi(k)), v \to s)$, i.e. for all $H \in \mathcal{N}_k(\mathcal{C})$ we have:*

$$\Delta_k(C, v \to s)[H] =$$
$$N \cdot [ustat_k(\mathcal{N}(C,v,\phi(k))_{v \to s}))[H] - ustat_k((\mathcal{N}(C,v,\phi(k)))[H]]$$

We can generalize this fact to the transition function of a network of MDPs. In the following, given $C \in \mathcal{C}$, we write C' for the random configuration distributed accordingly to the probability distribution $P(-|C,a)$. The following proposition is a direct consequence of proposition 1.

Proposition 2. *Let $\mathcal{M}$ be a network of MDPs, $a \in \Sigma$, $C \in \mathcal{C}$, $k \in \mathbb{N}$, and let Δ_k be a k-statistics shift vector. Then:*

$$P(ustat_k(C') = ustat_k(C) + \tfrac{\Delta_k}{N} \mid C, a) =$$
$$P(ustat_k(C') = ustat_k(C) + \tfrac{\Delta_k}{N} \mid ustat_{\phi(k)}(C), a)$$

In other words, the distribution of the k-order statistics of the configurations after a transition depends only on the $\phi(k)$-th order statistics of the configuration C before the transition.

4 Structured Population Protocols with Decisions

Let $\mathcal{G} = (V, E)$ be a graph network, and let $N = |V|$. Let S be a finite set of labels, and let $\mathcal{C}$ be the set of S-labeled graphs on $\mathcal{G}$. Given $k \in \mathbb{N}$, recall that $\mathcal{N}_k(\mathcal{C})$ is the set of all possible k-neighborhoods which can appear in S-labeled graphs on the structure $\mathcal{G}$. We define a *Structured Population Protocol* which will induce an MDP on the set of statistics vectors. Our model generalizes classical Population Protocols in two ways:

- it uses statistics on graphs neighborhoods, i.e. on a structured domain, as opposed to statistics on sets,
- decisions can be taken on states, using the same decision space Σ as the original device MDPs.

A *Population* of statistics of order k, or k-Population, on $\mathcal{G}$, is a vector $A \in \mathbb{N}^{\mathcal{N}_k(\mathcal{C})}$ whose components sum to N. We write $\mathcal{A}_k$ for the set of k-Populations. A Population can be seen as a *soup* of neighborhoods, i.e. a multiset of neighborhoods with no structure. Given a neighborhood H in $\mathcal{N}_k(\mathcal{C})$, $A[H]$ is equal to the number of times the neighborhood H appears in the soup of neighborhoods A. A k-Population A induces a distribution $\frac{A}{N}$ on $\mathcal{N}_k(\mathcal{C})$, with $\frac{A}{N}(H) = A[H]/N$. Reciprocally, an $ustat_k$ vector $x = ustat_k(C)$ induces a k-Population $A = N \cdot x$. Typically, a k-Population counts the different k-neighborhoods which appear in an S-labeled graph C on $\mathcal{G}$. In that case, the probability distribution $\frac{A}{N}(-)$ is equal to $ustat_k(C)(-)$. Notice however that there may exist k-Populations A such that for no S-labeled graph C on $\mathcal{G}$ we have $\frac{A}{N}(-) = ustat_k(C)(-)$. Notice also that if $L = |\mathcal{N}_k(\mathcal{C})|$, then $|\mathcal{A}_k| \leq N^L$.

As in [3], in our approach of Population Protocols, the devices, i.e. the nodes of the graph, will interact locally. The associated transition probabilities will be given by a transition function δ. A *k-Structured Population Protocols with Decisions* on $\mathcal{G}$ is given by a *transition function* δ and a *reconstruction function* R_k. The function R_k will impose the updates to depend on the structure of the underlying graph $\mathcal{G}$.

Definition 2 (k-Structured Population Protocols with Decisions). *Given $k \in \mathbb{N}$, a* k-Structured Population Protocols with Decisions, *or k-SPPD, on $\mathcal{G}$, is a triple $\mathcal{O}_k = (\delta, R_k, \Sigma)$ where $\delta : \mathcal{N}_1(\mathcal{C}) \times \Sigma \to \mathcal{D}(S)$ and R_k : $\mathcal{A}_k \times \mathcal{N}_{\phi(k)}(\mathcal{G}) \to \mathcal{D}(\mathcal{N}_{\phi(k)}(\mathcal{C}))$.*

The function $\delta : \mathcal{N}_1(\mathcal{C}) \times \Sigma \to \mathcal{D}(S)$ is the *transition function*. When $|\Sigma| = 1$, the domain of δ is $\mathcal{N}_1(\mathcal{C})$ and the system is called a k-Structured Population Protocol, or k-SPP. In that case, our model is close to the standard model of Population Protocol, [4]. Because the interaction is local only $\mathcal{N}_1(\mathcal{C})$ is considered. Given $k \geq 0$ and $\overline{H}_v \in \mathcal{N}_{\phi(k)}(\mathcal{C})$, δ induces a distribution $\delta(-|\overline{H}_v, a)$ on the set of k-shift vectors. If Δ_k is a k-shift vector, $\delta(\Delta_k|\overline{H}_v, a)$ is the probability that an update of the label of the center node v of $\overline{H}_v$ according to the transition function δ induces a change Δ_k in the k-statistics. Formally, let $\delta(\Delta_k|\overline{H}_v, a)$ be defined as:

$$\delta(\Delta_k|\overline{H}_v, a) = \sum_{s \in S \ s.t. \ \Delta_k(\overline{H}_v, v \to s) = \Delta_k} \delta(\mathcal{N}(\overline{H}_v, v, 1), a)(s)$$

The function $R_k : \mathcal{A}_k \times \mathcal{N}_{\phi(k)}(\mathcal{G}) \to \mathcal{D}(\mathcal{N}_{\phi(k)}(\mathcal{C}))$ is a *reconstruction function*: given a k-Population A and H_v a $\phi(k)$-neighborhood of the graph $\mathcal{G}$, it outputs randomly a valuation in S for the nodes of H_v. The distribution $R_k(A, H_v)(-)$ assigns probabilities to the labellings of H_v. A k-SPPD is an MDP whose domain is $\mathcal{A}_k$, action set is Σ and probabilities depend on δ and R_k.

4.1 The k-SPPD Associated to a Network of MDPs

Let $\mathcal{M} = (\mathcal{G}, \mathcal{S})$ be a network of MDPs, with $\mathcal{G} = (V, E)$ and $\mathcal{S} = (S, \Sigma, P_D)$. Let $\mathcal{S}(\mathcal{M}) = (\mathcal{C}, \Sigma, P)$ be the MDP associated to $\mathcal{M}$. Given $k \in \mathbb{N}$, we want to define a k-SPPD $\mathcal{O}_k(\mathcal{M}) = (\delta, R_k, \Sigma')$ on $\mathcal{G}$ such that the associated MDP $\mathcal{S}_{\mathcal{O}_k}$ mimics the transitions of $\mathcal{S}(\mathcal{M})$ on the set of k-statistics vectors.

The set of actions on $\mathcal{O}_k(\mathcal{M})$ will be Σ, the same as for $\mathcal{M}$. The state space of $\mathcal{O}_k(\mathcal{M})$ will be $\mathcal{A}_k$, the set of k-Populations on $\mathcal{G}$, which can also be seen as the set of k-statistics of configurations on $\mathcal{M}$. The transition function δ of $\mathcal{O}_k(\mathcal{M})$ will be equal to the transition function P_D of the device MDPs of $\mathcal{M}$. The point is to define a relevant reconstruction function R_k. The role of the function R_k is, given a k-Population A and a $\phi(k)$-neighborhood $H \in \mathcal{N}_k(\mathcal{G})$, to guess valuations for the nodes in H. Ideally, we would like, given $C \in \mathcal{C}$ and $H \in \mathcal{N}_{\phi(k)}(\mathcal{G})$, the distributions $R_k(N \cdot ustat_k(C), H)(-)$ and $(ustat_{\phi(k)}(C)|H)(-)$ to be equal. That is, the reconstruction of size $\phi(k)$ of the k-statistics of a configuration is the $\phi(k)$ statistics of the configuration. This is not possible in general, but we give an algorithm to compute the function R_k which will give good approximations on a restricted class of *mixed configuration*, defined in the next subsection. We use the following algorithm to sample the function R_k.

Algorithm 1 (Sampling from R_k)
Input: *A Population* $A \in \mathcal{A}_k$, $H \in \mathcal{N}_{\phi(k)}(\mathcal{G})$.
Output: $\overline{H} = (H, C) \in \mathcal{N}_{\phi(k)}(\mathcal{C})$ *an S-valuation of the nodes of* H.
Method: *We define* C *incrementally on the set of nodes of* H. *Until a valuation for all the node of* H *is defined:*

1: Sample a node v *uniformly at random among the nodes in* H *whose* k*-neighborhood contains unlabeled nodes, and which is at distance at most* k *from the center of* H.

2: Sample $K \in \mathcal{N}_k(\mathcal{C})$ *according to distribution* $(\frac{A}{N}|C)(-)$. *That is, sample* K *according to the distribution* $\frac{A}{N}$ *conditioned to the partial valuation* C *defined so far. This corresponds to sampling labels for a neighborhood of size* k *in* H. *For all* $w \in K$, *define* $C(w) = K(w)$.
Return $\overline{H} = (H, C)$.

Finally, given $\mathcal{M} = (\mathcal{G}, \mathcal{S})$ the network of MDPs, using the algorithm 1 for the construction of R_k, we have a k-SPPD $\mathcal{O}_k(\mathcal{M}) = (\delta, R_k, \Sigma)$ on $\mathcal{G}$, with state space $\mathcal{A}_k$.

Given a policy on the MDP $\mathcal{M}$, how can we build a related policy on $\mathcal{O}_k$? Since the state space of $\mathcal{M}$ has size exponential in the state space of $\mathcal{O}_k$, we cannot associate a policy on $\mathcal{O}_k$ to each policy on $\mathcal{M}$. We will have to restrict the class of policies that we consider on $\mathcal{M}$: a policy on $\mathcal{M}$ must satisfy certain compatibility properties to be transferable on $\mathcal{O}_k$. A natural condition is the fact that it depends only on the k*-statistics* of the configurations. We call *statistical policies* such policies on $\mathcal{M}$:

Definition 3 (Statistical Policies). *A policy* σ *on* $\mathcal{M}$ *is* k-statistical *if:*

$$\forall C, C' \in \mathcal{C},\ ustat_k(C) = ustat_k(C') \Rightarrow \sigma(C) = \sigma(C')$$

Let $SR^k(\mathcal{M})$ be the set of k-Statistical and Randomized policies. For instance, a policy which takes its decisions according to the 0-statistic of the configurations, i.e. according to the proportions of the different states among the devices, is k-statistical for all $k \in \mathbb{N}$. A policy $\sigma \in SR^k(\mathcal{M})$ induces trivially a policy σ on $\mathcal{O}_k$, since σ can be defined on the set of $ustat_k$ vectors, hence on $\mathcal{A}_k$.

5 Approximations on Networks of MDPs.

Let $\mathcal{M}$ be a network of MDPs as before, with state space $\mathcal{C}$ and transition function P, and let $\mathcal{O}_k$ be the associated k-SPPD, with state space $\mathcal{A}_k$ and transition function $P_{\mathcal{O}_k}$. In this section we show that we can bound the difference in the evolutions of the statistics of the real process induced by $\mathcal{M}$, and the evolution of the approximation process induced by $\mathcal{O}_k$. More precisely, we show that we can define a notion of *mixed configurations*, quantified by a *mixing parameter*, such that the reconstruction function R_k defined by algorithm 1 on the Population Protocol approximates $\mathcal{M}$.

5.1 Mixed Configurations

A *partially labeled graph* is a graph such that labels are associated only to a subset of nodes. In particular, a graph $H = (V, E)$ can be seen as a partially labeled graph, where no valuation is defined for any node. We write $\mathcal{N}_k(\mathcal{C}_p)$ for the set of k-neighborhoods of partially labeled graphs on $\mathcal{G}$. Given F, F' two partially labeled graphs on the same domain V, F and F' are said to be *compatible* if there exists no node of V to which F and F' assign different labels. Given C_H a partially labeled graph on a graph H, we define $\mathcal{L}(C_H)$ as the set of labeled graphs on H compatible with C_H. We need to condition probability distributions by a structure: given a distribution μ on $\mathcal{N}_k(\mathcal{C})$, given $H \in \mathcal{N}_k(\mathcal{G})$ and given C_H a partially labeled graph on H such that $\mu(\mathcal{L}(C_H)) > 0$, the distribution $(\mu|C_H)(-)$ on $\mathcal{N}_k(\mathcal{C})$ is defined as follows: for all $K \in \mathcal{N}_k(\mathcal{C})$,

$$(\mu|C_H)(K) = 0 \; if \; K \notin \mathcal{L}(C_H) \; and \; else \; (\mu|C_H)(K) = \frac{\mu(K)}{\mu(\mathcal{L}(C_H))}$$

We now want to quantify the quality of the reconstruction function R_k. As we said before, ideally, given a configuration $C \in \mathcal{C}$ and $H \in \mathcal{N}_{\phi(k)}(\mathcal{G})$, the distributions $R_k(N \cdot ustat_k(C), H)(-)$ and $(ustat_{\phi(k)}(C)|H)(-)$ should be equal. However, this is not possible in general, since there may exist configurations $C, C' \in \mathcal{C}$ such that $ustat_k(C) = ustat_k(C')$ but $ustat_{\phi(k)}(C) \neq ustat_{\phi(k)}(C')$. We present a class of configurations for which there exist good reconstruction functions, i.e. functions R_k such that the distributions $R_k(N \cdot ustat_k(C), H)(-)$ and $(ustat_{\phi(k)}(C)|H)(-)$ are close. Such configurations can be seen as "mixed" configurations, and we define a *mixing coefficient.*

Let $C \in \mathcal{C}$, let v be a node in C and let $H = \mathcal{N}(C, v, k)$. Let K be a partial labeling of $\mathcal{N}(\mathcal{G}, v, \phi(k))$ such that the partial labeling K_H induced by K on $\mathcal{N}(\mathcal{G}, v, k)$ is compatible with H. Let P_C be the probability distribution

$ustat_k(C)(-)$. We define the following conditional probabilities: $P_C[H|H \cap K]$ is the probability, among the k-neighborhoods of C, of the neighborhood H, given the partial valuation K_H is given. $P_C[H|K]$ is the probability, among the $\phi(k)$-neighborhoods of C, of the neighborhood which contains H around its center, given the partial valuation K is given. Formally:

$$P_C[H|H \cap K] = \frac{|\{u \in V \ s.t. \ \mathcal{N}(C,u,k) \simeq H\}|}{|\{u \in V \ s.t. \ \mathcal{N}(C,u,k) \in \mathcal{L}(K_H)\}|}$$

$$P_C[H|K] = \frac{|\{u \in V \ s.t. \ \mathcal{N}(C,u,k) \simeq H \wedge \mathcal{N}(C,u,\phi(k)) \in \mathcal{L}(K)\}|}{|\{u \in V \ s.t. \ \mathcal{N}(C,u,\phi(k)) \in \mathcal{L}(K)\}|}$$

If K is not compatible with C, let $P_C[H|H \cap K] = P_C[H|K] = 0$.

Definition 4 (Mixing coefficient ϵ_k). *Let $C \in \mathcal{C}$ be a configuration. The* k-mixing coefficient of C *is defined as:*

$$\epsilon_k(C) = Max_{H=\mathcal{N}(C,u,k),\ K \in \mathcal{N}_{\phi(k)}(\mathcal{C}_p)}\{|P_C[H|H \cap K] - P_C[H|K]|\}$$

Intuitively, $\epsilon_k(C)$ is small if the distribution of the k-neighborhoods does not depend on their environment. We say that C is *well mixed* if $\epsilon_k(C)$ is small. The following proposition, shows that if configuration C is well mixed, the function R_k defined by the algorithm 1 is a good reconstruction function. We measure the distance between the distributions using the $\| \ \|_\infty$-norm: given $v \in \mathbb{R}^n$, $\|v\|_\infty = max_{i \in [1;n]}|v_i|$. As a consequence, if C is well mixed, we can find a good approximation of $ustat_{\phi(k)}(C)$ from $ustat_k(C)$. This is exactly what the function R_k is supposed to do.

Proposition 3. *Let $C \in \mathcal{C}$ be a configuration, and let R_k be defined by algorithm 1. Then for all $H \in \mathcal{N}_{\phi(k)}(\mathcal{G})$ we have:*

$$\|R_k(N \cdot ustat_k(C), H)(-) - (ustat_{\phi(k)}(C)|H)(-)\|_\infty \leq \epsilon_k(C)$$

We now use the mixing coefficient to give bounds on approximation of the behavior of networks of MDPs by our k-SPPDs. In [12], the authors approximate the short term evolution of large Markov chains by using a "sliding windows" approach. As in [12], we try to bound the deviation between our approximation and the real process as time goes on. Given $C \in \mathcal{C}$ and $a \in \Sigma$, we write C' for the random configuration induced by the probability distribution $P(-|C,a)$. Given $C \in \mathcal{C}$ and $A = N \cdot ustat_k(C)$, define the distributions $\mu^k_{C,a}$ and $\nu^k_{A,a}$ on $\mathcal{A}_k$ as follows: given $A' \in \mathcal{A}_k$, let:

$$\mu^k_{C,a}(A') = P(N \cdot ustat_k(C') = A' \mid C, a), \quad and \quad \nu^k_{A,a}(A') = P_{\mathcal{O}_k}(A' \mid A, a)$$

In other words, $\mu^k_{C,a}(-)$ is the distribution of the k-statistics of configurations after a transition from the configuration C, on the network of MDPs $\mathcal{M}$. On the other hand, $\nu^k_{A,a}(-)$ is the distribution of the k-statistics after a transition from the population $N \cdot ustat_k(C)$, on $\mathcal{O}_k$. The following theorem measures the quality of the approximation of the network of MDPs by the k-SPPD, on the set of k-statistics.

Theorem 1. $\|\mu^k_{C,a} - \nu^k_{A,a}\|_\infty \leq \|ustat_k(C)(-) - \frac{A}{N}(-)\|_\infty + \epsilon_k(C)$

We first study the Evaluation problem for Markov chains, in the paper's final version.

5.2 Approximations for Markov Decision Processes.

Now, we extend our approach to networks of MDPs with non-determinism. Given $\mathcal{M}$, the associated polytope H (see section 2.1) is a subset of $\mathbb{R}^{\mathcal{C}}$. The k-SPPD $\mathcal{O}_k$ associated to $\mathcal{M}$ is also an MDP, and its polytope lies in $\mathbb{R}^{\mathcal{A}_k}$. How can we relate these two polytopes?

We consider the set of the limit points associated to stationary statistics policies on $\mathcal{M}$, and we prove that it is also a polytope. We obtain a natural approximation of $H^k_{stat}(\mathcal{M})$ by the polytope $H(\mathcal{O}_k)$ associated to all the stationary policies on $\mathcal{O}_k$.

Theorem 2. *The set* $H^k_{stat}(\mathcal{M}) = \{N \cdot ustat_k(x^\infty_{\sigma,\alpha}) \mid \sigma \in SR^k\}$ *is a polytope of* $\mathbb{R}^{\mathcal{A}_k}$*, with a number of extremal points polynomial in* N.

Proof. First, notice that a convex combination of k-statistical policy is clearly a statistical policy. Thus, $H^k_{stat}(\mathcal{M})$ is convex. Next, any statistical policy is a convex combination of "deterministic" statistical policies which assign Dirac distribution to statistic vectors of configurations. (i.e. policies σ such that, given $A \in \mathcal{A}_k$, $\sigma(A) \in \Sigma$). This proves that $H^k_{stat}(\mathcal{M})$ is a polytope, and it is the convex hull of the limit frequency vectors associated to deterministic statistical policies. We can conclude using the fact that since there exists only a polynomial number of $ustat_k$ vectors of configurations in $\mathcal{A}_k$, there exists only a polynomial number of "deterministic" statistical policies, hence of extremal points of the polytope.

6 Conclusion

We studied how to approximate the evolution of large probabilistic networks of MDPs. Given a network $\mathcal{M}$ of N device MDPs, we defined a k-Structured Population Protocol with Decisions $\mathcal{O}_k$, which is also an MDP, whose states are statistics vectors. From an exponential number of configurations, we obtain a polynomial number of statistics. This allows the use of standard evaluation methods on the approximate system, and we gave a sufficient condition, using a mixing parameter ϵ_k, to guarantee a good approximation. If we discretize the statistics vectors up to a coefficient γ, the size of the set of configurations of the approximate process becomes independent of N: it depends only on δ, k, and the degree of the underlying graph. In the paper's final version, we will present the values of the sizes of the state spaces of the real and the approximate processes for various parameters, underlying the efficiency of the discretization model. The application to the Ising model is described by an applet at *http://www.up2.fr/Ising* which visualizes both the real process and its approximate version.

References

1. Agapie, A., Höns, R., ühlenbein, H.: Markov Chain Analysis for One-Dimensional Asynchronous Cellular Automata. Methodology and Computing in Applied Probability 6(2), 181–201 (2004)
2. Angluin, D., Aspnes, J., Chan, M., Fischer, M.J., Jiang, H., Peralta, R.: Stably computable properties of network graphs. In: Prasanna, V.K., Iyengar, S.S., Spirakis, P.G., Welsh, M. (eds.) DCOSS 2005. LNCS, vol. 3560, pp. 63–74. Springer, Heidelberg (2005)
3. Angluin, D., Aspnes, J., Diamadi, Z., Fischer, M.J., Peralta, R.: Computation in networks of passively mobile finite-state sensors. Distributed Computing 18(4), 235–253 (2006)
4. Aspnes, J., Ruppert, E.: An introduction to population protocols. Chemistry 314, 315 (2007)
5. Barrett, C.L., Hunt, H.B., Marathe, M.V., Ravi, S.S., Rosenkrantz, D.J., Stearns, R.E.: Complexity of reachability problems for finite discrete dynamical systems. Journal of Computer and System Sciences 72(8), 1317–1345 (2006)
6. Courcoubetis, C., Yannakakis, M.: The complexity of probabilistic verification. JACM 42(4), 857–907 (1995)
7. de Rougemont, M., Tracol, M.: Statistical analysis for probabilistic processes. In: Proc. IEEE Logic in Computer Science, pp. 299–308 (2009)
8. Derman, C.: Finite State Markovian Decision Processes. Academic Press, Inc., Orlando (1970)
9. Durrett, R.: Stochastic spatial models. Siam Review 41(4), 677–718 (1999)
10. Esparza, J.: Decidability and complexity of Petri net problems-an introduction. In: Reisig, W., Rozenberg, G. (eds.) APN 1998. LNCS, vol. 1491, pp. 374–385. Springer, Heidelberg (1998)
11. Fischer, E., Magniez, F., de Rougemont, M.: Approximate satisfiability and equivalence. SIAM J. Comput. 39(6), 2251–2281 (2010)
12. Henzinger, T.A., Mateescu, M., Wolf, V.: Sliding window abstraction for infinite markov chains. In: Bouajjani, A., Maler, O. (eds.) CAV 2009. LNCS, vol. 5643, pp. 337–352. Springer, Heidelberg (2009)
13. Kahn, J.M., Katz, R.H., Pister, K.S.J.: Next century challenges: mobile networking for Smart Dust. In: Proceedings of the 5th Annual ACM/IEEE International Conference on Mobile Computing and Networking, pp. 271–278. ACM, New York (1999)
14. Peyronnet, S., De Rougemont, M., Strozecki, Y.: Approximate verification and enumeration problems. In: Roychoudhury, A., D'Souza, M. (eds.) ICTAC 2012. LNCS, vol. 7521, pp. 228–242. Springer, Heidelberg (2012)
15. Puterman, M.L.: Markov Decision Processes: Discrete Stochastic Dynamic Programming. John Wiley & Sons, Inc., New York (1994)
16. Sutner, K.: On the computational complexity of finite cellular automata. Journal of Computer and System Sciences 50(1), 87–97 (1995)
17. Vardi, M.Y.: Automatic verification of probabilistic concurrent finite state programs. In: FOCS 1984, pp. 327–338 (1985)
18. Wolfram, S.: Cellular automata. Los Alamos Science 9, 2–21 (1983)

Model-Checking by Infinite Fly-Automata

Bruno Courcelle and Irène Durand

Université Bordeaux-1, LaBRI, CNRS
351, Cours de la Libération
33405, Talence, France
{courcell,idurand}@labri.fr

Abstract. We present logic based methods for constructing XP and FPT graph algorithms, parameterized by tree-width or clique-width. We will use *fly-automata* introduced in a previous article. They make it possible to check properties that are *not* monadic second-order expressible because their states may include counters, so that their set of states may be infinite. We equip these automata with *output functions*, so that they can compute values associated with terms or graphs. We present tools for constructing easily algorithms by combining predefined automata for basic functions and properties.

1 Introduction

Finite automata on terms that denote graphs of bounded tree-width or clique-width can be used to check monadic second-order properties of the denoted graphs. However, these automata have in most cases so many states that their transition tables cannot be built [13,15]. In the article [4] we have introduced automata called *fly-automata* whose states are described (but not listed) and whose transitions are computed on the fly (and not tabulated). Fly-automata can have infinite sets of states. For example, a state can record, among other things, the (unbounded) number of occurrences of a particular symbol. We exploit this feature in the construction of fly-automata that check properties that are *not monadic second-order (MS) expressible*. Furthermore, we equip automata with *output functions*, which map accepting states to some effectively given domain $\mathcal{D}$ (e.g., the set of integers, or of pairs of integers, or the set of words over a fixed alphabet). Hence, a fly-automaton $\mathcal{A}$ defines a mapping from $T(F)$ (the set of terms over the signature F) to $\mathcal{D}$, and we construct automata that yield polynomial-time algorithms for these mappings. The height $ht(t)$ of a term t and the number $|t|$ of its positions are obviously computable in this way. The *uniformity* of a term, *i.e.*, the property that all maximal branches of its syntactic tree have the same length, can be checked by a polynomial-time fly-automaton (but not by a finite automaton). (*Symbolic automata* [16] have "small" sets of states and "large" sets of symbols. Symbols are described by properties rather than listed. Fly-automata have, to the opposite, "small" sets of symbols and "large" sets of states).

Our main interest is actually in the case where F is the signature F_∞ of "clique-width graph operations", and for fly-automata that define mappings from

T. Muntean, D. Poulakis, and R. Rolland (Eds.): CAI 2013, LNCS 8080, pp. 211–222, 2013.

the graphs defined by terms in $T(F_\infty)$ to $\mathcal{D}$. We construct fly-automata that yield FPT and XP algorithms [10,12] for clique-width as parameter. Since the clique-width $cwd(G)$ of a simple graph G is bounded in terms of its tree-width $twd(G)$ (we have $cwd(G) \leq 2^{2twd(G)+2}+1$, [7], Proposition 2.114, and [3]), all our results for graphs of bounded clique-width apply immediately to graphs of bounded tree-width. The graphs of clique-width at most k are those denoted by the terms in $T(F_k)$ where F_k is a finite subset of F_∞. As in [4], we construct elementary fly-automata for basic functions and properties, e.g., the degree of a vertex or the regularity of graph. Then, we consider more complex functions and properties written with these functions and properties (and the basic MS properties of [4]) and functional and logical constructors. For example, $\exists X, Y.(Partition(X,Y) \wedge Reg[X] \wedge Reg[Y])$ expresses that the graph is the union of two disjoint regular graphs with possibly some edges between them. Here are some typical examples of questions and functions that we can handle in this way:

(1) Is it possible to cover a graph with s cliques?

(2) Does there exist an equitable s-coloring? *Equitable* means that the sizes of any two color classes differ by at most 1 (see [11]). We express this property by: $\exists X_1,\ldots,X_s.(Partition(X_1,\ldots,X_s) \wedge St[X_1] \wedge \ldots \wedge St[X_s]$
$\wedge |X_1| = \ldots = |X_{i-1}| \geq |X_i| = \ldots = |X_s| \geq |X_1| - 1)$
where $St[X]$ means that $G[X]$ is *stable*, *i.e.*, has no edge.

(3) Assuming that the graph is s-colorable, what is the minimum size of X_1 in an s-coloring $(X_1,\ldots,X_s)$?

(4) Which sets X such that $G[X]$ and $G[V_G - X]$ are connected, minimize the number of edges between X and $V_G - X$?

More generally, let $P(X_1, ..., X_s)$ be a property of vertex sets $X_1, ..., X_s$. Everywhere in the sequel, we denote $(X_1, ..., X_s)$ by $\overline{X}$ and $t \models P(\overline{X})$ means that $\overline{X}$ satisfies P in the graph $G(t)$ defined by t; this writing does not assume that P is written in any particular logical language. We are interested, not only to check the validity of $\exists\overline{X}.P(\overline{X})$ in some term t, but also to compute from t the following objects:

$\#\overline{X}.P(\overline{X})$, defined as the number of assignments $\overline{X}$ such that $t \models P(\overline{X})$,
$\mathrm{Sp}\overline{X}.P(\overline{X})$, the *spectrum* of $P(\overline{X})$, defined as the set of tuples of the form $(|X_1|,\ldots,|X_s|)$ such that $t \models P(\overline{X})$,
$\mathrm{MSp}\overline{X}.P(\overline{X})$, the *multispectrum* of $P(\overline{X})$, defined as the multiset of tuples $(|X_1|,\ldots,|X_s|)$ such that $t \models P(\overline{X})$,
$\mathrm{Sat}\overline{X}.P(\overline{X})$ as the set of assignments $\overline{X}$ such that $t \models P(\overline{X})$.

Each prefix $\#\overline{X}, \mathrm{Sp}\overline{X}.$ etc. can be considered as a generalized quantifier that binds the variables of $\overline{X}$. The associated values (numbers or sets of tuples of numbers) can be computed from $\mathrm{Sat}\overline{X}.P(\overline{X})$, a set of s-tuples of subsets of $Pos(t)$ (the set of *positions* of t, i.e., of nodes of the syntactic tree of t) that may be of exponential cardinality $2^{s.|t|}$, hence, not computable by a polynomial-time algorithm.

We provide logic based methods for proving the existence of FPT and XP algorithms for terms and graphs. We generalize constructions of [1,2,8]. These constructions have been implemented and tested [4,5,6].

Lacking of space (see [5] for details and proofs), we do not review MS logiç and clique-width. We only recall that edges are introduced by means of operations on vertex labeled graphs. A vertex labeled by a is an *a-port*. Notation is as in [4,7]. If $t \in T(F)$, i.e., is a term over a signature F, we let $Sig(t)$ be the set of symbols of F that occur in t.

Tuples of Sets of Positions in Terms
Let F be a signature and s be a positive integer. We let $F^{(s)}$ be the set $F \times \{0,1\}^s$ made into the signature such that the arity $\rho((f,w))$ of (f,w) is $\rho(f)$. We let $pr_s : F^{(s)} \to F$ be the relabeling that deletes the second component of a symbol (f,w). To every term $t \in T(F^{(s)})$ corresponds the term $pr_s(t)$ in $T(F)$ and the s-tuple $\nu(t) = (X_1, ..., X_s)$ of subsets of $Pos(t) = Pos(pr_s(t))$ such that $u \in X_i$ if and only if $w[i] = 1$ where the symbol at position u in t is (f,w). Conversely, if $t \in T(F)$ and $(X_1, ..., X_s)$ is an s-tuple of sets of positions of t, then there is a unique term $t' \in T(F^{(s)})$ such that $pr_s(t') = t$ and $\nu(t') = (X_1, ..., X_s)$.We will denote this term by $t * (X_1, ..., X_s)$ or by $t * \overline{X}$.

A property $P(\overline{X})$ of sets of positions of terms over a signature F is characterized by the language $T_{P(\overline{X})}$ over $F^{(s)}$ defined as $\{t * \overline{X} \mid t \models P(\overline{X})\}$. A key fact about pr_s is that $T_{\exists \overline{X}.P(\overline{X})} = pr_s(T_{P(\overline{X})})$. A function α whose arguments are t and $\overline{X}$ such that $t \in T(F)$ and $\overline{X}$ is an s-tuple of positions of t, and whose values are in a set $\mathcal{D}$ corresponds to a function $\overline{\alpha} : T(F^{(s)}) \to \mathcal{D}$ such that $\overline{\alpha}(t * \overline{X}) = \alpha(t, \overline{X})$.

Tuples of sets of vertices
The operations defining clique-width form a countably infinite signature F_∞. Those using only labels in $[k]$ form F_k. The nullary symbols $\mathbf{a}$ (for vertex labels a) denote the vertices of the graph $G(t)$ defined by t. The same technique as above applies to tuples of sets of vertices of graphs defined by terms in $T(F_\infty)$. In particular, we define $F_\infty^{(s)}$ from F_∞ by replacing all nullary symbols $\mathbf{a}$ by the nullary symbols $(\mathbf{a}, w)$ for all $w \in \{0,1\}^s$. (The other symbols are not changed).

2 Polynomial-Time Fly-Automata

All automata run bottom-up (or frontier-to-root) on terms without ε-transitions.

Definitions 1: *Fly-automata recognizing languages.*
(a) Let F be a finite or infinite (effectively given) signature. A *fly-automaton* over F (in short, an FA over F) is a 4-tuple $\mathcal{A} = \langle F, Q_\mathcal{A}, \delta_\mathcal{A}, Acc_\mathcal{A} \rangle$ such that $Q_\mathcal{A}$ is the finite or infinite, effectively given set of *states*, $Acc_\mathcal{A}$ is a computable mapping $Q_\mathcal{A} \to \{True, False\}$ so that $Acc_\mathcal{A}^{-1}(True)$ is the set of *accepting states*, and $\delta_\mathcal{A}$ is a computable function that defines the *transition rules*: for each tuple $(f, q_1, \ldots, q_m)$ with $q_1, \ldots, q_m \in Q_\mathcal{A}$, $f \in F$, $\rho(f) = m \geq 0$, $\delta_\mathcal{A}(f, q_1, \ldots, q_m)$

is a finite set of states. We will write $f[q_1,\ldots,q_m] \to_{\mathcal{A}} q$ (and $f \to_{\mathcal{A}} q$ if f is nullary) to mean that $q \in \delta_{\mathcal{A}}(f,q_1,\ldots,q_m)$. Each set $\delta_{\mathcal{A}}(f,q_1,\ldots,q_m)$ is linearly ordered for some fixed (say lexicographic) linear order on Z^* where Z is the alphabet used to encode states. We say that $\mathcal{A}$ is *finite* if F and $Q_{\mathcal{A}}$ are finite. If furthermore, $Q_{\mathcal{A}}$, its accepting states and its transitions are listed in tables, it is called a *table-automaton.*

(b) *Runs* and *recognized languages* are defined as usual. A *deterministic* FA $\mathcal{A}$ ("deterministic" will mean "deterministic and complete") has a unique run on each term t, denoted by $run_{\mathcal{A},t}$; we let also $q_{\mathcal{A}}(t) := run_{\mathcal{A},t}(root_t)$. The mapping $q_{\mathcal{A}}$ is computable and the membership in $L(\mathcal{A})$ of a term t is decidable.

Every fly-automaton $\mathcal{A}$ over F can be *determinized* as follows. For every term $t \in T(F)$, we denote by $run^*_{\mathcal{A},t}$ the mapping: $Pos(t) \to \mathcal{P}_f(Q_{\mathcal{A}})$ that associates with every position u, the finite set of states of the form $r(root_{t/u})$ for some run r on the subterm t/u of t issued from u. The run of $\det(\mathcal{A})$ on t is called *the determinized run* of $\mathcal{A}$ and we define $ndeg_{\mathcal{A}}(t)$, the *nondeterminism degree of* $\mathcal{A}$ *on* t, as the maximal cardinality of $run^*_{\mathcal{A},t}(u)$ for u in $Pos(t)$. The mapping $run^*_{\mathcal{A},t}$ is computable and the membership in $L(\mathcal{A})$ of a term in $T(F)$ is decidable: clearly, $t \in L(\mathcal{A})$ if and only if the set $run^*_{\mathcal{A},t}(root_t)$ contains an accepting state.

Definitions 2: *Fly-automata computing functions.*
A fly-automaton over F *with output function* is a 4-tuple $\mathcal{A} = \langle F, Q_{\mathcal{A}}, \delta_{\mathcal{A}}, Out_{\mathcal{A}}\rangle$ as above except that $Acc_{\mathcal{A}}$ is replaced by a total and computable *output function* $Out_{\mathcal{A}}$: $Q_{\mathcal{A}} \to \mathcal{D}$ where $\mathcal{D}$ is an effectively given domain. If $\mathcal{A}$ is deterministic, the *function computed by* $\mathcal{A}$ is $Comp(\mathcal{A}) : T(F) \to \mathcal{D}$ such that $Comp(\mathcal{A})(t) := Out_{\mathcal{A}}(q_{\mathcal{A}}(t))$. If $\mathcal{A}$ is not deterministic, we let $\mathcal{B}$ be $\det(\mathcal{A})$ equipped with output function $Out_{\mathcal{B}}:\mathcal{P}_f(Q_{\mathcal{A}}) \to \mathcal{P}_f(\mathcal{D})$ such that $Out_{\mathcal{B}}(R) := \{Out_{\mathcal{A}}(q) \mid q \in R\}$. Then, we define $Comp(\det(\mathcal{A}))$ as $Comp(\mathcal{B})$. (In some cases, we may take a computable function $Out_{\mathcal{B}} : \mathcal{P}_f(Q_{\mathcal{A}}) \to \mathcal{D}'$ where $\mathcal{D}'$ is another effectively given domain).

Definitions 3: *Polynomial-time fly-automata and related notions*
(a) A fly-automaton $\mathcal{A}$ over a signature F, possibly with output, is a *polynomial-time fly-automaton* (a *P-FA*) if it is deterministic and there is a polynomial p such that its computation time on any term $t \in T(F)$ is at most $p(\|t\|)$, where $\|t\|$ is the size of t, written as a word; the operation symbols are encoded by words of non constant length. This time includes the time taken by the output function. We call p a *bounding polynomial* for $\mathcal{A}$.

(b) A fly-automaton $\mathcal{A}$ as above is an *XP fly-automaton* (an *XP-FA* in short) if, for each finite subsignature F' of F, $\mathcal{A} \upharpoonright F'$ (the subautomaton of $\mathcal{A}$ induced by F') is a P-FA. It is an *FPT fly-automaton* (an *FPT-FA* in short) if, for each finite subsignature F' of F, $\mathcal{A} \upharpoonright F'$ is a P-FA with bounding polynomial whose degree does not depend on F'. We have the inclusions of classes of automata:

P-FA $\subseteq$ FPT-FA $\subseteq$ XP-FA with equalities for finite signatures.

Lemma 4: Let $\mathcal{A}$ be a nondeterministic fly-automaton over a signature F.

(1) The fly-automaton $\det(\mathcal{A})$ is a P-FA if and only if there are polynomials $p_1, ..., p_4$ such that, in the determinized computation of $\mathcal{A}$ on any term $t \in T(F)$, $p_1(\|t\|)$ bounds the time for firing the next transition (and recognizing that there is no next transition), $p_2(\|t\|)$ bounds the size of a state, $p_3(\|t\|)$ bounds the time for checking if a state is accepting or for computing the output and $p_4(\|t\|)$ bounds the nondeterminism degree of $\mathcal{A}$ on t.

(2) The fly-automaton $\det(\mathcal{A})$ is an XP-FA if and only if, for each finite subsignature F' of F, there are polynomials $p_1, ..., p_4$ that bound as above the computations on terms in $T(F')$. It is an FPT-FA if and only if, for each finite subsignature F' of F, there are polynomials $p_1, ..., p_4$ that bound as above the computations on terms in $T(F')$ and whose degrees are independent of F'.

Definition 5: *Functions computable by fly-automata.*

A function $\alpha : T(F) \to \mathcal{D}$ *is P-FA computable* (or is a *P-FA function* for short) if it is computable by a P-FA over F that we have constructed or that we know how to construct by an algorithm. For a property P, we say that it is *P-FA decidable.* In this definition, F can be $H^{(s)}$ for some signature H, hence, a P-FA computable function or property can take as arguments, not only a term, but also a tuple of sets of positions or of vertices.

It is well-known that every MS property P of a term over a finite signature is P-FA decidable. The cardinality of a set and the height of a term are P-FA functions. We will construct an FPT-FA to check if a graph is regular (this not an MS property).

The mapping $\mathrm{Sat}X.P(X)$ is not P-FA computable, and not even XP-FA computable in general for the obvious reason that its output is not always of polynomial size (take $P(X)$ always true).

Proposition 6: Let F be a signature. Every **P**-computable (resp. **FPT**- computable or **XP**-computable) function α on $T(F)$ is computable by a P-FA (resp. by an FPT-FA or an XP-FA).

Hence, our three notions of FA may look trivial. Actually, we will be interested by giving effective constructions of P-FA, FPT-FA and XP-FA from logical expressions of functions and properties. These constructions will apply to properties that are *not MS expressible* but are decidable in polynomial time on graphs of bounded tree-width or clique-width.

3 Fly-Automata for Logically Defined Functions and Properties

Proposition 7: (1) If $\alpha_1, ..., \alpha_r$ are P-FA functions of same type and g is a **P**-computable function (or relation) of appropriate type, the function (or the property) $g \circ (\alpha_1, ..., \alpha_r)$ is P-FA computable (or decidable).

(2) If P and Q are P-FA properties of same type, then, so are $\neg P$, $P \vee Q$ and $P \wedge Q$.

(3) The same properties hold with FPT-FA and XP-FA.

The proof is based on easy constructions like taking a product of automata.

First-order constructions
We now consider the more delicate case of existential quantifications. We define one more construction: if $\alpha(\overline{X})$ is a function (relative to a term t), we define $\mathrm{SetVal}\overline{X}.\alpha(\overline{X})$ as the set of values $\alpha(\overline{X}) \neq \bot$ ($\bot$ stands for undefined) for all relevant tuples $\overline{X}$. We let $\exists x_1, ..., x_s.P(x_1, ..., x_s)$ (also written $\exists\overline{x}.P(\overline{x})$) abbreviate $\exists X_1, ..., X_s.(P(X_1, ..., X_s) \wedge Sgl(X_1) \wedge ... \wedge Sgl(X_s))$ (where $Sgl(X)$ means that X is singleton) and similarly, $\mathrm{SetVal}(x_1, ..., x_s).\alpha(x_1, ..., x_s)$ (also written $\mathrm{SetVal}\overline{x}.\alpha(\overline{x})$) is the set of well-defined values of $\alpha(X_1, ..., X_s)$ such that: $Sgl(X_1) \wedge ... \wedge Sgl(X_s)$.

Proposition 8: (1) If $P(\overline{X})$ is a P-FA property, then the property $\exists\overline{x}.P(\overline{x})$ is P-FA decidable and the functions $\mathrm{Sat}\overline{x}.P(\overline{x})$ and $\#\overline{x}.P(\overline{x})$ are P-FA computable.
(2) If $\alpha(\overline{X})$ is a P-FA function, then the function $\mathrm{SetVal}\overline{x}.\alpha(\overline{x})$ is P-FA computable.
(3) The same implications hold for the classes FPT-FA and XP-FA.

Proof Sketch: If $\mathcal{A}$, deterministic, checks $P(\overline{x})$, then the nondeterministic automaton $pr_s(\mathcal{A})$ defines $\exists\overline{x}.P(\overline{x})$ with nondeterminism degree bounded by the polynomial $p(n) = 1 + (n+1)^s$ that does not depend on $Sig(t)$. Hence $\det(pr_s(\mathcal{A}))$ is a P-FA, an FPT-FA or an XP-FA by Lemma 4 if $\mathcal{A}$ is so. □

These results remain valid if each condition $Sgl(X_i)$ is replaced by the condition $Card(X_i) = c_i$ or $Card(X_i) \leq c_i$ for fixed integers c_i. For example, we can compute

$$\#(X_1, ..., X_s).P(X_1, ..., X_s) \wedge Card(X_1) \leq c_1 \wedge ... \wedge Card(X_s) \leq c_s.$$

The exponents in the bounding polynomial become larger, but they still depend only on the numbers $c_1, ..., c_s$. This does not work for $Card(X_i) \geq c_i$ because the bound would not be polynomial.

Monadic second-order constructions
We recall that for finite signatures, the notions of P-FA, FPT-FA and XP-FA coincide. We let $P(\overline{X})$ be a property of terms in $T(F)$ with s set arguments and $\alpha(\overline{X})$ be similarly a function. The relabeling pr: $F^{(s)} \rightarrow F$ has a computable inverse. We consider infinite signatures F. Our main application will be to the infinite signature F_∞ that generates all finite graphs. We will use $Sig(t)$, the set of symbols that occur in a term t, as a parameter for FPT and XP algorithms. If $t \in T(F_\infty)$, this is equivalent to taking as parameter the minimal k such that $t \in T(F_k)$, hence, the clique-width of the considered graph because clique-width can be approximated in cubic time.

Definitions 9: *Multisets of tuples of numbers; a semi-ring.*

(a) If μ and μ' are two mappings $\mathbb{N}^s \to \mathbb{N}$, we define $\mu+\mu'$ and $\mu*\mu' : \mathbb{N}^s \to \mathbb{N}$ by:

$$(\mu + \mu')(n_1, ..., n_s) := \mu(n_1, ..., n_s) + \mu'(n_1, ..., n_s), \text{ and}$$
$$(\mu * \mu')(n_1, ..., n_s) := \sum_{0\leq p_i \leq n_i} \mu(p_1, ..., p_s).\mu'(n_1 - p_1, ..., n_s - p_s).$$

$[\mathbb{N}^s \to \mathbb{N}]_f$ is the set of *finite* mappings: $\mathbb{N}^s \to \mathbb{N}$, i.e., with value 0 almost everywhere. The functions $\mu + \mu'$ and $\mu * \mu'$ are finite if μ and μ' are. The operations + and $*$ are associative and commutative. The constant mapping: $\mathbb{N}^s \to \mathbb{N}$ with value 0 is denoted by $\mathbf{0}$. If $w \in \{0,1\}^s$, we let M_w: $\mathbb{N}^s \to \mathbb{N}$ be such that $M_w(\overline{n})$:=$\texttt{if}$ $\overline{n} = w$ $\texttt{then}$ 1 $\texttt{else}$ 0. We have $\mu + \mathbf{0} = \mu$, $\mu * \mathbf{0} = \mathbf{0}$ and $\mu * M_{0...0} = \mu$. Since $*$ is distributive over +, we get that $\langle[\mathbb{N}^s \to \mathbb{N}]_f, +, *, \mathbf{0}, M_{0...0}\rangle$ is a semi-ring; $\mu \in [\mathbb{N}^s \to \mathbb{N}]_f$ is (represents) a finite multiset of s-tuples of integers.

(b) If E is a set and $Z \subseteq \mathcal{P}_f(E)^s$, we define MSp$(Z)$ as the mapping: $\mathbb{N}^s \to \mathbb{N}$ such that MSp$(Z)(n_1, ..., n_s)$ is the number of tuples $(X_1, \dots, X_s) \in Z$ such that $n_i = |X_i|$ for each i, hence is the multiset $\{(|X_1|, ..., |X_s|) \mid (X_1, \dots, X_s) \in Z\}$. If Z and $Z' \subseteq \mathcal{P}_f(E)^s$ are disjoint, then MSp$(Z \cup Z')$ = MSp(Z) + MSp(Z'). If $Z \subseteq \mathcal{P}_f(E)^s$ and $Z' \subseteq \mathcal{P}_f(E')^s$ with $E \cap E' = \emptyset$, and if $W = \{(X_1 \cup Y_1, \dots, X_s \cup Y_s) \mid (X_1, \dots, X_s) \in Z, (Y_1, \dots, Y_s) \in Z'\}$, then MSp$(W)$ = MSp$(Z) *$ MSp(Z').

Definition 10: A fly-automaton $\mathcal{A}$ over F has an *FPT-bounded nondeterminism degree* (cf. p_4 in Lemma 4) if, for every $t \in T(F)$, $ndeg_{\mathcal{A}}(t) \leq f(Sig(t)) \cdot \|t\|^a$ for some fixed function f and constant a. It has an *XP-bounded nondeterminism degree* if $ndeg_{\mathcal{A}}(t) \leq f(Sig(t)) \cdot \|t\|^{g(Sig(t))}$ for some fixed functions f and g, equivalently, if $\mathcal{A} \restriction H$ has a polynomially bounded nondeterminism degree for each finite subsignature H of F.

Proposition 11: (1) If $P(\overline{X})$ is decided by a P-FA (resp. FPT-FA, resp. XP-FA) $\mathcal{A}$ over $F^{(s)}$ such that the FA $pr(\mathcal{A})$ has a polynomially bounded (resp. FPT-bounded, resp. XP-bounded) nondeterminism degree, then the property $\exists\overline{X}.P(\overline{X})$ is P-FA (resp. FPT-FA, resp. XP-FA) decidable, and the function MSp$\overline{X}.P(\overline{X})$ is P-FA (resp. FPT-FA, resp. XP-FA) computable. These results also hold for Sp$\overline{X}.P(\overline{X})$, $\#\overline{X}.P(\overline{X})$, MinCard$X.P(X)$ and MaxCard$X.P(X)$.

(2) If $\alpha(\overline{X})$ is computed by a P-FA (resp. FPT-FA, resp. XP-FA) $\mathcal{A}$ such that $pr(\mathcal{A})$ has a polynomially bounded (resp. FPT-bounded, resp. XP-bounded) nondeterminism degree, then the function SetVal$\overline{X}.\alpha(\overline{X})$ is P-FA (resp. FPT-FA, resp. XP-FA) computable.

Proof Sketch: We start from a deterministic automaton $\mathcal{A}$ over $F^{(s)}$ that defines $P(\overline{X})$. Let $t \in T(F)$. For each state q and position u of t, we let $Z(q, u)$ be the set of s-tuples $\overline{X} \in (\mathcal{P}_f(Pos(t)/u))^s$ (where $Pos(t)/u$ is the set of positions of t below or equal to u; to be distinguished from $Pos(t/u)$) such that $run_{\mathcal{A},t*\overline{X}}(u) = q$. At the root, these sets define Sat$\overline{X}.P(\overline{X})$. We extract information from $Z(q, u)$ and make it into an *attribute* of q. Depending on the case, this attribute may be a Boolean for emptiness of $Z(q, u)$ (for $\exists\overline{X}$), its cardinality (for $\#\overline{X}$), the multiset MSp$(Z(q, u))$ (for MSp$\overline{X}$). We focus on the last case.

The operation + sums the multisets coming from the different runs of $pr(\mathcal{A})$ that reach q at u. The operation $*$ combines the attributes at the sons of u in each run that reaches q at u. We obtain a nondeterministic automaton $\mathcal{B}$ whose states are pairs (q, α) where α is an attribute. Then $\det(\mathcal{B})$ is a deterministic FA that computes $\mathrm{MSp}\overline{X}.P(\overline{X})$ which is equal to the sum of the multisets $\mathrm{MSp}(Z(q, root_t))$ for q accepting.

We consider the case where $P(\overline{X})$ is MS expressible and F is finite. Then $\mathcal{A}$ is finite. A state of $\det(\mathcal{B})$ can be implemented as the finite set of tuples $(q, n_1, ..., n_s, m)$ such that $q \in Q_\mathcal{A}$, $m = \alpha(n_1, ..., n_s) \neq 0$ where α is the attribute of q (at some position). Then $\det(\mathcal{B})$ is a P-FA because its states (on a term with n nodes) can be encoded by words of length $\leq |Q_\mathcal{A}|.(n+1)^s. \log(2^{s.n}) = O(n^{s+1})$ (the numbers $\alpha(n_1, ..., n_s)$ being written in binary). Computing the transitions and the output takes polynomial time. The proof extends to infinite F as stated. The cases of $\mathrm{Sp}\overline{X}.P(\overline{X})$ etc. are even simpler because we have less information from $Z(q, u)$ to encode. The computation of $\mathrm{SetVal}\overline{X}.\alpha(\overline{X})$, the set of all values of α, is based on $\det(\mathcal{A})$.

If $\mathrm{Sat}\overline{X}.P(\overline{X}))$ is not empty, a P-FA (resp. an FPT-FA, resp. an XP-FA) can compute one of its tuples (but not all of them in general). □

4 Properties of Terms and Functions on Terms

The height of a term t can be computed by a P-FA $\mathcal{A}_{ht}$ whose states are positive integers (the state at u is $ht(t/u)$ where $ht(a) = 1$ for a nullary symbol a). A term t is *uniform* (this property is denoted by $Unif(t)$) if and only if any two leaves of its syntactic tree are at same distance to the root. This is equivalent to the condition that for every position u with sons u' and u", the subterms t/u' and t/u" have same height. The automaton $\mathcal{A}_{ht}$ can be modified into a P-FA $\mathcal{A}_{Unif}$ that decides uniformity. Its set of states is $\mathbb{N}_+ \cup \{Error\}$ and $q_{\mathcal{A}_{Unif}}(t) = ht(t)$ if t is uniform, $= Error$ if t is not uniform.

Definition 12: *An extension of MS logic on terms.*

We consider properties and functions constructed in the following way:

(a) We use free set variables $X_1, ..., X_s$ (that will not be quantified), first-order (FO) variables, $y_1, ..., y_m$ and set terms over $X_1, ..., X_s, \{y_1\}, ..., \{y_m\}$.

(b) As basic properties, we use *Unif* and all properties P expressible by MS formulas (that can use other bound variables than $X_1, ..., X_s, y_1, ..., y_m$). As basic functions, we use ht, $Card$ (that yields the cardinality of a set of positions).

(c) We construct properties from already constructed properties P, Q, ... and from functions $\alpha, \alpha_1, ..., \alpha_r, ...$ by the following compositions:

$P \wedge Q, P \vee Q, \neg P$,
$R \circ (\alpha_1, ..., \alpha_r)$ where R is an r-ary **P**-decidable relation on $\mathcal{D}$,
$P(S_1, ..., S_p)$ where $S_1, ..., S_p$ are set terms over $X_1, ..., X_s, \{y_1\}, ..., \{y_m\}$ (*set terms* are built with union, intersection and complementation; see [4]).
$\exists\overline{y}.P(\overline{y})$ where $\overline{y}$ is a tuple of variables among $y_1, ..., y_m$.

(d) Similarly, we construct functions in the following ways:

$g \circ (\alpha_1, ..., \alpha_r)$ where g is **P**-computable: $\mathcal{D} \to \mathcal{D}^r$,
$\alpha(S_1, ..., S_p)$ where $S_1, ..., S_p$ are set terms over $X_1, ..., X_s, \{y_1\}, ..., \{y_m\}$,
SetVal$\overline{y}.\alpha(\overline{y})$ (the set of values of α), $\#\overline{y}.P(\overline{y})$ and Sat$\overline{y}.P(\overline{y})$ where $\overline{y}$ is a tuple of variables among $y_1, ..., y_m$.

We assume that we have for R in (c) and g in (d) a certified polynomial-time algorithm. This is necessary to build automata. We denote by $\mathcal{PF}(F)$ the set of all these formulas.

Theorem 13: Every property (or function) defined by a formula of $\mathcal{PF}(F)$ is decidable (or computable) by a P-FA over F. Such an automaton can be constructed from automata for the basic properties and functions.

Our language $\mathcal{PF}(F)$ does not exhaust the possibilities of extension of MS logic that yield P-FA computable properties and functions. We can for example introduce a *relativized height* $ht(t, X)$ for $t \in T(F)$ and $X \subseteq Pos(t)$, defined as the maximal number of elements of X on a branch of the syntactic tree of t. However, we cannot use set quantifications.

5 Properties and Functions on Graphs

1. *Degrees of vertices*
For a directed graph G, we generalize the notion of *outdegree* by defining $e(X_1, X_2)$ as the number of edges from X_1 to X_2 if X_1 and X_2 are disjoint sets of vertices and as $\bot$ otherwise. Hence $e(\{x\}, V_G - \{x\})$ is the outdegree of x in G (all graphs are loop-free). Note that $e(X_1, X_2)$ *is not* of the form $\#Y.P(Y, X_1, X_2)$ for an MS property P as we do not allow edge set quantification.

We can define a deterministic FA $\mathcal{A}_k$ over $F_k^{(2)}$, intended to run on *irredundant terms* (such that no edge is defined twice, see [4]) written with labels in $C := [k]$. Its set of states is $(\mathbb{N} \times [C \to \mathbb{N}] \times [C \to \mathbb{N}]) \cup \{Error\}$. If $X \subseteq V_G$, we denote by λ_X the mapping that gives, for each a, the number of a-ports in X; if $X = V_G$, we denote it by λ_G. We want that $q_{\mathcal{A}_k}(t * (X_1, X_2)) = Error$ if $X_1 \cap X_2 \neq \emptyset$ and $q_{\mathcal{A}_k}(t * (X_1, X_2)) = (e(X_1, X_2), \lambda_{X_1}, \lambda_{X_2})$ otherwise. The transitions are easy to write. For example $\overrightarrow{add}_{a,b}[(m, \lambda_1, \lambda_2)] \to (m + \lambda_1(a).\lambda_2(b), \lambda_1, \lambda_2)$, is correct because t is assumed irredundant.

On a term that denotes a graph with n vertices, each state belongs to the set $([0, n^2] \times [C \to [0, n]] \times [C \to [0, n]]) \cup \{Error\}$ of cardinality less than $(n+1)^{2+2k}$, hence, has size $O(k.\log(n))$ (the integers m and the values of λ_1 and λ_2 are written in binary notation). Transitions and outputs can be computed in time $O(k.\log(n))$. Hence, $\mathcal{A}_k$ is a P-FA. We represent a function $\lambda : C \to \mathbb{N}$ by the set $\{(a, \lambda(a)) \mid \lambda(a) \neq 0\}$. This implies that $\mathcal{A}_k$ is a subautomaton of $\mathcal{A}_{k'}$ if $k < k'$. Hence, the union of the automata $\mathcal{A}_k$ is a P-FA $\mathcal{A}_\infty$ over $F_\infty^{(2)}$. For an undirected graph, we define $e(X_1, X_2)$ as the number of edges between X_1 and X_2 if X_1 and X_2 are disjoint and $\bot$ otherwise. The construction is similar.

2. *Regularity of a graph*
The regularity of an undirected graph is not MS expressible because the complete bipartite graph $K_{n,m}$ is regular if and only if $n = m$ and we apply the arguments of Proposition 5.13 of [7]. That a graph is not regular can be checked by a FA constructed from the formula $\exists X, Y.(P(X,Y) \wedge Sgl(X) \wedge Sgl(Y))$ where $P(X,Y)$ is the property $e(X, X^c) \neq e(Y, Y^c)$. By previous constructions, this property is P-FA decidable, and we can apply Proposition 14(1) to get a P-FA for checking regularity. However, we can construct directly a simpler P-FA without using an intermediate nondeterministic automaton. Its state at position u is *Error* if two a-ports of $G(t/u)$ have different degrees, and otherwise indicates, for each a, the number of a-ports and their common degree. In its run on a term t such that $G(t)$ has n vertices, less than $(n+1)^{2k}$ states occur and these states have size $O(k.\log(n))$. We get a P-FA $\mathcal{A}_{Reg[X]}$. The nondeterminism degree of $pr(\mathcal{A}_{Reg[X]})$ is bounded by $O((n+1)^{2k})$ where the exponent depends on the bound k to the clique-width.

The property $\exists X.(Card_{\leq p}(X) \wedge Reg[X^c])$ expressing that the considered graph becomes regular if we remove at most p vertices, is P-FA decidable by Proposition 11(1) and the remark following it. The function MaxCard$X.Reg[X]$ that defines the maximal cardinality of a regular induced subgraph of the considered graph is XP-FA computable. So is the property that the graph can be partitioned into two regular subgraphs, expressed by $\exists X.(Reg[X] \wedge Reg[X^c])$ (the proof uses the same propositions).

3. *Graph partition problems with numerical constraints*
Many partition problems (cf. also [14]) consist in finding $(X_1, \cdots, X_s)$, an s-tuple satisfying:

$$Partition(X_1, ..., X_s) \wedge P_1(X_1) \wedge ... \wedge P_s(X_s) \wedge R(|X_1|, ..., |X_s|),$$

where, $P_1, ..., P_s$ are properties of sets and R is a **P**-computable arithmetic condition. We may also wish to count the number of such partitions, or to find one that maximizes or minimizes the number $Ext(\overline{X}) := \Sigma_{1 \leq i < j \leq s} e(X_i, X_j)$ of *external edges*, *i.e.*, of edges not in the induced subgraphs $G[X_1], ..., G[X_s]$. This number is P-FA computable.

We can handle partitions in s planar induced subgraphs with an FPT-FA, however, its implementation does not seem doable (planarity is MS expressible, but the formula is complicated).

If $P_i(X_i)$ is stability for each i, (*i.e.*, the induced subgraphs have no edge), we get a *constrained coloring problem* of the form:

$$\exists X_1, ..., X_s.(Partition(X_1, ..., X_s) \wedge St[X_1] \wedge \; ... \; \wedge St[X_s] \wedge R(|X_1|, ..., |X_s|)).$$

An example is the notion of *equitable s-coloring*: condition $R(|X_1|, ..., |X_s|)$ is $\exists i \in [s].(|X_1| = ... = |X_{i-1}| \geq |X_i| = ... = |X_s| \geq |X_1| - 1)$, which means that any two color classes have same cardinality up to 1. The existence of an equitable 3-coloring is not trivial: it holds for the cycles but not for the graphs $K_{n,n}$ for large n. The existence of an equitable s-coloring is W[1]-hard for the parameter defined as s plus the tree-width [11], hence presumably not FPT for this parameter. Our constructions yield, for each integer s, an FPT-FA for checking the existence of an equitable s-coloring for clique-width as parameter.

6 Implementations

Let $\mathcal{A}_{P(\overline{X})}$ be an automaton recognizing graphs with assignments of sets to the variables of $\overline{X}$. From $\mathcal{A}_{P(\overline{X})}$, we can obtain the automaton $\mathcal{A}_{\Gamma.P(\overline{X})}$ for $\Gamma \in \{\#\overline{X}, \mathrm{Sp}\overline{X}, \mathrm{MSp}\overline{X}, \ \mathrm{MinCard}X, \mathrm{MaxCard}X\}$ for graphs with no assignments. This is done in two steps. The first step is to associate to the automaton an *attribute mechanism* such that the automaton, instead of computing a state q, computes a state $[q, a]$ where a is the attribute to be computed according to Γ, for instance the number of runs yielding q for $\#\overline{X}.P(\overline{X})$. The attribute mechanism is composed of two functions: the first function applies to symbols and yields a function for computing the attribute obtained at $t = f(t_1, \ldots, t_p)$ from the ones obtained for $t_1, \ldots, t_p$ in the deterministic case; sometimes this function is the same for all symbols as for the counting case. The second function is for combining several attributes of identical states accessed with different runs. For $\#\overline{X}.P(\overline{X})$, the first function is the addition function for all symbols and the second is the multiplication function. (The case of $\exists\overline{X}.P(\overline{X})$ is handled by determinizing $pr(\mathcal{A}_{P(\overline{X})})$.)

This can be applied for counting the s-colorings of a graph or for constructing "special" colorings. From an appropriate $\mathcal{A}$, we can obtain an automaton that computes the number of s-colorings as the number of runs of $\mathcal{A}$ on the representing term. This is done by using the attribute mechanism for counting runs. For classic graphs such as Petersen's, with known chromatic polynomials, we can verify the computation. We can also count acyclic-colorings [4]. The number of 4-acyclic colorings of Petersen's graph is 10800. The number of 3-acyclic-colorings of McGee's graph is 57024. We also provide a mechanism for enumerating colorings (more generally, satisfying assignments) [9]. It is also useful for determining "quickly" the existence a coloring (but not for counting them).

7 Conclusion and References

In this communication, we have given logic based methods for proving the existence of FPT and XP algorithms that check properties or compute functions on terms and on graphs defined by terms. These constructions are currently under implementation. They are quite general and flexible and so, they do not give necessarily the best possible time complexities. They generalize constructions of [1,2,8]. Detailed definitions and proofs are in [5]. Implementation issues are described in [6,9].

References

1. Arnborg, S., Lagergren, J., Seese, D.: Easy problems for tree-decomposable graphs. J. Algorithms 12, 308–340 (1991)
2. Courcelle, B., Mosbah, M.: Monadic second-order evaluations on tree-decomposable graphs. Theor. Comput. Sci. 109, 49–82 (1993)
3. Courcelle, B.: On the model-checking of monadic second-order formulas with edge set quantifications. Discrete Applied Mathematics 160, 866–887 (2012)
4. Courcelle, B., Durand, I.: Automata for the verification of monadic second-order graph properties. J. Applied Logic 10, 368–409 (2012)
5. Courcelle, B., Durand, I.: Computations by fly-automata beyond monadic second-order logic (preprint, June 2013)
6. Courcelle, B., Durand, I.: Infinite transducers on terms denoting graphs In: Proceedings of the 6th European Lisp Symposium, Madrid (June 2013)
7. Courcelle, B., Engelfriet, J.: Graph structure and monadic second-order logic, a language theoretic approach. Encyclopedia of mathematics and its application, vol. 138. Cambridge University Press (June 2012)
8. Courcelle, B., Makowsky, J., Rotics, U.: Linear-time solvable optimization problems on graphs of bounded clique-width. Theory Comput. Syst. 33, 125–150 (2000)
9. Durand, I.: Object enumeration. In: Proceedings of the 5th European LISP Conference, Zadar, Croatia, pp. 43–57 (May 2012)
10. Downey, R., Fellows, M.: Parameterized complexity. Springer (1999)
11. Fellows, M., et al.: On the complexity of some colorful problems parameterized by treewidth. Inf. Comput. 209, 143–153 (2011)
12. Flum, J., Grohe, M.: Parametrized complexity theory. Springer (2006)
13. Frick, M., Grohe, M.: The complexity of first-order and monadic second-order logic revisited. Ann. Pure Appl. Logic 130, 3–31 (2004)
14. Rao, M.: MSOL partitioning problems on graphs of bounded treewidth and clique-width. Theor. Comput. Sci. 377, 260–267 (2007)
15. Reinhardt, K.: The complexity of translating logic to finite automata. In: Grädel, E., Thomas, W., Wilke, T. (eds.) Automata, Logics, and Infinite Games. LNCS, vol. 2500, pp. 231–238. Springer, Heidelberg (2002)
16. Veanes, M., Bjørner, N.: Symbolic automata: The toolkit. In: Flanagan, C., König, B. (eds.) TACAS 2012. LNCS, vol. 7214, pp. 472–477. Springer, Heidelberg (2012)

A Selection-Quotient Process for Packed Word Hopf Algebra

Gérard H.E. Duchamp[1], Nguyen Hoang-Nghia[1], and Adrian Tanasa[1,2]

[1] LIPN, UMR 7030 CNRS, Institut Galilée - Université Paris 13, Sorbonne Paris Cité, 99 avenue J.-B. Clément, 93430 Villetaneuse, France, EU
[2] Horia Hulubei National Institute for Physics and Nuclear Engineering, P.O.B. MG-6, 077125 Magurele, Romania, EU
{ghed,hoang}@lipn.univ-paris13.fr,
adrian.tanasa@ens-lyon.org

Abstract. In this paper, we define a Hopf algebra structure on the vector space spanned by packed words using a selection-quotient coproduct. We show that this algebra is free on its irreducible packed words. its primitive elements. Finally, we give some brief explanations on the Maple codes we have used.

Keywords: Hopf algebras, free algebras.

1 Introduction

In computer science, one is led to the study of algebraic structures based on trees, graphs, tableaux, matroids, words and other discrete structures. Hopf algebras are also shown to play an important role in quantum field theory [1], non-commutative QFT [4], [5], (see also the review articles [6] [8]), or in quantum gravity spin-foam models [3], [7]. These algebras use a selection-quotient rule for the coproduct.

$$\Delta(S) = \sum_{\substack{A\subseteq S \\ + Conditions}} S[A] \otimes S/_A, \qquad (1)$$

where S is some (general) combinatorial object (tree, graph, matroid, *etc.*), $S[A]$ is a substructure of S and $S/_A$ is a quotient.

The present article introduces a new Hopf algebraic structure, which we call WMat, on the set of packed words. The product is given by the shifted concatenation and the coproduct is given by such a selection-quotient principle.

2 Algebra Structure

2.1 Definitions

Let X be an infinite totally ordered alphabet $\{x_i\}_{i\geq 0}$ and X^* be the set of words with letters in the alphabet X.

T. Muntean, D. Poulakis, and R. Rolland (Eds.): CAI 2013, LNCS 8080, pp. 223–234, 2013.

A word w of length $n = |w|$ is a mapping $i \mapsto w[i]$ from $[1..|w|]$ to X. For a letter $x_i \in X$, the partial degree $|w|_{x_i}$ is the number of times the letter x_i occurs in the word w. One has:

$$|w|_{x_i} = \sum_{j=1}^{|w|} \delta_{w[j],x_i}. \tag{2}$$

For a word $w \in X^*$, one defines the alphabet $Alph(w)$ as the set of its letters, while $IAlph(w)$ is the set of indices in $Alph(w)$.

$$Alph(w) = \{x_i | \, |w|_{x_i} \neq 0\} \; ; \; IAlph(w) = \{i \in \mathbb{N} | \, |w|_{x_i} \neq 0\}. \tag{3}$$

The upper bound $sup(w)$ is the supremum of $IAlph(w)$, i. e.

$$sup(w) = sup_{\mathbb{N}}(IAlph(w)). \tag{4}$$

Note that $sup(1_{X^*}) = 0$.

Let us define the substitution operators. Let $w = x_{i_1} \dots x_{i_m}$ and $\phi : IAlph(w) \longrightarrow \mathbb{N}$, with $\phi(0) = 0$. One then defines:

$$S_\phi(x_{i_1} \dots x_{i_m}) = x_{\phi(i_1)} \dots x_{\phi(i_m)}. \tag{5}$$

Let us define the pack operator of a word w. Let $\{j_1, \dots, j_k\} = IAlph(w) \setminus \{0\}$ with $j_1 < j_2 < \dots < j_k$ and define ϕ_w as

$$\phi_w(i) = \begin{cases} m \text{ if } i = j_m \\ 0 \text{ if } i = 0 \end{cases}. \tag{6}$$

The corresponding packed word, denoted by $pack(w)$, is $S_{\phi_w}(w)$. This means that if the word w has (one or several) "gap(s)" between the indices of its letters, then in the word $pack(w)$ these gaps have vanished (the indices of the respective letters being modified accordingly).

Example 1. Let $w = x_1x_1x_5x_0x_4$. One then has $pack(w) = x_1x_1x_3x_0x_2$.

A word $w \in X^*$ is said to be *packed* if $w = pack(w)$.

Example 2. The packed words of weight 2 are $x_0^{k_1}x_1x_0^{k_2}x_1x_0^{k_3}$, with $k_1, k_2, k_3 \geq 0$.

The operator $pack : X^* \longrightarrow X^*$ is idempotent ($pack \circ pack = pack$). It defines, by linear extension, a projector. The image, $pack(X^*)$, is the set of packed words.

Let u, v be two words; one defines the shifted concatenation $*$ by

$$u * v = uT_{sup(u)}(v), \tag{7}$$

where, for $t \in \mathbb{N}$, $T_t(w)$ denotes the image of w by S_ϕ for $\phi(n) = n + t$ if $n > 0$ and $\phi(0) = 0$ (in general, all letters can be reindexed except x_0). It is straightforward to check that, in the case the words are packed, the result of a shifted concatenation is a packed word.

Definition 1. *Let k be a field. One defines a vector space $\mathcal{H} = span_k(pack(X^*))$ and endows this space with a product (on the words) given by*

$$\mu : \mathcal{H} \otimes \mathcal{H} \longrightarrow \mathcal{H},$$
$$u \otimes v \longmapsto u * v.$$

Remark 1. The product above is similar to the shifted concatenation for permutations. Moreover, if u, v are two words in X^*, then $sup(u*v) = sup(u) + sup(v)$.

Proposition 1. *$(\mathcal{H}, \mu, 1_{X^*})$ is an associative algebra with unit (AAU).*

Proof. Let u, v, w be three words in $\mathcal{H}$. One then has:

$$(u * v) * w = u(T_{sup(u)}(v)(T_{sup(u)+sup(v)})(w)) = u * (v * w). \tag{8}$$

Thus, the algebra $(\mathcal{H}, \mu)$ is associative. On the other hand, for all $u \in pack(X^*)$, one can easily check that

$$u * 1_{X^*} = u = 1_{X^*} * u.$$

Now remark that $pack(1_{X^*}) = 1_{X^*}$. This is clear from the fact that $1_{X^*} = 1_{\mathcal{H}}$. One concludes that $(\mathcal{H}, \mu, 1_{X^*})$ is an AAU. □

As already announced in the introduction, we call this algebra WMat.

Remark 2. The product is non-commutative, for example: $x_1 * x_1x_1 \neq x_1x_1 * x_1$.

Let $w = x_{k_1} \dots x_{k_n}$ be a word and $I \subseteq [1 \dots n]$. A sub-word $w[I]$ is defined as $x_{k_{i_1}} \dots x_{k_{i_l}}$, where $i_j \in I$.

Lemma 1. *Let u, v be two words. Let $I \subset [1 \dots |u|]$ and $J \subset [|u|+1 \dots |u|+|v|]$. One then has*

$$pack(u * v[I + J]) = pack(u[I]) * pack(v[J']), \tag{9}$$

where J' is the set $\{i - |u|\}_{i \in J}$.

Proof. By direct computation, one has:

$$\begin{aligned} pack(u * v[I + J]) &= pack(uT_{sup(u)}(v)[I + J]) = pack(u[I]T_{sup(u)}(v)[J]) \\ &= pack(u[I]T_{sup(u[I])}(v[J'])) = pack(u[I]) * pack(v[J']). \end{aligned} \tag{10}$$

□

Theorem 1. *Let $k < X >$ be equipped with the shifted concatenation. The mapping pack*

$$k < X > \xrightarrow{pack} \mathcal{H} \tag{11}$$

is then a morphism AAU.

Proof. By using the Lemma 1 and taking $I = [1 \dots |u|]$ and $J = [|u| + 1 \dots |u| + |v|]$, one gets the conclusion. □

2.2 WMat Is a Free Algebra

WMat is, by construction, the algebra of the monoid $pack(X^*)$, therefore to check that WMat is a free algebra, it is sufficient to show that $pack(X^*)$ is a free monoid on its atoms. a free monoid is a pair $(F(X), j_X)$ where $F(X)$ is a monoid, $j_X : X \to F(X)$ is a mapping such that $(\forall M \in Mon)$ $(\forall f : X \to M)$ $(\exists! f_X \in Mor(F(X), M))$ $f = f_x \circ j_X$.

Here we will use an "internal" characterization of free monoids in terms of irreducible elements.

Definition 2. *A packed word w in $pack(X^*)$ is called an irreducible word if and only if it can not be written under the form $w = u * v$, where u and v are two non trivial packed words.*

Example 3. The word $x_1x_1x_1$ is an irreducible word. The word $x_1x_1x_2$ is a reducible word because it can be written as $x_1x_1x_2 = x_1x_1 * x_1$.

Proposition 2. *If w is a packed word, then w can be written uniquely as $w = v_1 * v_2 * \cdots * v_n$, where v_i, $1 \le i \le n$, are non-trivial irreducible words.*

Proof. The ith position of word w is called an admissible cut if $sup(w[1 \ldots i]) = inf(w[i+1 \ldots |w|]) - 1$ or $sup(w[i+1 \ldots |w|]) = 0$, where $inf(w)$ is infimum of $IAlph(w)$.

Because the length of word is finite, one can get $w = v_1 * v_2 * \cdots * v_n$, with n maximal and v_i non trivial, $\forall 1 \le i \le n$.

One assumes that one word can be written in two ways

$$w = v_1 * v_2 * \cdots * v_n \tag{12}$$

and

$$w = v'_1 * v'_2 * \cdots * v'_m. \tag{13}$$

Denoting by k the first number such that $v_k \neq v'_k$, without loss of generality, one can suppose that $|v_k| < |v'_k|$. From equation (12), the kth position is an admissible cut of w. From equation (13), the kth position is not an admissible cut of w. One thus has a contradiction. One has $n = m$ and $v_i = v'_i$ for all $1 \le i \le n$. □

One can thus conclude that $pack(X^*)$ is free as monoid with the packed words as a basis.

3 Bialgebra Structure

Let us give the definition of the coproduct and prove that the coassociativity property holds.

Definition 3. *Let $A \subset X$, one defines $w/A = S_{\phi_A}(w)$ with*
$$\phi_A(i) = \begin{cases} i \text{ if } x_i \notin A, \\ 0 \text{ if } x_i \in A. \end{cases}$$
Let u be a word. One defines ${}^w/_u = {}^w/_{Alph(u)}$.

Definition 4. *The coproduct of $\mathcal{H}$ is given by*
$$\Delta(w) = \sum_{I+J=[1\ldots|w|]} pack(w[I]) \otimes pack(w[J]/_{w[I]}), \forall w \in \mathcal{H}, \tag{14}$$
where this sum runs over all partitions of $[1\ldots|w|]$ divided into two blocks, $I \cup J = [1\ldots|w|]$ and $I \cap J = \emptyset$.

Example 4. One has:
$$\Delta(x_1x_2x_1) = x_1x_2x_1 \otimes 1_{X^*} + x_1 \otimes x_1x_0 + x_1 \otimes x_1^2 + x_1 \otimes x_0x_1 + x_1x_2 \otimes x_0 \\ + x_1^2 \otimes x_1 + x_2x_1 \otimes x_0 + 1_{X^*} \otimes x_1x_2x_1.$$
Let us now prove the coassociativity.

Let $I = [i_1, \ldots, i_n]$, and α be a mapping:
$$\alpha : I \longrightarrow [1\ldots n], \\ i_s \longmapsto s. \tag{15}$$

Lemma 2. *Let $w \in X^*$ be a word, I be a subset of $[1\ldots|w|]$ and $I_1 \subset [1\ldots|I|]$. One then has*
$$pack(w[I])[I_1] = S_{\phi_{w[I]}}(w[I_1']), \tag{16}$$
where I_1' is $\alpha^{-1}(I_1)$ and $\phi_{w[I]}$ is the packing map of $w[I]$ that is given in (6) .

Proof. Using the definition of packing map $\phi_{w[I]}$, one can directly check that equation 16 holds. □

Lemma 3. *Let $w \in X^*$ be a word and ϕ be a strictly increasing map from $IAlph(w)$ to $\mathbb{N}$. One then has:*

1)
$$pack(S_\phi(w)) = pack(w). \tag{17}$$
2)
$$S_\phi({}^{w_1}/_{w_2}) = {}^{S_\phi(w_1)}/_{S_\phi(w_2)}. \tag{18}$$

Proof. 1) Using the definition (6) of the packing map, one has the following identity:
$$pack(S_\phi(w)) = S_{\phi_{S_\phi(w)}}(S_\phi(w)) = S_{\phi_{S_\phi(w)}\circ\phi}(w) = S_{\phi_w}(w) = pack(w).$$

2) Let $I_2 = Alph(w_2)$ and $I_2' = Alph(S_\phi(w_2))$. One can directly check that $\phi \circ \phi_{I_2}(i) = \phi_{I_2'} \circ \phi(i)$, for all $i \in IAlph(w_1)$. One thus has
$$S_\phi({}^{w_1}/_{w_2}) = S_\phi(S_{\phi_{I_2}}(w_1)) = S_{\phi\circ\phi_{I_2}}(w_1) = \\ S_{\phi_{I_2'}\circ\phi}(w_1) = S_{\phi_{I_2'}}(S_\phi(w_1)) = {}^{S_\phi(w_1)}/_{S_\phi(w_2)}. \tag{19}$$
□

Lemma 4. *Let w be a word in $\mathcal{H}$, and I, J, K be three disjoint subsets of the set $\{1 \dots |w|\}$. One then has:*

$$\frac{{}^{w[K]}/_{w[I]}}{{}^{w[J]}/_{w[I]}} = {}^{w[K]}/_{w[I+J]}. \tag{20}$$

Proof. Using Lemma 3, one has:

$$\frac{{}^{w[K]}/_{w[I]}}{{}^{w[J]}/_{w[I]}} = {}^{S_{\phi_I}(w[K])}/_{S_{\phi_I}(w[J])} = S_{\phi_I}({}^{w[K]}/_{w[J]}) = S_{\phi_I}(S_{\phi_J}(w[K]))$$
$$= S_{\phi_I \circ \phi_J}(w[K]) = {}^{w[K]}/_{w[I+J]}. \tag{21}$$

□

Proposition 3. *The vector space $\mathcal{H}$ endowed with the coproduct (14) is a coassociative coalgebra with co-unit (c-AAU). The co-unit is given by:*

$$\epsilon(w) = \begin{cases} 1 \textit{ if } w = 1_{\mathcal{H}}, \\ 0 \textit{ otherwise.} \end{cases}$$

Proof. Using the lemmas 2, 3 and 4, one has the following identity:

$$(\Delta \otimes Id) \circ \Delta(w) = \textstyle\sum_{I+J+K=[1\dots|w|]} pack(w[I]) \otimes pack({}^{w[J]}/_{w[I]})$$
$$\otimes pack({}^{w[K]}/_{w[I+J]}) = (Id \otimes \Delta) \circ \Delta(w). \tag{22}$$

Thus, one can conclude that the coproduct (14) is coassociative.

One can easily check that

$$(\epsilon \otimes Id) \circ \Delta(w) = (Id \otimes \epsilon) \circ \Delta(w), \tag{23}$$

for all word $w \in \mathcal{H}$.

One thus concludes that $(\mathcal{H}, \Delta, \epsilon)$ is a c-AAU. □

Remark 3. This coalgebra is not cocommutative, for example:

$$T_{12} \circ \Delta(x_1^2) = T_{12}(x_1^2 \otimes 1_{\mathcal{H}} + 2x_1 \otimes x_0 + 1_{\mathcal{H}} \otimes x_1^2)$$
$$= x_1^2 \otimes 1_{\mathcal{H}} + 2x_0 \otimes x_1 + 1_{\mathcal{H}} \otimes x_1^2 \neq \Delta(x_1^2),$$

where the operator T_{12} is given by $T_{12}(u \otimes v) = v \otimes u$.

Lemma 5. *Let u, v be two words. Let $I_1 + J_1 = [1 \dots |u|]$ and $I_2 + J_2 = [|u| + 1 \dots |u| + |v|]$. One then has*

$$pack({}^{u*v[J_1+J_2]}/_{u*v[I_1+I_2]}) = pack({}^{u[J_1]}/_{u[I_1]}) * pack({}^{v[J_2']}/_{v[I_2']}), \tag{24}$$

where I_2' is the set $\{k - |u|, k \in I_2\}$ and J_2' is the set $\{k - |u|, k \in J_2\}$.

Proof. One has:

$$pack({}^{u*v[J_1+J_2]}/_{u*v[I_1+I_2]}) = pack(S_{\phi_{I_1+I_2}}(u * v[J_1 + J_2]))$$
$$= pack(S_{\phi_{I_1}+\phi_{I_2}}(u[J_1]T_{sup(u)}(v)[J_2])) = pack(S_{\phi_{I_1}}(u[J_1])S_{\phi_{I_2}}(T_{sup(u)}(v[J_2'])))$$
$$= pack({}^{u[J_1]}/_{u[I_1]})T_{sup({}^{u[J_1]}/_{u[I_1]})}pack(S_{\phi_{I_2'}}(v[J_2']))$$
$$= pack(({}^{u[J_1]}/_{u[I_1]}) * pack({}^{u[J_2']}/_{u[I_2']}). \quad (25)$$

□

Proposition 4. *Let u, v be two words in $\mathcal{H}$. One has:*

$$\Delta(u * v) = \Delta(u) *^{\otimes 2} \Delta(v). \quad (26)$$

Proof. Using the Lemma 5, the proof can be done by a direct check.

□

Since $\mathcal{H}$ is graded by the word's length, one has the following theorem:

Theorem 2. *$(\mathcal{H}, *, 1_{\mathcal{H}}, \Delta, \epsilon)$ is a Hopf algebra.*

Proof. The proof follows from the above results. □

For $w \neq 1_{\mathcal{H}}$, the antipode is given by the recursion:

$$S(w) = -w - \sum_{I+J=[1...|w|], I,J\neq\emptyset} S(pack(w[I])) * pack({}^{w[J]}/_{w[I]}). \quad (27)$$

4 Hilbert Series of the Hopf Algebra WMat

In this section, we compute the number of packed words with length n and supremum k. Using the formula of Stirling numbers of the second kind, one can get the explicit formula for the number of packed words with length n, number which we denote by d_n.

Definition 5. *The Stirling numbers of the second kind count the number of set unordered partitions of an n-element set into precisely k non-void parts (or blocks). The Stirling numbers, denoted by $S(n, k)$ are given by the recursive definition:*

1. $S(n, n) = 1 (n \geq 0)$,
2. $S(n, 0) = 0 (n > 0)$,
3. $S(n + 1, k) = S(n, k - 1) + kS(n, k)$, *for* $0 < k \leq n$.

One can define a word without x_0 by its positions, this means that if a word $w = x_{i_1}x_{i_2}\dots x_{i_n}$ has length n and alphabet $IAlph(w) = \{1, 2, \dots, k\}$, then this word can be determine from the list $[S_1, S_2, \dots, S_k]$, where S_i is the set of positions of x_i in the word w, with $1 \le i \le k$. It is straightforward to check that $(S_i)_{0\le i\le k}$ is a partition of $[1\dots n]$.

One can divide the set of packed words with length n and supremum k in two parts: *"pure"* packed words (which have no x_0 in their alphabet), denote $pack^+_{n,k}(X)$ and packed words which have x_0 in their alphabet, denote $pack^0_{n,k}(X)$. It is clear that:

$$d(n,k) = \#pack^+_{n,k}(X) + \#pack^0_{n,k}(X). \tag{28}$$

Let us now compute the cardinal of these two sets $pack^+_{n,k}(X)$ and $pack^0_{n,k}(X)$.

Consider a word $w \in pack^+_{n,k}(X)$, then $IAlph(w) = \{1, 2, \dots, k\}$. This word is determined by $[S_1, S_2, \dots, S_k]$, in which S_i is a set of positions of x_i, for $1 \le i \le k$. One can see that:

1. $S_i \neq \emptyset$, $\forall i \in [1, k]$;
2. $\sqcup_{1\le i\le k} S_i = \{1, 2, \dots, n\}$.

Note that 1-2 hold even with $w = 1_{\mathcal{H}}$.

Thus, one has the cardinal of packed words with length n and supremum k:

$$d^+(n,k) = \#pack^+_{n,k}(X) = S(n,k)k!. \tag{29}$$

Similarly, a word $w \in \#pack^0_{n,k}(X)$ can be determined by $[S_0, S_1, S_2, \dots, S_k]$ where S_i is the set of positions of x_i, for all $0 \le i \le k$. One then has:

$$d^0(n,k) = \#pack^0_{n,k}(X) = S(n,k+1)(k+1)!. \tag{30}$$

From the two equations above, one can get the number of packed word with length n, supremum k:

$$d(n,k) = d^+(n,k) + d^0(n,k) = S(n,k)k! + S(n,k+1)(k+1)! = S(n+1,k+1)k!. \tag{31}$$

From this formula, using Maple, one can get some values of $d(n,k)$. We give in the Table 1 the first values.

Note that the values of Table 1 correspond to those of the triangular array *A*028246 of Sloane [9].

Remark 4. Formulas (29) and (30) imply that the packed words of length n and supremum k without, and respectively with, x_0 are in bijection with the circularly ordered partitions of $[n]$ in k parts and respectively in $k+1$ parts. Therefore (31) implies that the set of packed words of length n with supremum k is in bijection with the circularly ordered partitions of $n+1$ elements in $k+1$ parts.

Table 1. Values of $d(n,k)$ given by the explicit formula (31) and computed with Maple

		k								
		0	1	2	3	4	5	6	7	8
	0	1	0	0	0	0	0	0	0	0
	1	1	1	0	0	0	0	0	0	0
	2	1	3	2	0	0	0	0	0	0
	3	1	7	12	6	0	0	0	0	0
n	4	1	15	50	60	24	0	0	0	0
	5	1	31	180	390	360	120	0	0	0
	6	1	63	602	2100	3360	2520	720	0	0
	7	1	127	1932	10206	25200	31920	20160	5040	0
	8	1	255	6050	46620	166824	317520	332640	181440	40320

The formula for the number of packed words of length n, d_n $(n \geq 1)$, is then given by

$$d_n = \sum_{k=0}^{n} d(n,k) = \sum_{k=0}^{n} S(n+1,k+1)k!. \tag{32}$$

Using again Maple, one can get the values listed in Table 2.

Table 2. Some values of d_n obtained from formula (32)

n	0	1	2	3	4	5	6	7	8	9	10
d_n	1	2	6	26	150	1082	9366	94586	1091670	14174522	204495126

The number of packed words is the sequence $A000629$ of Sloane [9], where it is also mentioned that this sequence corresponds to the ordered Bell numbers sequence times two (except for the 0th order term).

The ordinary and exponential generating function of our sequence are also given in [9]. The ordinary one is given by the formula: $\sum_{n\geq 0} \frac{2^n n! x^n}{\prod_{k=0}^{n}(1+kx)}$. The exponential one is given by: $\frac{e^x}{2-e^x}$. Let us give the proof of this.

Firstly, recall that the exponential generating function of the ordered Bell numbers (see, for example, page 109 of Philippe Flajolet's book [2]) is:

$$\frac{1}{2-e^x} = \sum_{n\geq 0}\sum_{k=0}^{n} S(n,k)k!\frac{x^n}{n!}. \tag{33}$$

By deriving both side of equation (33) with respect to x , one obtains:

$$\frac{e^x}{2-e^x} = \sum_{n\geq 1}\sum_{k=1}^{n} S(n,k)k!\frac{x^{n-1}}{(n-1)!}. \tag{34}$$

From equations (32) and (34), one gets the exponential generating function of our sequence:

$$\frac{e^x}{2-e^x} = \sum_{n\geq 0}\sum_{k=0}^{n} S(n+1,k+1)k!\frac{x^n}{n!} = \sum_{n\geq 0} d_n \frac{x^n}{n!}. \tag{35}$$

Let us now investigate the combinatorics of irreducible packed words (see Definition 2). Firstly, we notice that one still has an infinity of irreducible packed words of weight m, which are again obtained by adding multiple copies of the letter x_0.

Example 5. The word $x_1 x_0^k x_1 x_0^k x_1$ (with k an arbitrary integer) is an irreducible packed word of weight 3.

Let us denote by i_n the number of irreducible packed words of length n. Then one has:

$$i_n = \sum_{\substack{j_1+\cdots+j_k=n \\ j_l\neq 0}} (-1)^{k+1} d_{j_1}\ldots d_{j_k}. \tag{36}$$

Using Maple, one can get the values of i_n, which we give in Table 3 below.

Table 3. Ten first values of the number of irreducible packed words.

n	0	1	2	3	4	5	6	7	8	9	10
i_n	1	2	2	10	66	538	5170	59906	704226	9671930	145992338

Note that this sequence does not appear in Sloane's On-Line Encyclopedia of Integer Sequences [9].

5 Primitive Elements of WMat

Let us emphasize that this Hopf algebra, although graded, is not cocommutative and thus the primitive elements do not generate the whole algebra but only a sub Hopf algebra on which Δ is cocommutative (the biggest on which CQMM theorem holds).

We denote by *Prim*(WMat) the algebra generated by the primitive elements of $\mathcal{H}$.

Let us recall the following result:

Lemma 6. *Let $V^{(1)}$ and $V^{(2)}$ be two graded vector space.*

$$V^{(i)} = \oplus_{n\geq 0} V_n^{(i)},\ i=1,2. \tag{37}$$

Let $\phi \in Hom^{gr}(V^{(1)}, V^{(2)})$, that means $(\forall n\geq 0)(\phi(V_n^{(1)}) \subseteq V_n^{(2)})$. Then, $Ker(\phi)$ is graded.

One then has:

Proposition 5. *Prim*(WMat) *is a Lie subalgebra of* WMat *graded by the word's length.*

Proof. Let us define the mapping

$$\Delta^+ : \quad \mathrm{WMat} \longrightarrow \mathrm{WMat} \otimes \mathrm{WMat} \qquad (38)$$
$$\begin{cases} 1_{\mathcal{H}} & \longmapsto 0 \\ h & \longmapsto \Delta(h) - 1_{\mathcal{H}} \otimes h - h \otimes 1_{\mathcal{H}} \end{cases}.$$

This mapping is graded. Using Lemma 6, one has $Prim(\mathrm{WMat}) = Ker(\Delta^+)$. Thus, the subalgebra $Prim(\mathrm{WMat})$ is graded. □

Let WMat_n be the subalgebra generated by the packed words of length n, $n \geq 0$. One can see an element P of this subalgebra as a polynomial of packed words of length n.

$$P = \sum_{w \in \mathrm{WMat}_n} \langle P|w \rangle \, w. \qquad (39)$$

Let us now compute the dimensions of the first few spaces $Prim(\mathrm{WMat})_n$. For $n = 1$, one has a basis formed by the primitive elements x_0 and x_1. Then one can check that the primitive elements of length 1 have the form $ax_0 + bx_1$, with a and b scalars. For $n = 2$, one has a basis formed by the primitive elements: $x_0x_1 - x_1x_0$ and $x_1x_2 - x_2x_1$. Then one can check that all the primitive elements of the length 2 have the form $a(x_0x_1 - x_1x_0) + b(x_1x_2 - x_2x_1)$, with a and b scalars. This comes from explicitly solving a system of 4 equations with $d_2 = 6$ variables.

Nevertheless, the explicit calculations quickly become lengthy. Thus, for $n = 3$, one has to solve a system of 22 equations with 26 variables.

6 Maple Coding

To test our results with Maple, we implement a random word in the following way. To each word we associate a certain monomial which encodes, using a given alphabet the position of any letter and its value. For example, to the word x_2x_3 we associate the monomial $a_1^2a_2^3$ where the powers (2 and respectively 3) correspond to the values of the letters (x_2 and respectively x_3) and the indices (1 and 2) correspond to the positions of the respective letters.

One has to keep in mind that the letter x_0 can also be present in the words. which is encoded with a supplementary word length variable.

Using this idea we can then implement in Maple packed words (obtained with a Maple function taking as an argument a general word).

We have also implemented the LHS (left hand side) and the RHS (right hand side) of the coproduct formula. For this purpose, one needs to refine the above function by considering two distinct alphabets to "build up" the words, such that one can easily separate - as function of the different alphabets - the LHS from the RHS.

Finally, using all of the above, we have checked the coassociativity condition for random words up to length 7, with maximal power 7.

Acknowledgements. We acknowledge Jean-Yves Thibon for various discussions and suggestions. The authors also acknowledge a Univ. Paris 13, Sorbonne Paris Cité BQR grant.

A. Tanasa further acknowledges the grants PN 09 37 01 02 and CNCSIS Tinere Echipe 77/04.08.2010.

G. H. E. Duchamp acknowledges the grants ANR BLAN08-2_332204 (Physique Combinatoire) and PAN-CNRS 177494 (Combinatorial Structures and Probability Amplitudes).

References

1. Connes, A., Kreimer, D.: Renormalization in quantum field theory and the Riemann-Hilbert problem I: The Hopf algebra structure of graphs and the main theorem. Commun. Math. Phys. 210(1), 249–273 (2000)
2. Flajolet, P., Sedgewick, R.: Analytic Combinatorics. Cambridge University Press (2009)
3. Markopoulou, F.: Coarse graining in spin foam models. Class. Quant. Grav. 20, 777–800 (2003), gr-qc/0203036
4. Tanasa, A., Kreimer, D.: Combinatorial Dyson-Schwinger equations in noncommutative field theory. J. Noncomm. Geom. (in press) arXiv:0907.2182 [hep-th]
5. Tanasa, A., Vignes-Tourneret, F.: Hopf algebra of non-commutative field theory. J. Noncomm. Geo. 2(1), 125–139 (2008)
6. Tanasa, A.: Combinatorial Hopf Algebras in (Noncommutative) Quantum Field Theory. Rom. J. Phys. 55, 1142–1155 (2010) [arXiv:1008.1471 [math.CO]]
7. Tanasa, A.: Algebraic structures in quantum gravity. Class. Quant. Grav. 27, 095008 (2010) [arXiv:0909.5631 [gr-qc]]
8. Tanasa, A.: Some Combinatorial Aspects of Quantum Field Theory. Sem. Lothar. Combin. 65, B65g (2012)
9. Sloane, N.J.A.: The On-Line Encyclopedia of Integer Sequences (OEIS), `http://www.oeis.org`

Synchronous Forest Substitution Grammars

Andreas Maletti*

Universität Stuttgart, Institute for Natural Language Processing
Pfaffenwaldring 5b, 70569 Stuttgart, Germany
andreas.maletti@ims.uni-stuttgart.de

Abstract. The expressive power of synchronous forest (tree-sequence) substitution grammars (SFSG) is studied in relation to multi bottom-up tree transducers (MBOT). It is proved that SFSG have exactly the same expressive power as compositions of an inverse MBOT with an MBOT. This result is used to derive complexity results for SFSG and the fact that compositions of an MBOT with an inverse MBOT can compute tree translations that cannot be computed by any SFSG, although the class of tree translations computable by MBOT is closed under composition.

1 Introduction

Synchronous forest substitution grammars (SFSG) [19] or the rational binary tree relations [17] computed by them received renewed interest recently due to their applications in Chinese-to-English machine translation [21,22]. The fact that [19] and [17] arrived independently and with completely different backgrounds at the same model shows that SFSG are a natural, practically relevant, and theoretically interesting model for tree translations. Roughly speaking, SFSG are a synchronous grammar formalism [2] that utilizes only first-order substitution (as in a regular tree grammar [7,8]), but allows several components that develop simultaneously for both the input and the output side. This feature allows them to model linguistic discontinuity on both the source and target language. The rational binary tree relations (or tree translations computed by SFSG) can also be characterized by rational expressions [17] and automata [16].

Multi bottom-up tree transducers (MBOT) [1,4] are restricted SFSG, in which only the output side is allowed to have several components. They were rediscovered in [5,6], but were studied extensively by [3,11,1] already in the 70s and 80s. Their properties [13] are desirable in statistical syntax-based machine translation [10]. This led to a closer inspection [4,15,9] of their properties in recent years. Overall, their expressive power is rather well-understood by now.

In this contribution, we investigate the expressive power of SFSG in terms of MBOT. We show that the expressive power of SFSG coincides exactly with that of compositions of an inverse MBOT followed by an MBOT. This characterization is natural in terms of bimorphisms and shows that the input and the

* The author gratefully acknowledges the financial support by the German Research Foundation (DFG) grant MA / 4959 / 1-1.

T. Muntean, D. Poulakis, and R. Rolland (Eds.): CAI 2013, LNCS 8080, pp. 235–246, 2013.

output tree are independently obtained by a full MBOT from an intermediate tree language (which is always regular [7,8]). This paves the way to complementary results. In particular, we derive the first complexity results for SFSG and we demonstrate that the composition in the other order (first an MBOT followed by an inverse MBOT) contains tree translations that cannot be computed by any SFSG. This shows a limitation of MBOT, which are closed under composition [4]. Overall, we can thus also characterize the expressive power of SFSG by an arbitrary chain of inverse MBOT followed by an arbitrary chain of MBOT.

2 Preliminaries

The set of nonnegative integers is $\mathbb{N}$. We write $[k]$ for the set $\{i \in \mathbb{N} \mid 1 \leq i \leq k\}$, and we treat functions (or maps) as special relations. For all relations $R \subseteq A \times B$ and subsets $A' \subseteq A$, we let $R(A') = \{b \in B \mid \exists a \in A' \colon (a, b) \in R\}$. Moreover,

$$R^{-1} = \{(b, a) \mid (a, b) \in R\} \qquad \mathrm{dom}(R) = R^{-1}(B) \qquad \mathrm{ran}(R) = \mathrm{dom}(R^{-1}) \ ,$$

which are called the *inverse* of R, the *domain* of R, and the *range* of R, respectively. Given $R_1 \subseteq A \times B$ and $R_2 \subseteq B \times C$, the *composition* $R_1 \,;\, R_2 \subseteq A \times C$ of R_1 and R_2 is $R_1 \,;\, R_2 = \{(a, c) \in A \times C \mid \exists b \in B \colon (a, b) \in R_1, (b, c) \in R_2\}$. These notions and notations are lifted to sets of relations as usual. Given a set Σ, the set of all words over Σ is Σ^*, of which ε is the empty word. The concatenation of two words $u, w \in \Sigma^*$ is denoted by uw. The length of a word $w = \sigma_1 \cdots \sigma_k$ with $\sigma_i \in \Sigma$ for all $i \in [k]$ is $|w| = k$. We simply write w_i for the ith letter of w (i.e., $w_i = \sigma_i$) for all $i \in [k]$. For every $k \in \mathbb{N}$, we let $\Sigma^k = \{w \in \Sigma^* \mid k = |w|\}$.

A ranked alphabet (Σ, rk) consists of an alphabet Σ and a map $\mathrm{rk} \colon \Sigma \to \mathbb{N}$. The symbol $\sigma \in \Sigma$ has rank $\mathrm{rk}(\sigma)$, and we let $\Sigma_k = \{\sigma \in \Sigma \mid \mathrm{rk}(\sigma) = k\}$ for all $k \in \mathbb{N}$. We usually denote the ranked alphabet (Σ, rk) by just Σ and write $\sigma^{(k)}$ to indicate that $\mathrm{rk}(\sigma) = k$. The set $T_\Sigma(N)$ of all *Σ-trees indexed by* the set N is the smallest set T such that $N \subseteq T$ and $\sigma(\boldsymbol{t}) \in T$ for all $\sigma \in \Sigma$ and $\boldsymbol{t} \in T^{\mathrm{rk}(\sigma)}$. Such a sequence $\boldsymbol{t}$ of trees is also called *forest*. Consequently, a tree t is either an element of N or it consists of a root node labeled σ followed by a forest $\boldsymbol{t}$ of $\mathrm{rk}(\sigma)$ children. To improve the readability, we often write a forest $t_1 \cdots t_k$ as $t_1, \ldots, t_k$. The *positions* $\mathrm{pos}(t), \mathrm{pos}(\boldsymbol{u}) \subseteq \mathbb{N}^*$ of a tree $t \in T_\Sigma(N)$ and a forest $\boldsymbol{u} \in T_\Sigma(N)^*$ are inductively defined by (i) $\mathrm{pos}(n) = \{\varepsilon\}$, (ii) $\mathrm{pos}(\sigma(\boldsymbol{t})) = \{\varepsilon\} \cup \mathrm{pos}(\boldsymbol{t})$, and (iii) $\mathrm{pos}(\boldsymbol{u}) = \bigcup_{i=1}^{|\boldsymbol{u}|} \{ip \mid p \in \mathrm{pos}(u_i)\}$ for every $n \in N$, $\sigma \in \Sigma_k$, and $\boldsymbol{t} \in T_\Sigma(N)^k$. This yields an undesirable difference between $\mathrm{pos}(t)$ and $\mathrm{pos}(\boldsymbol{u})$ with $\boldsymbol{u} = (t)$. Note that positions are totally ordered via the (standard) lexicographic ordering on $\mathbb{N}^*$. Let $t, t' \in T_\Sigma(N)$ and $p \in \mathrm{pos}(t)$. The label of t at position p is $t(p)$, the subtree rooted at position p is $t|_p$, and the tree obtained by replacing the subtree at position p by t' is denoted by $t[t']_p$. Formally, they are defined by $n(\varepsilon) = n|_\varepsilon = n$ and $n[t']_\varepsilon = t'$ for every $n \in N$ and

$$t(p) = \begin{cases} \sigma & \text{if } p = \varepsilon \\ \boldsymbol{t}(p) & \text{if } p \neq \varepsilon \end{cases} \qquad t|_p = \begin{cases} t & \text{if } p = \varepsilon \\ \boldsymbol{t}|_p & \text{if } p \neq \varepsilon \end{cases} \qquad t[t']_p = \begin{cases} t' & \text{if } p = \varepsilon \\ \boldsymbol{t}[t']_p & \text{if } p \neq \varepsilon \end{cases}$$

$$\boldsymbol{u}(ip') = u_i(p') \qquad \boldsymbol{u}|_{ip'} = u_i|_{p'} \qquad \boldsymbol{u}[t']_{ip'} = u_i[t']_{p'}$$

for all $t = \sigma(\boldsymbol{t})$ with $\sigma \in \Sigma_k$ and $\boldsymbol{t} \in T_\Sigma(N)^k$, $\boldsymbol{u} \in T_\Sigma(N)^*$, $1 \leq i \leq |\boldsymbol{u}|$, and $p' \in \mathrm{pos}(u_i)$. As demonstrated, these notions are also defined for forests $\boldsymbol{u}$. A position $p \in \mathrm{pos}(t)$ is a *leaf* (in t) if $p1 \notin \mathrm{pos}(t)$. For every $S \subseteq N \cup \Sigma$, we let $\mathrm{pos}_S(t) = \{p \in \mathrm{pos}(t) \mid t(p) \in S\}$ and $\mathrm{pos}_s(t) = \mathrm{pos}_{\{s\}}(t)$ for every $s \in N \cup \Sigma$. The tree $t \in T_\Sigma(N)$ is *linear* in $S \subseteq N$ if $|\mathrm{pos}_s(t)| \leq 1$ for every $s \in S$. The *variables* of t are $\mathrm{var}(t) = \{n \in N \mid \mathrm{pos}_n(t) \neq \emptyset\}$, and $\mathrm{var}(\boldsymbol{u}) = \bigcup_{i=1}^{|\boldsymbol{u}|} \mathrm{var}(\boldsymbol{u}_i)$ for all $\boldsymbol{u} \in T_\Sigma(N)^*$. Given $S \subseteq N$, $\boldsymbol{u} \in T_\Sigma(N)^*$, and $\theta\colon S \to T_\Sigma(N)^*$ such that $|\theta(s)| = |\mathrm{pos}_s(\boldsymbol{u})|$ for every $s \in S$, the forest $\boldsymbol{u}\theta$ is obtained from $\boldsymbol{u}$ by replacing for every $s \in S$ the occurrences $\mathrm{pos}_s(\boldsymbol{u}) = \{p_1, \ldots, p_k\}$ with $p_1 < \cdots < p_k$ of (the leaf) s in $\boldsymbol{u}$ by the trees $\theta(s)_1, \ldots, \theta(s)_k$, respectively.

Given ranked alphabets Σ and Δ, a mapping $d\colon \bigcup_{k \in \mathbb{N}} \Sigma_k \to (\Delta_k \cup \{\Box\})$ is a *delabeling* if $d(\sigma) \in \Delta_k$ for all $\sigma \in \Sigma_k$ with $k \neq 1$. Thus, a delabeling is similar to a relabeling [7,8], but it can also erase unary symbols. It induces a mapping $d\colon T_\Sigma \to T_\Delta$ such that $d(\sigma(\boldsymbol{t})) = d(t_1)$ if $d(\sigma) = \Box$ and $d(\sigma)(d(t_1), \ldots, d(t_k))$ otherwise for all $\sigma \in \Sigma_k$ and $\boldsymbol{t} \in T_\Sigma^k$. Finally, let us recall the regular tree languages [7,8]. A *regular tree grammar* (RTG) is a tuple $G = (N, \Sigma, I, R)$ such that N is a finite set of *nonterminals*, Σ is a ranked alphabet of symbols, $I \subseteq N$ is a set of *initial nonterminals*, and $R \subseteq N \times T_\Sigma(N)$ is a finite set of *rules*. A rule $(n, r) \in R$ is typically written $n \to r$, and for every $n \in N$, we let $R_n = \{n \to r \mid n \to r \in R\}$. Given $\xi, \zeta \in T_\Sigma(N)$ we write $\xi \Rightarrow_G \zeta$ if there exists a a rule $n \to r \in R$ and a position $p \in \mathrm{pos}_n(\xi)$ such that $\zeta = \xi[r]_p$. The regular tree grammar G generates the tree language $L(G) = \{t \in T_\Sigma \mid \exists n \in I\colon n \Rightarrow_G^* t\}$, where $\Rightarrow_G^*$ is the reflexive and transitive closure of $\Rightarrow_G$. A tree language $L \subseteq T_\Sigma$ is *regular* if there exists a regular tree grammar G such that $L = L(G)$. The class of regular tree languages is denoted by Reg. Moreover, FTA denotes the class of partial identities computed by the regular tree languages; i.e., FTA $= \{\mathrm{id}_L \mid L \in \mathrm{Reg}\}$, where $\mathrm{id}_L = \{(t, t) \mid t \in L\}$.

3 Synchronous Forest Substitution Grammars

The *(stateful) synchronous forest substitution grammars* (SFSG) are a natural generalization of the non-contiguous synchronous tree sequence substitution grammars of [19] to include full grammar nonterminals (or states). They naturally coincide with the binary rational relations studied by [17,16]. To keep the presentation simple, we assume a global ranked alphabet Σ of input and output terminal symbols. Moreover, we immediately present it in a form inspired by tree bimorphisms [1] and tree grammars with multi-variables [17].

Definition 1. *A* (stateful) synchronous forest substitution grammar *(SFSG) is a tuple* $G = (N, \Sigma, I, R, B)$, *where*

- (N, Σ, I, R) *is a regular tree grammar, and*
- $B \subseteq (\bigcup_{n \in I} R_n \times R_n) \cup (\bigcup_{n \in N \setminus I} R_n^* \times R_n^*)$ *is a finite set of* aligned rules.

It is a multi bottom-up tree transducer *(MBOT) if* $B \subseteq \bigcup_{n \in N} R_n \times R_n^*$.

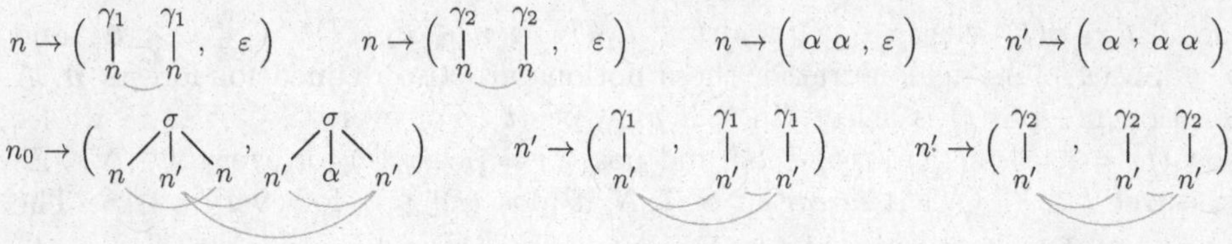

Fig. 1. Aligned example rules of the SFSG of Example 1

Roughly speaking, we have a regular tree grammar containing all the potentially used rules. However, potentially several rules with the same left-hand side are applied at the same time on both the input and the output side. This dependence is expressed by the set B of aligned rules. For all initial nonterminals, only one rule is applied to the input and output side as we want to compute a tree translation. For the remaining nonterminals we can use arbitrarily many rules on the input and the output side. The alignment in the rules is established implicitly by occurrences of the same nonterminal in the right-hand sides. To make aligned rules more readable, we also write $n \to (\ell_1 \cdots \ell_k, r_1 \cdots r_{k'})$ or $n \to (\boldsymbol{\ell}, \boldsymbol{r})$ for a rule $(n \to \ell_1 \cdots n \to \ell_k, n \to r_1 \cdots n \to r_{k'}) \in B$, where $n \to \ell_1, \ldots, n \to \ell_k, n \to r_1, \ldots, n \to r_{k'} \in R_n$ are rules for the same nonterminal $n \in N$. In short, we write the common nonterminal only once on the left-hand side and then group all the right-hand sides of the rules of R_n. We assume that the nonterminals N of each SFSG are totally ordered by $\leq_N$. Finally, we let $\text{var}(\chi) = \text{var}(\boldsymbol{\ell}) \cup \text{var}(\boldsymbol{r})$ for every rule $\chi = n \to (\boldsymbol{\ell}, \boldsymbol{r})$, where $\boldsymbol{\ell}$ and $\boldsymbol{r}$ contain only the right-hand sides of rules of R (as per the previous declaration).

Example 1. Let $(N, \Sigma, \{n_0\}, R)$ be the regular tree grammar such that

- $N = \{n_0, n, n'\}$ with $n_0 <_N n <_N n'$ and $\Sigma = \{\alpha^{(0)}, \gamma_1^{(1)}, \gamma_2^{(1)}, \sigma^{(3)}\}$, and
- the following rules are in R:

$$\begin{array}{llll} \rho_0 \colon n_0 \to \sigma(n, n', n) & \rho_2 \colon n \to \gamma_1(n) & \rho_4 \colon n \to \gamma_2(n) & \rho_6 \colon n \to \alpha \\ \rho_1 \colon n_0 \to \sigma(n', \alpha, n') & \rho_3 \colon n' \to \gamma_1(n') & \rho_5 \colon n' \to \gamma_2(n') & \rho_7 \colon n' \to \alpha \ . \end{array}$$

Based on this RTG we construct the SFSG $G = (N, \Sigma, \{n_0\}, R, B)$ with

$$B = \{(\rho_0, \rho_1), (\rho_2\rho_2, \varepsilon), (\rho_4\rho_4, \varepsilon), (\rho_6\rho_6, \varepsilon), (\rho_3, \rho_3\rho_3), (\rho_5, \rho_5\rho_5) (\rho_7, \rho_7\rho_7)\} \ .$$

We illustrate these aligned rules in Fig. 1, where we indicate the implicit links by splines. Clearly, the SFSG G is (syntactically) not an MBOT.

Next, we introduce the (bottom-up) semantics of an SFSG G. It works on pre-translations, which are pairs of input and output tree sequences together with a governing nonterminal. The pre-translations computed by G are inductively defined, and each pre-translation is obtained from an aligned rule $\chi = n \to (\boldsymbol{\ell}, \boldsymbol{r})$ of G by replacing each nonterminal $n \in \text{var}(\chi)$ by a pre-translation computed by G that is governed by n. Alongside, we introduce the derivation tree, which records how the aligned rules combined.

Definition 2. *Let $G = (N, \Sigma, I, R, B)$ be an SFSG. A* pre-translation *for G is a triple $\langle \boldsymbol{t}, n, \boldsymbol{u}\rangle$ consisting of a nonterminal $n \in N$ and input and output tree sequences $\boldsymbol{t}, \boldsymbol{u} \in T_\Sigma^*$. The set* PT($G$) *of* pre-translations generated by G *is the smallest set T such that (†): $\langle \boldsymbol{\ell}\theta\,,\, n\,,\, \boldsymbol{r}\theta'\rangle \in \mathrm{PT}(G)$ for all aligned rules $\chi = n \to (\boldsymbol{\ell}, \boldsymbol{r}) \in B$, all mappings $\theta, \theta' \colon \mathrm{var}(\chi) \to T_\Sigma^*$, and for all $n' \in \mathrm{var}(\chi)$*

- *$|\theta(n')| = |\mathrm{pos}_{n'}(\boldsymbol{\ell})|$ and $|\theta'(n')| = |\mathrm{pos}_{n'}(\boldsymbol{r})|$, and*
- *$\langle \theta(n'), n', \theta'(n')\rangle \in T$ is a pre-translation generated by G.*

The derivation tree corresponding to the pre-translation (†) is $\chi(d_{n_1}, \ldots, d_{n_k})$, where $\mathrm{var}(\chi) = \{n_1, \ldots, n_k\}$ with $n_1 <_N \cdots <_N n_k$ and d_n is the derivation tree corresponding to the pre-translation $\langle \theta(n), n, \theta'(n)\rangle$ for every $n \in \mathrm{var}(\chi)$.

Example 2. Recall the SFSG G of Example 1. The aligned rules $\chi_6 = (\rho_6\rho_6, \varepsilon)$ and $\chi_7 = (\rho_7, \rho_7\rho_7)$ immediately yield the pre-translations $\langle(\alpha,\alpha)\,,\, n\,,\, \varepsilon\rangle$ and $\langle \alpha\,,\, n'\,,\, (\alpha,\alpha)\rangle$ with derivation trees χ_6 and χ_7, respectively. The former pre-translation (and the pre-translations obtained) can be used with the aligned rules $\chi_2 = (\rho_2\rho_2, \varepsilon)$ and $\chi_4 = (\rho_4\rho_4, \varepsilon)$ to obtain the pre-translations

$\langle(\gamma_1(\alpha), \gamma_1(\alpha))\,,\, n\,,\, \varepsilon\rangle$ with derivation tree $\chi_2(\chi_6)$, or more generally,
$\{\langle(t,t)\,,\, n\,,\, \varepsilon\rangle \mid t \in T_{\{\gamma_1,\gamma_2,\alpha\}}\}$ with derivation trees $d \in T_{\{\chi_2,\chi_4,\chi_6\}}$,

where the rules χ_2 and χ_4 have rank 1 in the derivation trees. Similarly, with the help of the rules $\chi_3 = (\rho_3, \rho_3\rho_3)$ and $\chi_5 = (\rho_5, \rho_5\rho_5)$ we can obtain the pre-translations $\{\langle(t,t)\,,\, n'\,,\, t\rangle \mid t \in T_{\{\gamma_1,\gamma_2,\alpha\}}\}$ with derivation trees $d \in T_{\{\chi_3,\chi_5,\chi_7\}}$. Plugging those pre-translations into the rule $\chi_1 = (\rho_0, \rho_1)$, we obtain

$$\{\langle \sigma(t,u,t)\,,\, n_0\,,\, \sigma(u,\alpha,u)\rangle \mid t, u \in T_{\{\gamma_1,\gamma_2,\alpha\}}\} \subseteq \mathrm{PT}(G)$$

with derivation trees $\{\chi_1(d_1, d_2) \mid d_1 \in T_{\{\chi_2,\chi_4,\chi_6\}}, d_2 \in T_{\{\chi_3,\chi_5,\chi_7\}}\}$. We illustrate the last step of the process in Fig. 2.

Now we are ready to define the tree translation computed by an SFSG. Intuitively all pre-translations governed by initial nonterminals are translations.

Definition 3. *Let $G = (N, \Sigma, I, R, B)$ be an SFSG. It computes the tree translation $\tau_G \subseteq T_\Sigma \times T_\Sigma$ defined by $\tau_G = \bigcup_{n \in I}\{(t,u) \mid \langle t, n, u\rangle \in \mathrm{PT}(G)\}$. The derivation tree language $D(G)$ contains all derivation trees for the pre-translations $\langle t, n, u\rangle \in \mathrm{PT}(G)$ with $n \in I$. As usual, two SFSG are equivalent if their computed tree translations coincide. Finally, we denote the classes of tree translations computable by SFSG and MBOT by* SFSG *and* MBOT, *respectively.*

In the rest of this section, we present a normal form for MBOT, which allows us to relate our notion of MBOT to that of [4]. Moreover, we present some simple properties of SFSG. Let us start with classic MBOT [4].

Definition 4. *The MBOT (N, Σ, I, R, B) is* classic *if ℓ is linear in N and $\mathrm{var}(\boldsymbol{r}) \subseteq \mathrm{var}(\ell)$ for every $n \to (\ell, \boldsymbol{r}) \in B$.*

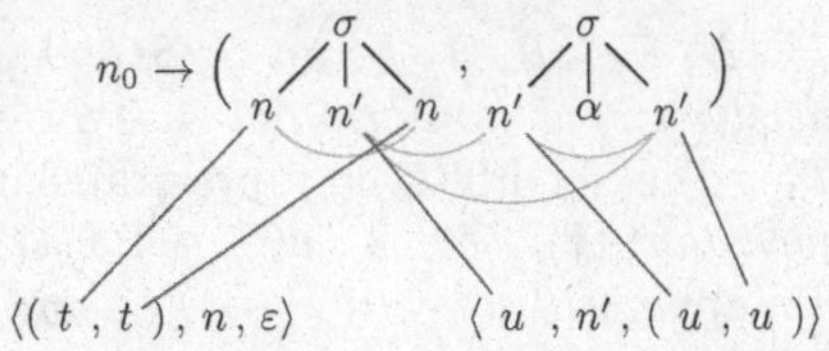

Fig. 2. Illustration of the combination of an aligned rule with pre-translations

Proposition 1. *For every MBOT there exists an equivalent classic MBOT.*

Proof. Let $G = (N, \Sigma, I, R, B)$ be the given MBOT. We construct the MBOT $G' = (N, \Sigma, I, R, B')$ with $B' = \{n \to (\ell, \boldsymbol{r}) \in B \mid \ell \text{ linear in } N, \text{var}(\boldsymbol{r}) \subseteq \text{var}(\ell)\}$ that is obviously classic. It remains to prove that G and G' are equivalent. To this end, we observe that $|\boldsymbol{t}| = 1$ for all $\langle \boldsymbol{t}, n, \boldsymbol{u} \rangle \in \text{PT}(G)$ due to the rule shape of G. Now, let $\chi = n \to (\ell, \boldsymbol{r}) \in B$ be a rule and $n' \in \text{var}(\boldsymbol{r}) \setminus \text{var}(\ell)$. To build a pre-translation of $\text{PT}(G)$ with χ, we need an existing pre-translation $\langle \varepsilon, n', \boldsymbol{u} \rangle \in \text{PT}(G)$ because $n' \in \text{var}(\chi)$, but $n' \notin \text{var}(\ell)$. Such pre-translations do not exist, hence the rule χ is useless (i.e., there are no derivation trees that contain χ), which proves that deleting it does not affect the semantics. In the same manner, rules whose left-hand side is not linear in N can be deleted (because they would require a pre-translation $\langle \boldsymbol{t}, n, \boldsymbol{u} \rangle \in \text{PT}(G)$ with $|\boldsymbol{t}| \geq 2$). □

Consequently, our class MBOT coincides the standard notion [4], so we can freely use the known properties of MBOT. Already in [12,4] the MBOT were transformed into a special normal form before composition. In this normal form, at most one (input or output) symbol is allowed per aligned rule. For our purposes, a slightly less restricted variant, in which at most one input symbol may occur per aligned rule is sufficient since we compose the input parts of two MBOT. Let us recall the property and the associated normalization result [4].

Definition 5. *The classic MBOT* (N, Σ, I, R, B) *is in* one-symbol (input) normal form *if* $|\text{pos}_\Sigma(\ell)| \leq 1$ *for every aligned rule* $n \to (\ell, \boldsymbol{r})$.

Lemma 1 (see [4, Lemma 14]). *For every MBOT there exists an equivalent classic MBOT in one-symbol (input) normal form.*

Proof. By Proposition 1 we can construct an equivalent classic MBOT for every MBOT. With the help of [4, Lemma 14] we can then construct an equivalent MBOT in one-symbol normal form. □

Given one-symbol normal form, we can now define deterministic MBOT, which we use instead of k-morphisms [1] to avoid another concept. It should be noted that deterministic MBOT are slightly more expressive than k-morphisms.

Definition 6. *A classic MBOT* (N, Σ, I, R, B) *in one-symbol normal form is* deterministic *if (i)* I *is a singleton, (ii)* $\ell \notin N$ *for every* $n \to (\ell, \boldsymbol{r}) \in B$, *and (iii) for every* $n \in N$ *and* $\sigma \in \Sigma$ *there exists at most one aligned rule* $n \to (\ell, \boldsymbol{r}) \in B$ *such that* $\ell(\varepsilon) = \sigma$.

Theorem 1. *The following simple properties can easily be observed:*

1. SFSG = SFSG^{-1}.
2. *The domain* $\text{dom}(\tau)$ *and the range* $\text{ran}(\tau)$ *of a tree translation* $\tau \in$ SFSG *are not necessarily regular.*
3. MBOT $\subsetneq$ SFSG.

Proof. The first property is immediate because the syntactic definition of SFSG is completely symmetric. For the second property we observe that the tree translation τ_G computed by the SFSG G of Example 1 is such that both its domain and its range are not regular. Finally, the inclusion in the third item is obvious. Moreover, we know that $\text{dom}(\tau)$ is regular for every $\tau \in$ MBOT by Proposition 1 and [4, Theorem 25], so the tree translation τ_G is not in MBOT. □

4 Composition and Decomposition

In this section, we develop a characterization of SFSG in terms of MBOT in order to better understand the expressive power of SFSG. Since we already showed MBOT $\subsetneq$ SFSG in Theorem 1, we will use compositions of MBOT to characterize the expressive power of SFSG. To this end, we need a decomposition (see Theorem 2) and a composition (see Theorem 4) result.

Theorem 2 (see [17, Proposition 4.5]). *For every SFSG* G*, there exist two deterministic MBOT* G_1 *and* G_2 *such that* $\tau_G = \tau_{G_1}^{-1} ; \tau_{G_2}$.

Proof. Let $G = (N, \Sigma, I, R, B)$ be the original SFSG. Without loss of generality, we can assume that I is a singleton. Whenever we explicitly list nonterminals like $\{n_1, \ldots, n_k\}$, we assume that $n_1 <_N \cdots <_N n_k$. We construct the two MBOT $G_1 = (N, \Sigma \cup B, I, R \cup R', B')$ and $G_2 = (N, \Sigma \cup B, I, R \cup R', B'')$ with

- $R' = \{n \to \chi(n_1, \ldots, n_k) \mid \chi = n \to (\boldsymbol{\ell}, \boldsymbol{r}) \in B, \text{var}(\chi) = \{n_1, \ldots, n_k\}\}$,
- $B' = \{n \to (\chi(n_1, \ldots, n_k), \boldsymbol{\ell}) \mid \chi = n \to (\boldsymbol{\ell}, \boldsymbol{r}) \in B, \text{var}(\chi) = \{n_1, \ldots, n_k\}\}$, and
- $B'' = \{n \to (\chi(n_1, \ldots, n_k), \boldsymbol{r}) \mid \chi = n \to (\boldsymbol{\ell}, \boldsymbol{r}) \in B, \text{var}(\chi) = \{n_1, \ldots, n_k\}\}$.

Obviously, both G_1 and G_2 are classic MBOT in one-symbol normal form, and moreover, they are deterministic. It only remains to prove that $\tau_G = \tau_{G_1}^{-1} ; \tau_{G_2}$. A straightforward induction can be used to prove that G_1 and G_2 translate derivation trees of $D(G)$ to the corresponding input and output tree, respectively. Since each derivation tree $d \in D(G)$ uniquely determines the corresponding input and the output tree, we immediately obtain the statement. A more detailed proof can be found in [17]. □

Corollary 1 (of Theorem 2). *The derivation tree language* $D(G)$ *of an SFSG* G *is regular.*

Proof. By the proof of Theorem 2, there exist classic MBOT that translate the derivation trees to the corresponding input and output tree. Moreover, by [4, Theorem 25] the domain of each MBOT is regular, which yields the result. □

Note that in the proof of Theorem 2 the rule χ uniquely determines the nonterminal n. Nevertheless, the constructed MBOT have (potentially) several nonterminals as we need to check that the behavior of the original SFSG is properly matched. In fact, it follows straightforwardly from the proof of Theorem 2 that each SFSG can be characterized by a regular derivation tree language and two deterministic MBOT mapping the derivation trees to the input and output trees. This view essentially coincides with the bimorphism approach of [1] (essentially, SFSG are equally expressive the bimorphisms of [1], in which both the input and output morphisms are allowed to be k-morphisms). We will reuse this characterization, so let us make it more explicit.

Theorem 3. $\text{SFSG} = \text{d-MBOT}^{-1} ; \text{FTA} ; \text{d-MBOT}$, *where* d-MBOT *is the class of all tree translations computed by deterministic MBOT.*

Now we are ready to state our composition result. We first prove it using several known results on decompositions and compositions together with a few new results. However, for the reader's benefit, we will present an fully integrated construction and an example after the next theorem.

Theorem 4. $\text{MBOT}^{-1} ; \text{MBOT} \subseteq \text{SFSG}$.

Proof. Let G_1 and G_2 be the given MBOT. By Lemma 1 we can assume without loss of generality that G_1 and G_2 are classic MBOT in one-symbol normal form. By the construction of [4, Lemma 6] applied to both G_1 and G_2 we obtain that

$$\tau_{G_1} = d_1^{-1} ; \text{id}_{L_1} ; \tau_{G'_1} \qquad \text{and} \qquad \tau_{G_1} = d_2^{-1} ; \text{id}_{L_2} ; \tau_{G'_2}$$

for some delabelings d_1 and d_2, regular tree languages $L_1, L_2 \in \text{Reg}$, and deterministic MBOT G'_1 and G'_2. Our approach is displayed in Fig. 3. Consequently,

$$\tau_{G_1}^{-1} ; \tau_{G_2} = (d_1^{-1} ; \text{id}_{L_1} ; \tau_{G'_1})^{-1} ; (d_2^{-1} ; \text{id}_{L_2} ; \tau_{G'_2}) = (\tau_{G'_1}^{-1} ; \text{id}_{L_1} ; d_1) ; (d_2^{-1} ; \text{id}_{L_2} ; \tau_{G'_2})$$

Now we show that $d_1 ; d_2^{-1} = e_2^{-1} ; e_1$ for some delabelings e_1 and e_2 in the spirit of [3, Sect. II-1-4-2-1]. Let $\Sigma' = \{\underline{\sigma} \mid \sigma \in \Sigma, d_1(\sigma) = \square\}$ be the ranked alphabet containing (same-rank) copies of the elements of Σ that are erased by d_1. Similarly, let $\Sigma'' = \{\overline{\sigma} \mid \sigma \in \Sigma, d_2(\sigma) = \square\}$ contain copies of those elements that are erased by d_2. Moreover, let

$$\Sigma''' = \{\langle \sigma, \sigma' \rangle \mid \sigma, \sigma' \in \Sigma, d_1(\sigma) = d_2(\sigma') \neq \square\}$$

and $\Delta = \Sigma' \cup \Sigma'' \cup \Sigma'''$. Then we construct delabelings $e_1, e_2 \colon T_\Delta \to T_\Sigma$ as follows:

$$\begin{array}{lll} e_2(\underline{\sigma}) = \sigma & \qquad e_2(\overline{\sigma}) = \square & \qquad e_2(\langle \sigma, \sigma' \rangle) = \sigma \\ e_1(\underline{\sigma}) = \square & \qquad e_2(\overline{\sigma}) = \sigma & \qquad e_2(\langle \sigma, \sigma' \rangle) = \sigma' \end{array}$$

for all $\sigma, \sigma' \in \Sigma$ provided that the listed elements belong to Σ', Σ'', and Σ''', respectively. We omit the formal proof of $d_1 ; d_2^{-1} = e_2^{-1} ; e_1$, but it can be achieved by a simple induction. So far we thus obtained

$$\tau_{G_1}^{-1} ; \tau_{G_2} = (\tau_{G'_1}^{-1} ; \text{id}_{L_1} ; d_1) ; (d_2^{-1} ; \text{id}_{L_2} ; \tau_{G'_2}) = (\tau_{G'_1}^{-1} ; \text{id}_{L_1} ; e_2^{-1}) ; (e_1 ; \text{id}_{L_2} ; \tau_{G'_2})$$

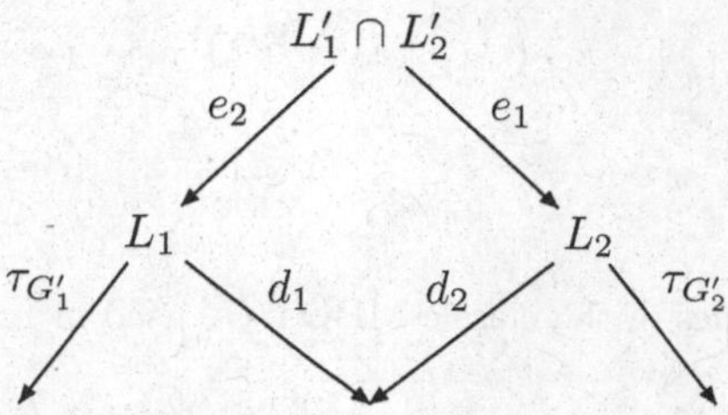

Fig. 3. Illustration of the approach used in the proof of Theorem 4

by the exchange of the delabelings. Now let $L'_1 = e_2^{-1}(L_1)$ and $L'_2 = e_1^{-1}(L_2)$. Clearly, both L'_1 and L'_2 are regular, and also $L'_1 \cap L'_2$ is regular [7,8]. Thus

$$\tau_{G_1}^{-1} ; \tau_{G_2} = (\tau_{G'_1}^{-1} ; e_2^{-1}) ; \mathrm{id}_{L'_1 \cap L'_2} ; (e_1 ; \tau_{G'_2}) \ ,$$

which can be simplified to $\tau_{G''_1}^{-1} ; \mathrm{id}_{L'_1 \cap L'_2} ; \tau_{G''_2}$ because we can compose the delabelings e_1 and e_2 with the deterministic MBOT G'_1 and G'_2 to obtain the deterministic MBOT G''_1 and G''_2, respectively, using [4, Theorem 23]. With this final step, we obtain a form suitable for Theorem 3, so $\tau_{G_1}^{-1} ; \tau_{G_2} \in$ SFSG. □

Corollary 2 (of Theorems 2 and 4). SFSG = MBOT^{-1} ; MBOT.

As mentioned, we provide an explicit construction for the composition of an inverse MBOT with an MBOT into an SFSG. Our construction follows the general approach of translating the output of the first MBOT with the help of the second MBOT as also demonstrated in [4].

Definition 7. *Let $G_1 = (N_1, \Sigma, I_1, R_1, B_1)$ and $G_2 = (N_2, \Sigma, I_2, R_2, B_2)$ be classic MBOT such that $N_1 \cap N_2 = \emptyset$. Moreover, let $G'_1 = (N_1, \Sigma, I_1, R_1)$ and $G'_2 = (N_2, \Sigma, I_2, R_2)$ be the underlying regular tree grammars, respectively. We construct the composed SFSG $(G_1^{-1} ; G_2) = (N_1 \times N_2, \Sigma, I_1 \times I_2, R, B)$ such that*

- *the set R of rules is given by:*
  - $\langle n_1, n_2 \rangle \to \langle n_1, n'_2 \rangle \in R$ *for every* $n_1 \in N_1$ *and* $n_2, n'_2 \in N_2$,
  - $\langle n_1, n_2 \rangle \to \langle n'_1, n_2 \rangle \in R$ *for every* $n_1, n'_1 \in N_1$ *and* $n_2 \in N_2$,
  - $\langle n_1, n_2 \rangle \to r(f_1)$ *with* $r(f_1) = r[n \leftarrow \langle n, f_1(n) \rangle \mid n \in \mathrm{var}(r)] \in R$ *for every rule* $\rho = n_1 \to r \in R_1$, $n_2 \in N_2$, *and injection* $f_1 \colon \mathrm{var}(r) \to N_2$,
  - $\langle n_1, n_2 \rangle \to r(f_2)$ *with* $r(f_2) = r[n \leftarrow \langle f_2(n), n \rangle \mid n \in \mathrm{var}(r)] \in R$ *for every rule* $\rho = n_2 \to r \in R_2$, $n_1 \in N_1$, *and injection* $f_2 \colon \mathrm{var}(r) \to N_1$,
  - *and no further rules are in R, and*
- *the set B of aligned rules is given by:*
  - $\langle n_1, n_2 \rangle \to (\boldsymbol{r}[n'_1 \leftarrow \langle n'_1, n_2 \rangle], \langle n'_1, n_2 \rangle) \in B$ *for every aligned rule* $n_1 \to (n'_1, \boldsymbol{r}) \in B_1$ *with* $n'_1 \in N_1$ *and* $n_2 \in N_2$,
  - $\langle n_1, n_2 \rangle \to (\langle n_1, n'_2 \rangle, \boldsymbol{r}[n'_2 \leftarrow \langle n_1, n'_2 \rangle]) \in B$ *for every aligned rule* $n_2 \to (n'_2, \boldsymbol{r}) \in B_2$ *with* $n'_2 \in N_2$ *and* $n_1 \in N_1$,

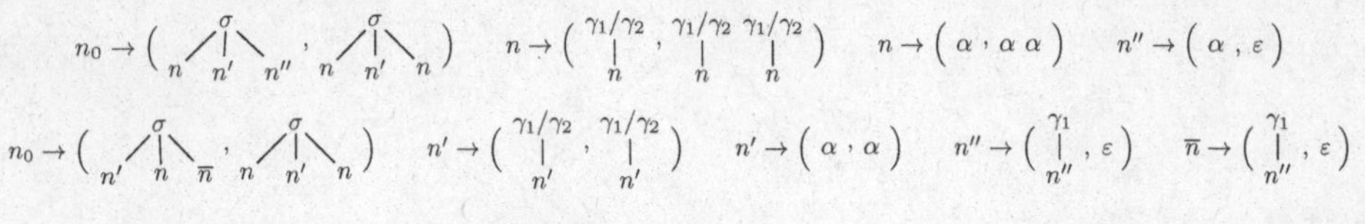

$$n_0 \to \left(\begin{array}{c} \sigma \\ \diagup | \diagdown \\ n \quad n' \quad n'' \end{array} , \begin{array}{c} \sigma \\ \diagup | \diagdown \\ n \quad n' \quad n \end{array} \right) \qquad n \to \left(\begin{array}{ccc} \gamma_1/\gamma_2 & \gamma_1/\gamma_2 & \gamma_1/\gamma_2 \\ | & | & | \\ n & n & n \end{array} \right) \qquad n \to \left(\alpha \cdot \alpha\,\alpha \right) \qquad n'' \to \left(\alpha \,,\, \varepsilon \right)$$

$$n_0 \to \left(\begin{array}{c} \sigma \\ \diagup | \diagdown \\ n' \quad n \quad \overline{n} \end{array} , \begin{array}{c} \sigma \\ \diagup | \diagdown \\ n \quad n' \quad n \end{array} \right) \qquad n' \to \left(\begin{array}{cc} \gamma_1/\gamma_2 & \gamma_1/\gamma_2 \\ | & | \\ n' & n' \end{array} \right) \qquad n' \to \left(\alpha \cdot \alpha \right) \qquad n'' \to \left(\begin{array}{c} \gamma_1 \\ | \\ n'' \end{array} , \varepsilon \right) \qquad \overline{n} \to \left(\begin{array}{c} \gamma_1 \\ | \\ n'' \end{array} , \varepsilon \right)$$

Fig. 4. Rules of the classic MBOT G_1 used in Example 3

$$m_0 \to \left(\begin{array}{c} \sigma \\ \diagup | \diagdown \\ m \quad m' \quad m'' \end{array} , \begin{array}{c} \sigma \\ \diagup | \diagdown \\ m' \quad \alpha \quad m' \end{array} \right) \qquad m' \to \left(\begin{array}{ccc} \gamma_1/\gamma_2 & \gamma_1/\gamma_2 & \gamma_1/\gamma_2 \\ | & | & | \\ m' & m' & m' \end{array} \right) \qquad m' \to \left(\alpha \cdot \alpha\,\alpha \right)$$

$$m \to \left(\begin{array}{c} \gamma_1/\gamma_2 \\ | \\ m \end{array} , \varepsilon \right) \qquad m \to \left(\alpha \,,\, \varepsilon \right) \qquad m'' \to \left(\begin{array}{c} \gamma_2 \\ | \\ n'' \end{array} , \varepsilon \right) \qquad m'' \to \left(\alpha \,,\, \varepsilon \right)$$

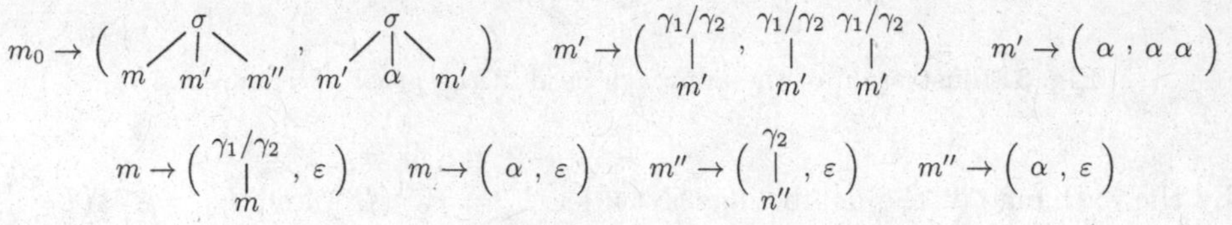

Fig. 5. Rules of the classic MBOT G_2 used in Example 3

- $\chi = \langle n_1, n_2 \rangle \to (\ell(f_1), r(f_2)) \in B$ *for all aligned rules* $n_1 \to (r, \ell) \in B_1$ *and* $n_2 \to (r', r) \in B_2$, *and injective mappings* $f_1 \colon \mathrm{var}(r) \to N_2$ *and* $f_2 \colon \mathrm{var}(r') \to N_1$ *such that* $r(f_1) = r'(f_2)$ *and* $L(G_1')_{n_1'} \cap L(G_2')_{n_2'} \neq \emptyset$ *for all omitted nonterminals* $\langle n_1', n_2' \rangle \in \mathrm{var}(r(f_1)) \setminus \mathrm{var}(\chi)$,[1]
- *and no further aligned rules are in* B.

Let us illustrate the construction on an example.

Example 3. Let $G_1 = (N, \Sigma, \{n_0\}, R_1, B_1)$ be the classic MBOT with nonterminals $N = \{n_0, n, n', n'', \overline{n}\}$, $\Sigma = \{\alpha^{(0)}, \gamma_1^{(1)}, \gamma_2^{(1)}, \sigma^{(3)}\}$, and the rules R_1 and aligned rules B_1 that are depicted in Fig. 4. Let $G_2 = (M, \Sigma, \{m_0\}, R_2, B_2)$ be the classic MBOT with nonterminals $M = \{m_0, m, m', m''\}$ and the rules R_2 and aligned rules B_2 depicted in Fig. 5. The SFSG $G_1^{-1} ; G_2$ is essentially the SFSG of Example 1, but we will explain the construction of two aligned rules. The aligned rule $\langle n_0, m_0 \rangle \to (\sigma(\langle n, m \rangle, \langle n', m' \rangle, \langle n, m \rangle), \sigma(\langle n', m' \rangle, \alpha, \langle n', m' \rangle))$ is constructed from the first aligned rule of G_1 (left, top row in Fig. 4) and the first aligned rule of G_2 (left, top row in Fig. 5). During the overlay of the left-hand sides also the state $\langle n'', m'' \rangle$ is created. Since the languages of n'' and m'' both contain the tree α, the previous aligned rule can be constructed. The process is illustrated in Fig. 6. However, if we want to use the left rule in the second row in Fig. 4 instead, then we can construct

$$\langle n_0, m_0 \rangle \to (\sigma(\langle n, m' \rangle, \langle n', m \rangle, \langle n, m' \rangle), \sigma(\langle n, m' \rangle, \alpha, \langle n, m' \rangle)) \ ,$$

but it is not in the composition because the state $\langle \overline{n}, m'' \rangle$ combines the states $\overline{n}$ and m'', which have an empty intersection.

We conclude with some further properties of SFSG and their consequences for MBOT using our main result of Corollary 2. In particular, it is known [9] that the

---

[1] As usual $\ell(f_1) = \ell_1(f_1) \cdots \ell_k(f_1)$ provided that $\ell = \ell_1 \cdots \ell_k$.

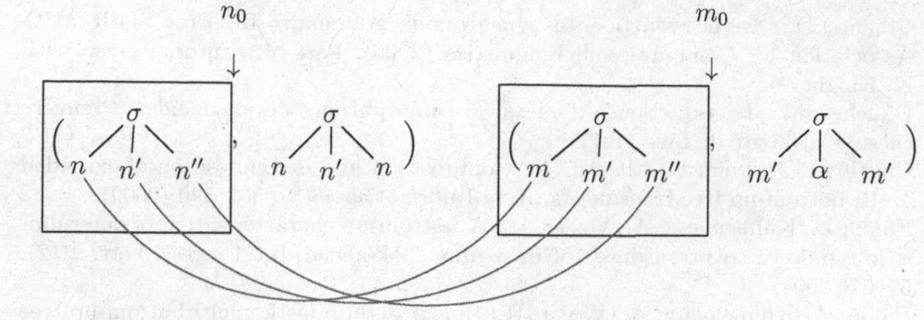

Fig. 6. Illustration of the composition construction (see Example 3). The matching happens inside the boxes and the obtained linked states are paired in the left-hand and right-hand side outside the box.

Table 1. Complexity results for a SFSG G and input strings (w_1, w_2) and trees (t_1, t_2), where $\text{rk}(G)$ is the length of the longest sequence in an aligned rule of G

problem	string level	tree level
Parsing	$\mathcal{O}(\lvert G\rvert \cdot (\lvert w_1\rvert \cdot \lvert w_2\rvert)^{2\text{rk}(G)+2})$	$\mathcal{O}(\lvert G\rvert \cdot \lvert t_1\rvert \cdot \lvert t_2\rvert)$
Translation	$\mathcal{O}(\lvert G\rvert \cdot \lvert w_1\rvert^{2\text{rk}(G)+2})$	$\mathcal{O}(\lvert G\rvert \cdot \lvert t_1\rvert)$

output string language of an MBOT is an LCFRS [20,18]. Using Corollary 2, we can conclude that both the input and the output string language of an SFSG are LCFRS. Moreover, we can import several complexity results from MBOT [14] to SFSG as indicated in Table 1.

Theorem 5 (see [16, Example 5]). SFSG *is not closed under composition.*

Corollary 3. MBOT ; $\text{MBOT}^{-1} \not\subseteq$ SFSG.

Proof. Let us assume that (†): MBOT ; $\text{MBOT}^{-1} \subseteq$ SFSG. Then

$$\begin{aligned} &\text{SFSG} ; \text{SFSG} \\ &\subseteq (\text{MBOT}^{-1} ; \text{MBOT}) ; (\text{MBOT}^{-1} ; \text{MBOT}) \subseteq \text{MBOT}^{-1} ; \text{SFSG} ; \text{MBOT} \\ &\subseteq \text{MBOT}^{-1} ; (\text{MBOT}^{-1} ; \text{MBOT}) ; \text{MBOT} \quad \subseteq \text{MBOT}^{-1} ; \text{MBOT} = \text{SFSG} \end{aligned}$$

using Corollary 2, (†), Corollary 2, the closure under composition for MBOT [4, Theorem 23], and Corollary 2 once more. However, the result contradicts Theorem 5, thus (†) is false, proving the result. □

References

1. Arnold, A., Dauchet, M.: Morphismes et bimorphismes d'arbres. Theor. Comput. Sci. 20(1), 33–93 (1982)

2. Chiang, D.: An introduction to synchronous grammars. In: Proc. 44th ACL. Association for Computational Linguistics (2006), Part of a tutorial given with K. Knight
3. Dauchet, M.: Transductions de forêts — Bimorphismes de magmoïdes. Première thèse, Université de Lille (1977)
4. Engelfriet, J., Lilin, E., Maletti, A.: Composition and decomposition of extended multi bottom-up tree transducers. Acta Informatica 46(8), 561–590 (2009)
5. Fülöp, Z., Kühnemann, A., Vogler, H.: A bottom-up characterization of deterministic top-down tree transducers with regular look-ahead. Inf. Process. Lett. 91(2), 57–67 (2004)
6. Fülöp, Z., Kühnemann, A., Vogler, H.: Linear deterministic multi bottom-up tree transducers. Theor. Comput. Sci. 347(1-2), 276–287 (2005)
7. Gécseg, F., Steinby, M.: Tree Automata. Akadémiai Kiadó, Budapest (1984)
8. Gécseg, F., Steinby, M.: Tree languages. In: Rozenberg, G., Salomaa, A. (eds.) Handbook of Formal Languages, vol. 3, pp. 1–68. Springer (1997)
9. Gildea, D.: On the string translations produced by multi bottom-up tree transducers. Computational Linguistics 38(3), 673–693 (2012)
10. Knight, K., Graehl, J.: An overview of probabilistic tree transducers for natural language processing. In: Gelbukh, A. (ed.) CICLing 2005. LNCS, vol. 3406, pp. 1–24. Springer, Heidelberg (2005)
11. Lilin, E.: Propriétés de clôture d'une extension de transducteurs d'arbres déterministes. In: Astesiano, E., Böhm, C. (eds.) CAAP 1981. LNCS, vol. 112, pp. 280–289. Springer, Heidelberg (1981)
12. Maletti, A.: Compositions of extended top-down tree transducers. Inform. and Comput. 206(9-10), 1187–1196 (2008)
13. Maletti, A.: Why synchronous tree substitution grammars? In: Proc. HLT-NAACL. Association for Computational Linguistics, pp. 876–884 (2010)
14. Maletti, A.: An alternative to synchronous tree substitution grammars. J. Nat. Lang. Engrg. 17(2), 221–242 (2011)
15. Maletti, A.: How to train your multi bottom-up tree transducer. In: Proc. 49th ACL, pp. 825–834. Association for Computational Linguistics (2011)
16. Radmacher, F.G.: An automata theoretic approach to rational tree relations. In: Geffert, V., Karhumäki, J., Bertoni, A., Preneel, B., Návrat, P., Bieliková, M. (eds.) SOFSEM 2008. LNCS, vol. 4910, pp. 424–435. Springer, Heidelberg (2008)
17. Raoult, J.C.: Rational tree relations. Bull. Belg. Math. Soc. Simon Stevin 4(1), 149–176 (1997)
18. Seki, H., Matsumura, T., Fujii, M., Kasami, T.: On multiple context-free grammars. Theoretical Computer Science 88(2), 191–229 (1991)
19. Sun, J., Zhang, M., Tan, C.L.: A non-contiguous tree sequence alignment-based model for statistical machine translation. In: Proc. 47th ACL, pp. 914–922. Association for Computational Linguistics (2009)
20. Vijay-Shanker, K., Weir, D.J., Joshi, A.K.: Characterizing structural descriptions produced by various grammatical formalisms. In: Proc. 25th ACL, pp. 104–111. Association for Computational Linguistics (1987)
21. Zhang, M., Jiang, H., Aw, A., Li, H., Tan, C.L., Li, S.: A tree sequence alignment-based tree-to-tree translation model. In: Proc. 46th ACL, pp. 559–567. Association for Computational Linguistics (2008)
22. Zhang, M., Jiang, H., Li, H., Aw, A., Li, S.: Grammar comparison study for translational equivalence modeling and statistical machine translation. In: Proc. 22nd CoLing, pp. 1097–1104. Association for Computational Linguistics (2008)

Characterizations of Weighted First-Order Logics over Semirings

Eleni Mandrali* and George Rahonis

Department of Mathematics
Aristotle University of Thessaloniki
54124, Thessaloniki, Greece
{elemandr,grahonis}@math.auth.gr

Abstract. We generalize, in the weighted setup over idempotent, zero-divisor free and totally commutative complete semirings, the expressive equivalence of linear temporal logic, first-order logic, star-free expressions and counter-free Büchi automata.

Keywords: Weighted LTL, weighted FO logic, star-free series, counter-free weighted Büchi automata.

1 Introduction

Linear temporal logic (*LTL* for short) and its several alternatives serve as specification languages in model checking for real world applications [1,12,15]. *LTL* definable languages have several characterizations, namely they coincide with *FO* (first-order) logic definable, aperiodic, star-free, and counter-free languages. It is well-known that all these characterizations hold for finitary and infinitary languages as well (cf. for instance [3]).

The aforementioned equivalence, excluding the counter-freeness characterization, has been established in [5] in the setup of arbitrary bounded lattices. A quantitative *LTL*, in the framework of regular cost functions over finite words, has been introduced in [10], and in [11], the expressive equivalence of that logic, cost *FO* logic, and very-weak alternating automata, over finite and infinite words, has been proved. Clearly, the last few years, there is an increasing interest in lifting fundamental results from classical to quantitative models. This is motivated by the need to create model checking tools which incorporate quantitative features. Recently, in [13], in the framework of infinitary series over the max-plus semiring with discounting, we established the coincidence of (a fragment of) *LTL* definability, of (a fragment of) *FO* logic definability, of star-freeness and of (a fragment of) counter-freeness.

* Research of the first author has been co-financed by the European Union (European Social Fund – ESF) and Greek national funds through the Operational Program "Education and Lifelong Learning" of the National Strategic Reference Framework (NSRF) - Research Funding Program: Heracleitus II. Investing in knowledge society through the European Social Fund.

T. Muntean, D. Poulakis, and R. Rolland (Eds.): CAI 2013, LNCS 8080, pp. 247–259, 2013.

In this paper, we develop the theory of [3] within the setup of arbitrary idempotent and zero-divisor free semirings which satisfy concrete completeness' axioms. More precisely, we consider a weighted *LTL*, a weighted *FO* logic, ω-star-free series, and counter-free weighted Büchi automata, and we prove the expressive equivalence of (fragments) of all these objects.

2 Preliminaries

Let A be an alphabet, i.e., a finite nonempty set. As usually, we denote by A^* (resp. A^ω) the set of all finite (resp. infinite) words over A and $A^+ = A^* \setminus \{\varepsilon\}$, where ε is the empty word. For every infinite word $w = a_0a_1\ldots$, which is written also as $w = w(0)w(1)\ldots$, we denote by $w_{\geq i}$ the suffix $w(i)w(i+1)\ldots$ of w, for every $i \geq 0$. Throughout the paper A will denote an alphabet.

Let $(K, +, \cdot, 0, 1)$ be a *semiring* which will be simply denoted by K if the operations and the constant elements are understood. The semiring K is called *commutative* iff $k \cdot k' = k' \cdot k$ for every $k, k' \in K$. It is called *additively idempotent* (or simply *idempotent*), if $k + k = k$ for every $k \in K$. Moreover, the semiring K is *zero-sum free* (resp. *zero-divisor free*) if $k + k' = 0$ implies $k = k' = 0$ (resp. $k \cdot k' = 0$ implies $k = 0$ or $k' = 0$) for every $k, k' \in K$. It is well known that every idempotent semiring is necessarily zero-sum free.

Next assume that the semiring K is equipped, for every index set I, with infinitary sum operations $\sum_I : K^I \to K$, such that for every family $(k_i \mid i \in I)$ of elements of K and $k \in K$ we have

$$\sum_{i\in\emptyset} k_i = 0, \quad \sum_{i\in\{j\}} k_i = k_j, \quad \sum_{i\in\{j,l\}} k_i = k_j + k_l \text{ for } j \neq l,$$

$$\sum_{j\in J}\Big(\sum_{i\in I_j} k_i\Big) = \sum_{i\in I} k_i, \text{ if } \textstyle\bigcup_{j\in J} I_j = I \text{ and } I_j \cap I_{j'} = \emptyset \text{ for } j \neq j',$$

$$\sum_{i\in I}(k \cdot k_i) = k \cdot \Big(\sum_{i\in I} k_i\Big), \quad \sum_{i\in I}(k_i \cdot k) = \Big(\sum_{i\in I} k_i\Big) \cdot k.$$

Then the semiring K together with the operations $\sum_I$ is called *complete* [6,9]. Moreover, a complete semiring is said to be *totally complete* [8], if it is endowed with a countably infinite product operation satisfying for every sequence $(k_i \mid i \geq 0)$ of elements of K the subsequent conditions:

$$\prod_{i\geq 0} 1 = 1, \quad \prod_{i\geq 0} k_i = \prod_{i\geq 0} k_i'$$

$$k_0 \cdot \prod_{i\geq 0} k_{i+1} = \prod_{i\geq 0} k_i, \quad \prod_{j\geq 1}\sum_{i\in I_j} k_i = \sum_{(i_1,i_2,\ldots)\in I_1\times I_2\times\ldots} \prod_{j\geq 1} k_{i_j},$$

where in the second equality $k_0' = k_0 \cdot \ldots \cdot k_{n_1}, k_2' = k_{n_1+1} \cdot \ldots \cdot k_{n_2}, \ldots$ for an increasing sequence $0 < n_1 < n_2 < \ldots$, and in the last equality $I_1, I_2, \ldots$ are arbitrary index sets.

Furthermore, we will call a totally complete semiring K *totally commutative complete* if it satisfies the statement:

$$\prod_{i\geq 0}(k_i \cdot k_i') = \left(\prod_{i\geq 0} k_i\right) \cdot \left(\prod_{i\geq 0} k_i'\right).$$

Obviously a totally commutative complete semiring is commutative.

Example 1. The following semirings are totally commutative complete, and all but the first one are idempotent. The semiring $(\mathbb{N}\cup\{\infty\},+,\cdot,0,1)$ of *extended natural numbers* [7], the *min-plus semiring* $(\mathbb{R}_+\cup\{\infty\},\min,+,\infty,0)$ where $\mathbb{R}_+ = \{r\in\mathbb{R} \mid r\geq 0\}$, the *max-plus semiring* $(\mathbb{R}_+\cup\{\pm\infty\},\max,+,-\infty,0)$, every completely distributive complete lattice with the operations supremum and infimum, in particular the *fuzzy semiring* $F = ([0,1],\sup,\inf,0,1)$.

In the rest of the paper K will denote a totally commutative complete, idempotent and zero-divisor free semiring.

Let Q be a set. A *formal series* (or simply *series*) *over* Q *and* K is a mapping $s : Q \to K$. For every $v\in Q$ we write (s,v) for the value $s(v)$ and refer to it as the *coefficient of* s *on* v. The *support of* s is the set $supp(s) = \{v\in Q \mid (s,v)\neq 0\}$. The *constant series* $\widetilde{k}$ $(k\in K)$ is defined, for every $v\in Q$, by $\left(\widetilde{k},v\right) = k$. We denote by $K\langle\langle Q\rangle\rangle$ the class of all series over Q and K.

Let $s,r\in K\langle\langle Q\rangle\rangle$ and $k\in K$. The *sum* $s+r$, the *scalar products* ks and sk as well as the *Hadamard product* $s\odot r$ are defined elementwise by $(s+r,v) = (s,v)+(r,v)$, $(ks,v) = k\cdot(s,v)$, $(sk,v) = (s,v)\cdot k$, and $(s\odot r,v) = (s,v)\cdot(r,v)$ for every $v\in Q$. Abusing notations, if $P\subseteq Q$, then we shall identify the restriction $s|_P$ of s on P with the series $s\odot 1_P$, where as usually we denote by 1_P the characteristic series of P. It is well-known that the structure $\left(K\langle\langle Q\rangle\rangle,+,\odot,\widetilde{0},\widetilde{1}\right)$ is a commutative semiring. In our paper, we work with the semirings $K\langle\langle A^*\rangle\rangle$ and $K\langle\langle A^\omega\rangle\rangle$ of finitary and infinitary series over A and K, respectively.

Let B be another alphabet and $h : A^*\to B^*$ be a nondeleting homomorphism, i.e., $h(a)\neq\varepsilon$ for each $a\in A$. Then h can be extended to a mapping $h : A^\omega\to B^\omega$ by letting $h(w) = (h(w(i)))_{i\geq 0}$ for every $w\in A^\omega$. Moreover, h is extended to a mapping $h : K\langle\langle A^*\rangle\rangle \to K\langle\langle B^*\rangle\rangle$ as follows. For every $s\in K\langle\langle A^*\rangle\rangle$ the series $h(s)\in K\langle\langle B^*\rangle\rangle$ is given by $(h(s),u) = \sum_{w\in h^{-1}(u)}(s,w)$ for every $u\in B^*$. Since K is complete, h is also extended to a mapping $h : K\langle\langle A^\omega\rangle\rangle \to K\langle\langle B^\omega\rangle\rangle$ in the same way as for finitary series. If $r\in K\langle\langle B^*\rangle\rangle$ (resp. $r\in K\langle\langle B^\omega\rangle\rangle$), then the series $h^{-1}(r)\in K\langle\langle A^*\rangle\rangle$ (resp. $h^{-1}(r)\in K\langle\langle A^\omega\rangle\rangle$) is determined by $(h^{-1}(r),w) = (r,h(w))$ for every $w\in A^*$ (resp. $w\in A^\omega$).

3 Weighted LTL and FO Logic

For every letter $a\in A$ we consider a proposition p_a and we let $AP = \{p_a \mid a\in A\}$. As usually, for every $p\in AP$ we identify $\neg\neg p$ with p.

Definition 1. *The syntax of formulas of the* weighted linear temporal logic (LTL *for short)* over A and K *is given by the grammar*

$$\varphi ::= k \mid p_a \mid \neg\varphi \mid \varphi \vee \varphi \mid \varphi \wedge \varphi \mid \bigcirc\varphi \mid \varphi U \varphi \mid \Box\varphi$$

where $k \in K$ and $p_a \in AP$. We denote by $LTL(K, A)$ the set of all such weighted LTL *formulas φ.*

We represent the semantics $\|\varphi\|$ of formulas $\varphi \in LTL(K, A)$ as infinitary series in $K\langle\langle A^\omega\rangle\rangle$.

Definition 2. *Let $\varphi \in LTL(K, A)$. The* semantics *of φ is a series $\|\varphi\| \in K\langle\langle A^\omega\rangle\rangle$ which is defined inductively as follows. For every $w \in A^\omega$ we set*

- $(\|k\|, w) = k$,
- $(\|p_a\|, w) = \begin{cases} 1 \text{ if } w(0) = a \\ 0 \text{ otherwise} \end{cases}$, - $(\|\neg\varphi\|, w) = \begin{cases} 1 \text{ if } (\|\varphi\|, w) = 0 \\ 0 \text{ otherwise} \end{cases}$,
- $(\|\varphi \vee \psi\|, w) = (\|\varphi\|, w) + (\|\psi\|, w)$, - $(\|\varphi \wedge \psi\|, w) = (\|\varphi\|, w) \cdot (\|\psi\|, w)$,
- $(\|\bigcirc\varphi\|, w) = (\|\varphi\|, w_{\geq 1})$, - $(\|\Box\varphi\|, w) = \prod_{i\geq 0} (\|\varphi\|, w_{\geq i})$,
- $(\|\varphi U \psi\|, w) = \sum_{i\geq 0} \left(\left(\prod_{0\leq j<i} (\|\varphi\|, w_{\geq j}) \right) \cdot (\|\psi\|, w_{\geq i}) \right)$.

The *eventually* operator is defined as in the classical LTL, i.e., by $\Diamond\varphi := 1U\varphi$, and then we have $(\|\Diamond\varphi\|, w) = \sum_{i\geq 0} (\|\varphi\|, w_{\geq i})$ for every $w \in A^\omega$.

The syntactic boolean fragment $bLTL(K, A)$ of $LTL(K, A)$ is given by the grammar

$$\varphi ::= 0 \mid 1 \mid p_a \mid \neg\varphi \mid \varphi \vee \varphi \mid \bigcirc\varphi \mid \varphi U \varphi.$$

For every formula $\varphi \in bLTL(K, A)$ it is easily obtained, by structural induction on φ and using idempotency of K, that $\|\varphi\|$ gets only values in $\{0, 1\}$.

We aim to define a further fragment of $LTL(K, A)$. For this we need some preliminary matter. More precisely, an *atomic-step formula* is an $LTL(K, A)$ formula of the form $\bigvee_{a\in A} (k_a \wedge p_a)$ where $k_a \in K$ and $p_a \in AP$ for every $a \in A$. An *LTL-step formula* is an $LTL(K, A)$ formula of the form $\bigvee_{1\leq i\leq n} (k_i \wedge \varphi_i)$ where $k_i \in K$ and $\varphi_i \in bLTL(K, A)$ for every $1 \leq i \leq n$. We shall denote by $stLTL(K, A)$ the class of *LTL*-step formulas over A and K. Furthermore, we shall denote by $abLTL(K, A)$ the class of *almost boolean LTL* formulas over A and K, i.e., formulas of the form $\bigwedge_{1\leq i\leq n} \varphi_i$ with $\varphi_i \in bLTL(K, A)$ or $\varphi_i = \bigvee_{a\in A} (k_a \wedge p_a)$, for every $1 \leq i \leq n$.

Definition 3. *The fragment $ULTL(K, A)$ of U-*nesting LTL formulas over A and K *is the least class of formulas in $LTL(K, A)$ which is defined inductively in the following way.*

- $k \in ULTL(K, A)$ *for every* $k \in K$.
- $abLTL(K, A) \subseteq ULTL(K, A)$.
- *If* $\varphi \in ULTL(K, A)$, *then* $\neg\varphi \in ULTL(K, A)$.
- *If* $\varphi, \psi \in ULTL(K, A)$, *then* $\varphi \wedge \psi, \varphi \vee \psi \in ULTL(K, A)$.
- *If* $\varphi \in ULTL(K, A)$, *then* $\bigcirc\varphi \in ULTL(K, A)$.
- *If* $\varphi \in bLTL(K, A)$ *or* φ *is an atomic-step formula, then* $\Box\varphi \in ULTL(K, A)$.
- *If* $\varphi \in abLTL(K, A)$ *and* $\psi \in ULTL(K, A)$, *then* $\varphi U \psi \in ULTL(K, A)$.

A series $r \in K\langle\langle A^\omega\rangle\rangle$ is called *ω-ULtl-definable* if there is a formula $\varphi \in ULTL(K, A)$ such that $r = \|\varphi\|$. We shall denote by ω-$ULtl(K, A)$ the class of all ω-$ULtl$-definable series over A and K.

In the sequel, we consider the weighted first-order logic (weighted FO logic, for short), and we show that the class of semantics in a fragment of this logic contains the class ω-$ULtl(K, A)$.

Definition 4. *The syntax of formulas of the* weighted FO logic over A and K *is given by the grammar*

$$\varphi ::= k \mid P_a(x) \mid x \le y \mid \neg\varphi \mid \varphi \vee \varphi \mid \varphi \wedge \varphi \mid \exists x \centerdot \varphi \mid \forall x \centerdot \varphi$$

where $k \in K$ *and* $a \in A$.

We shall denote by $FO(K, A)$ the set of all weighted FO formulas over A and K. In order to define the semantics of $FO(K, A)$ formulas, we recall the notions of extended alphabet and valid assignment (cf. for instance [14]). Let $\mathcal{V}$ be a finite set of first-order variables. For an infinite word $w \in A^\omega$ we let $dom(w) = \omega$. A *$(\mathcal{V}, w)$-assignment* σ is a mapping associating variables from $\mathcal{V}$ to elements of ω. For every $x \in \mathcal{V}$ and $i \in \omega$, we denote by $\sigma[x \to i]$ the $(\mathcal{V}, w)$-assignment which associates i to x and acts as σ on $\mathcal{V} \setminus \{x\}$. We encode pairs (w, σ) for every $w \in A^\omega$ and $(\mathcal{V}, w)$-assignment σ, by using the extended alphabet $A_\mathcal{V} = A \times \{0, 1\}^\mathcal{V}$. Each pair (w, σ) is a word in $A_\mathcal{V}^\omega$ where w is the projection over A and σ is the projection over $\{0, 1\}^\mathcal{V}$. Then σ is called a *valid* $(\mathcal{V}, w)$-assignment whenever for every $x \in \mathcal{V}$ the x-row contains exactly one 1. In this case, we identify σ with the $(\mathcal{V}, w)$-assignment so that for every first-order variable $x \in \mathcal{V}$, $\sigma(x)$ is the position of the 1 on the x-row. It is well-known (cf. [3]) that the set $\mathcal{N}_\mathcal{V} = \{(w, \sigma) \mid w \in A^\omega,\ \sigma \text{ is a valid } (\mathcal{V}, w)\text{-assignment}\}$ is an ω-star-free language over $A_\mathcal{V}$. The set $free(\varphi)$ of free variables in a formula $\varphi \in FO(K, A)$ is defined as usual.

Definition 5. *Let* $\varphi \in FO(K, A)$ *and* $\mathcal{V}$ *be a finite set of variables with* $free(\varphi) \subseteq \mathcal{V}$. *The semantics of* φ *is a series* $\|\varphi\|_\mathcal{V} \in K\langle\langle A_\mathcal{V}^\omega\rangle\rangle$. *Consider an element* $(w, \sigma) \in A_\mathcal{V}^\omega$. *If* σ *is not a valid assignment, then we put* $(\|\varphi\|_\mathcal{V}, (w, \sigma)) = 0$. *Otherwise, we inductively define* $(\|\varphi\|_\mathcal{V}, (w, \sigma)) \in K$ *as follows.*

- $(\|k\|_\mathcal{V}, (w, \sigma)) = k$, - $(\|\neg\varphi\|_\mathcal{V}, (w, \sigma)) = \begin{cases} 1 \text{ if } (\|\varphi\|_\mathcal{V}, (w, \sigma)) = 0 \\ 0 \text{ otherwise} \end{cases}$,
- $(\|P_a(x)\|_\mathcal{V}, (w, \sigma)) = \begin{cases} 1 \text{ if } w(\sigma(x)) = a \\ 0 \text{ otherwise} \end{cases}$,

- $(\|x \leq y\|_{\mathcal{V}}, (w,\sigma)) = \begin{cases} 1 \text{ if } \sigma(x) \leq \sigma(y) \\ 0 \text{ otherwise} \end{cases}$,
- $(\|\varphi \vee \psi\|_{\mathcal{V}}, (w,\sigma)) = (\|\varphi\|_{\mathcal{V}}, (w,\sigma)) + (\|\psi\|_{\mathcal{V}}, (w,\sigma))$,
- $(\|\varphi \wedge \psi\|_{\mathcal{V}}, (w,\sigma)) = (\|\varphi\|_{\mathcal{V}}, (w,\sigma)) \cdot (\|\psi\|_{\mathcal{V}}, (w,\sigma))$,
- $(\|\exists x \,.\, \varphi\|_{\mathcal{V}}, (w,\sigma)) = \sum_{i \geq 0} \left(\|\varphi\|_{\mathcal{V} \cup \{x\}}, (w, \sigma[x \to i]) \right)$,
- $(\|\forall x \,.\, \varphi\|_{\mathcal{V}}, (w,\sigma)) = \prod_{i \geq 0} \left(\|\varphi\|_{\mathcal{V} \cup \{x\}}, (w, \sigma[x \to i]) \right)$.

If $\mathcal{V} = free(\varphi)$, then we simply write $\|\varphi\|$ for $\|\varphi\|_{free(\varphi)}$. The syntactic boolean fragment $bFO(K,A)$ of $FO(K,A)$ is defined by the grammar

$$\varphi ::= 0 \mid 1 \mid P_a(x) \mid x \leq y \mid \neg\varphi \mid \varphi \vee \varphi \mid \exists x \,.\, \varphi.$$

For every formula $\varphi \in bFO(K,A)$ it is easily obtained, by structural induction on φ and using idempotency of K, that $\|\varphi\|$ gets only values in $\{0,1\}$.

Next, we define a fragment of our logic. For this, we recall the notion of an *FO*-step formula from [2]. More precisely, a formula $\varphi \in FO(K,A)$ is an *FO-step formula* if $\varphi = \bigvee_{1 \leq i \leq n} (k_i \wedge \varphi_i)$ with $\varphi_i \in bFO(K,A)$ and $k_i \in K$ for every $1 \leq i \leq n$. Moreover, a formula $\varphi \in FO(K,A)$ is called a *letter-step formula* whenever $\varphi = \bigvee_{a \in A} (k_a \wedge P_a(x))$ with $k_a \in K$ for every $a \in A$. We shall need also the following macros:

- $x = y := x \leq y \wedge y \leq x$,
- $x < y := x \leq y \wedge \neg(x = y)$,
- $z \leq x < y := z \leq x \wedge x < y$,
- $(y \leq x) \to \varphi := \neg(y \leq x) \vee ((y \leq x) \wedge \varphi)$,
- $(z \leq x < y) \to \varphi := \neg(z \leq x < y) \vee ((z \leq x < y) \wedge \varphi)$.

Definition 6. *A formula $\varphi \in FO(K,A)$ will be called* weakly quantified *if whenever φ contains a subformula of the form $\forall x \,.\, \psi$, then ψ is either a boolean or a letter-step formula with free variable x or a formula of the form $y \leq x \to \psi'$ or $z \leq x < y \to \psi'$ where ψ' is a letter-step formula with free variable x.*

We denote by $WQFO(K,A)$ the set of all weakly quantified $FO(K,A)$ formulas over A and K. A series $s \in K\langle\langle A^\omega \rangle\rangle$ is called *ω-wqFo-definable* if there is a sentence $\varphi \in WQFO(K,A)$ such that $s = \|\varphi\|$. We write ω-$wqFo(K,A)$ for the class of all ω-*wqFo*-definable series in $K\langle\langle A^\omega \rangle\rangle$.

By structural induction on $ULTL(K,A)$ formulas, we can show that for every $\varphi \in ULTL(K,A)$ we can construct a $WQFO(K,A)$ formula $\varphi'(x)$ such that $(\|\varphi'(x)\|, (w, [x \to i])) = (\|\varphi\|, w_{\geq i})$ for every $w \in A^\omega, i \geq 0$. By this, we deduce that for every $\varphi \in ULTL(K,A)$ there exists a sentence $\varphi' \in WQFO(K,A)$ such that $\|\varphi\| = \|\varphi'\|$. Therefore, we obtain the next first main result of our paper.

Theorem 1. *ω-$ULtl(K,A) \subseteq \omega$-$wqFo(K,A)$.*

4 Star-Free Series

In this section, we introduce the notions of star-free and ω-star-free series over A and K, and we show that the class of ω-*wqFo*-definable series is contained into the class of ω-star-free series.

Let $L \subseteq A^*$ (resp. $L \subseteq A^\omega$). If L is a singleton, i.e., $L = \{w\}$, then we simply write 1_w for the characteristic series $1_{\{w\}}$. Furthermore, we simply denote by k_L the series $k1_L$ for $k \in K$. The *monomials over A and K* are series of the form $(k_a)_a$ for $a \in A$ and $k_a \in K$. For simplicity, we shall consider also the series of the form k_ε with $k \in K$ as monomials. A series $s \in K\langle\langle A^*\rangle\rangle$ is called a *letter-step series* if $s = \sum_{a\in A} (k_a)_a$ where $a \in A$ and $k_a \in K$ for every $a \in A$. The *complement $\overline{s}$ of a series s* is given by $(\overline{s}, w) = 1$ if $(s, w) = 0$ and 0 otherwise. Let $r, s \in K\langle\langle A^*\rangle\rangle$. The *Cauchy product of r and s* is the series $r \cdot s \in K\langle\langle A^*\rangle\rangle$ defined for every $w \in A^*$ by

$$(r \cdot s, w) = \sum \{(r, u) \cdot (s, v) \mid u, v \in A^*, w = uv\}.$$

The nth-iteration $r^n \in K\langle\langle A^*\rangle\rangle$ $(n \geq 0)$ of a series $r \in K\langle\langle A^*\rangle\rangle$ is defined inductively by

$$r^0 = 1_\varepsilon \quad \text{and} \quad r^{n+1} = r \cdot r^n \quad \text{for } n \geq 0.$$

Then, we have $(r^n, w) = \sum \left\{\prod_{1\leq i\leq n}(r, u_i) \mid u_i \in A^*, w = u_1 \dots u_n\right\}$ for every $w \in A^*$. A series $r \in K\langle\langle A^*\rangle\rangle$ is called *proper* if $(r, \varepsilon) = 0$. If r is proper, then for every $w \in A^*$ and $n > |w|$ we have $(r^n, w) = 0$. The *iteration* $r^+ \in K\langle\langle A^*\rangle\rangle$ *of a proper series* $r \in K\langle\langle A^*\rangle\rangle$ is defined by $r^+ = \sum_{n>0} r^n$. Thus, for every $w \in A^+$ we have $(r^+, w) = \sum_{1\leq n\leq |w|} (r^n, w)$ and $(r^+, \varepsilon) = 0$.

Definition 7. *The class of* star-free series over A and K, *denoted by $sf(K, A)$, is the least class of series containing the monomials (over A and K) and being closed under sum, Hadamard product, complement, Cauchy product, and iteration restricted to letter-step series.*

Next, let $r \in K\langle\langle A^*\rangle\rangle$ be a finitary and $s \in K\langle\langle A^\omega\rangle\rangle$ an infinitary series. Then, the *Cauchy product of r and s* is the infinitary series $r \cdot s \in K\langle\langle A^\omega\rangle\rangle$ defined for every $w \in A^\omega$ by

$$(r \cdot s, w) = \sum \{(r, u) \cdot (s, v) \mid u \in A^*, v \in A^\omega, w = uv\}.$$

The *ω-iteration of a proper finitary series* $r \in K\langle\langle A^*\rangle\rangle$ is the infinitary series $r^\omega \in K\langle\langle A^\omega\rangle\rangle$ which is defined by

$$(r^\omega, w) = \sum \left\{\prod_{i\geq 1}(r, u_i) \mid u_i \in A^*, w = u_1u_2 \dots\right\}$$

for every $w \in A^\omega$.

Definition 8. *The class of* ω-star-free series over A and K, *denoted by ω-$sf(K, A)$, is the least class of infinitary series generated by the monomials (over A and K) by applying finitely many times the operations of sum, Hadamard product, complement, Cauchy product, iteration restricted to letter-step series, and ω-iteration restricted to letter-step series.*

Due to the idempotency of K one can easily show that $L \subseteq A^*$ (resp. $L \subseteq A^\omega$) is a star-free (resp. ω-star-free) language iff $1_L \in sf(K,A)$ (resp. $1_L \in \omega\text{-}sf(K,A)$). Next, we state properties of the classes $sf(K,A)$ and $\omega\text{-}sf(K,A)$.

Proposition 1 (Splitting lemma for finitary series). *Let $s \in sf(K,A)$ and $B, \Gamma \subseteq A$ with $B \cap \Gamma = \emptyset$. Then $s|_{B^*\Gamma B^*} = \sum_{1\le i\le n} \left(s_1^{(i)} \cdot \left(s_2^{(i)} \cdot s_3^{(i)}\right)\right)$ where for every $1 \le i \le n$, $s_1^{(i)}, s_3^{(i)} \in sf(K,B)$ and $s_2^{(i)} = (k_i)_{\gamma_i}$ with $\gamma_i \in \Gamma, k_i \in K$.*

Proposition 2 (Splitting lemma for infinitary series). *Let $s \in \omega\text{-}sf(K,A)$ and $B, \Gamma \subseteq A$ with $B \cap \Gamma = \emptyset$. Then $s|_{B^*\Gamma B^\omega} = \sum_{1\le i\le n} \left(s_1^{(i)} \cdot \left(s_2^{(i)} \cdot s_3^{(i)}\right)\right)$ where for every $1 \le i \le n$, $s_1^{(i)} \in sf(K,B)$, $s_3^{(i)} \in \omega\text{-}sf(K,B)$, and $s_2^{(i)} = (k_i)_{\gamma_i}$ with $\gamma_i \in \Gamma, k_i \in K$.*

Proposition 3. *Let A, B be two alphabets.*

(i) If $h : A \to B$ is a bijection and $s \in sf(K,A)$ (resp. $s \in \omega\text{-}sf(K,A)$), then $h(s) \in sf(K,B)$ (resp. $h(s) \in \omega\text{-}sf(K,B)$).

(ii) If $h : A \to B$ is a strict alphabetic epimorphism and $s \in sf(K,B)$ (resp. $s \in \omega\text{-}sf(K,B)$), then $h^{-1}(s) \in sf(K,A)$ (resp. $h^{-1}(s) \in \omega\text{-}sf(K,A)$).

Our next main result states that $\omega\text{-}wqFo(K,A) \subseteq \omega\text{-}sf(K,A)$. For this, we shall need the following auxiliary lemma.

Lemma 1. *Let $\varphi \in FO(K,A)$ and $\mathcal{V}$ be a finite set of first-order variables containing $free(\varphi)$. If $\|\varphi\|$ is an ω-star-free series, then $\|\varphi\|_{\mathcal{V}}$ is an ω-star-free series.*

Theorem 2. *$\omega\text{-}wqFo(K,A) \subseteq \omega\text{-}sf(K,A)$.*

Proof. (Sketch) The proof is by induction on the structure of weakly quantified FO formulas. For $k \in K$, atomic formulas, and the inductive steps of negation, conjunction, and disjunction we use standard arguments. We argue on the inductive step for the existential operator. To this end, let $\varphi \in FO(K,A)$ such that $\|\varphi\|$ is an ω-star-free series. Let $\mathcal{W} = free(\varphi) \cup \{x\}$ and $\mathcal{V} = free(\exists x.\varphi) = \mathcal{W} \setminus \{x\}$. We define $B, \Gamma \subseteq A_{\mathcal{W}}$, by $B = \{(a,f) \in A_{\mathcal{W}} \mid f(x) = 0\}$ and $\Gamma = \{(a,f) \in A_{\mathcal{W}} \mid f(x) = 1\}$. Since $\|\varphi\|_{\mathcal{W}} \in \omega\text{-}sf(K, A_{\mathcal{W}})$ (by Lemma 1, in case $x \notin free(\varphi)$), by Proposition 2 we get $\|\varphi\|_{\mathcal{W}}|_{B^*\Gamma B^\omega} = \sum\limits_{1\le i\le n} \left(s_1^{(i)} \cdot \left(s_2^{(i)} \cdot s_3^{(i)}\right)\right)$ with $s_1^{(i)} \in sf(K,B)$, $s_3^{(i)} \in \omega\text{-}sf(K,B)$, and $s_2^{(i)} = (k_i)_{\gamma_i}$, where $k_i \in K$, $\gamma_i \in \Gamma$ for every $1 \le i \le n$. It holds

$$\|\exists x.\varphi\| = \left(\sum_{1\le i\le n} \left(h|_B\left(s_1^{(i)}\right) \cdot \left((k_i)_{h(\gamma_i)} \cdot h|_B\left(s_3^{(i)}\right)\right)\right)\right) \odot 1_{\mathcal{N}_{\mathcal{V}}}$$

where $h : A_{\mathcal{W}} \to A_{\mathcal{V}}$ is the strict alphabetic epimorphism assigning $(a, f|_{\mathcal{V}})$ to (a,f) for every $(a,f) \in A_{\mathcal{W}}$. Clearly $h|_B$ is a bijection. By Proposition 3, for every $1 \leq i \leq n$, we get that $h|_B\left(s_1^i\right) \in sf\left(K, A_{\mathcal{V}}\right)$, $h|_B\left(s_3^{(i)}\right) \in \omega$-$sf\left(K, A_{\mathcal{V}}\right)$. Therefore $\|\exists x.\varphi\|$ is an ω-star-free series.

Now, let $\varphi \in FO(K, A)$ being a boolean, or a letter-step formula with free variable x, or $\varphi = (y \leq x) \to \psi$, or $\varphi = (y \leq x < z) \to \psi$ where ψ is a letter-step formula with free variable x. We will show that $\|\forall x.\varphi\|$ is an ω-star-free series. Due to space limitations, we only argue on the case where $\varphi = (y \leq x) \to \bigvee_{a\in A}(k_a \wedge P_a(x))$. We consider the subsets $F = \{(a,0) \mid a \in A\}$ and $F' = \{(a,1) \mid a \in A\}$ of $A_{\{y\}}$. The language F^+ is star-free, hence, the series 1_{F^+} is star-free. Consider the series $s = \sum_{a\in A}\left((k_a)_{(a,0)}\right)$ and $s' = \sum_{a\in A}\left((k_a)_{(a,1)}\right)$ over $A_{\{y\}}$ and K. Then, it holds $\|\forall x.\varphi\| = 1_{F^+} \cdot (s' \cdot s^\omega)$ which proves our claim.

5 Counter-Free Series

In this section, we consider the concept of counter-freeness within weighted (resp. weighted Büchi) automata over A and K. Our models will be nondeterministic. Firstly, we recall the notions of weighted automata and weighted Büchi automata over A and K.

A weighted automaton over A and K is a quadruple $\mathcal{A} = (Q, in, wt, F)$ where Q is the *finite state set*, $in : Q \to K$ is the *initial distribution*, $wt : Q \times A \times Q \to K$ is a mapping assigning *weights* to the transitions of the automaton and $F \subseteq Q$ is the *final state set*.

Given a word $w = a_0 \dots a_{n-1} \in A^*$, a path of $\mathcal{A}$ over w is a finite sequence of transitions $P_w := ((q_i, a_i, q_{i+1}))_{0\leq i\leq n-1}$. The *running weight of* P_w is the value
$rwt(P_w) := \prod_{0\leq i\leq n-1} wt\left((q_i, a_i, q_{i+1})\right)$
and the *weight of* P_w is given by
$weight(P_w) := in(q_0) \cdot rwt(P_w)$.

The path P_w is called *successful* if $q_n \in F$. Then, the *behavior of* $\mathcal{A}$ is the series $\|\mathcal{A}\| : A^* \to K$ which is defined, for every $w \in A^*$, by $(\|\mathcal{A}\|, w) = \sum_{P_w \text{ succ}} weight(P_w)$. A series $r \in K\langle\langle A^*\rangle\rangle$ is called *recognizable* if it is the behavior of a weighted automaton over A and K.

A weighted Büchi automaton $\mathcal{A} = (Q, in, wt, F)$ over A and K is defined as a weighted automaton. Given an infinite word $w = a_0a_1 \ldots \in A^\omega$, a path of $\mathcal{A}$ over w is an infinite sequence of transitions $P_w := \left((q_i, a_i, q_{i+1})\right)_{i\geq 0}$. The *running weight of* P_w is the value
$rwt(P_w) := \prod_{i\geq 0} wt\left((q_i, a_i, q_{i+1})\right)$
and the *weight of* P_w is given by
$weight(P_w) := in(q_0) \cdot rwt(P_w)$.

A path P_w is called *successful* if at least one final state occurs infinitely often along P_w. Then, the *behavior of* $\mathcal{A}$ is the infinitary series $\|\mathcal{A}\| : A^\omega \to K$ whose coefficients are given by $(\|\mathcal{A}\|, w) = \sum_{P_w \text{ succ}} weight(P_w)$, for every $w \in A^\omega$. An infinitary series $r \in K\langle\langle A^\omega\rangle\rangle$ is called *ω-recognizable* if it is the behavior of a weighted Büchi automaton over A and K.

We shall need the following notation. Given a weighted (resp. weighted Büchi) automaton $\mathcal{A} = (Q, in, wt, F)$, a word $w = a_0 \ldots a_{n-1} \in A^*$, and states $q, q' \in Q$, we shall denote by $P_{(q,w,q')}$ a path of $\mathcal{A}$ over w starting at state q and terminating at state q', i.e., $P_{(q,w,q')} = (q, a_0, q_1)\left((q_i, a_i, q_{i+1})\right)_{1 \leq i \leq n-2} (q_{n-1}, a_{n-1}, q')$. Then $rwt\left(P_{(q,w,q')}\right) = wt\left((q, a_0, q_1)\right) \cdot \prod_{1 \leq i \leq n-2} wt\left((q_i, a_i, q_{i+1})\right) \cdot wt\left((q_{n-1}, a_{n-1}, q')\right)$.

Now, we are ready to introduce our counter-free weighted and counter-free weighted Büchi automata.

Definition 9. *A weighted automaton (resp. weighted Büchi automaton) $\mathcal{A} = (Q, in, wt, F)$ over A and K is called* counter-free *(*cfwa, *resp.* cfwBa, *for short) if for every $q \in Q$, $w \in A^*$, and $n \geq 1$, the relation $\sum_{P_{(q,w^n,q)}} rwt\left(P_{(q,w^n,q)}\right) \neq 0$ implies* $\sum_{P_{(q,w^n,q)}} rwt\left(P_{(q,w^n,q)}\right) = \left(\sum_{P_{(q,w,q)}} rwt\left(P_{(q,w,q)}\right)\right)^n$.

A series $r \in K\langle\langle A^* \rangle\rangle$ (resp. $r \in K\langle\langle A^\omega \rangle\rangle$) is called *counter-free* (resp. *ω-counter-free*) if it is accepted by a cfwa (resp. cfwBa) over A and K. We shall denote by $cf(K, A)$ (resp. ω-$cf(K, A)$) the class of all counter-free (resp. ω-counter-free) series over A and K.

Proposition 4. *(i) The class $cf(K, A)$ contains the monomials and it is closed under sum, Hadamard product, complement, Cauchy product, and iteration restricted to letter-step series.*

(ii) The class ω-$cf(K, A)$ is closed under sum, complement, Cauchy product and ω-iteration restricted to letter-step series.

Note that the Cauchy product in Proposition 4(ii) is considered among series in $cf(K, A)$ and ω-$cf(K, A)$.

Next, we introduce the subclass of almost simple counter-free (resp. almost simple ω-counter-free) series and we show that it contains the class $sf(K, A)$ (resp. ω-$sf(K, A)$).

Definition 10. *A cfwa (resp. cfwBa) $\mathcal{A} = (Q, in, wt, F)$ over A and K is called* simple *if for every $q, q', p, p' \in Q$, and $a \in A$, $in(q) \neq 0 \neq in(q')$ implies $in(q) = in(q')$, and $wt((q, a, q')) \neq 0 \neq wt((p, a, p'))$ implies $wt((q, a, q')) = wt((p, a, p'))$. Furthermore, a series $r \in K\langle\langle A^* \rangle\rangle$ (resp. $r \in K\langle\langle A^\omega \rangle\rangle$) is* simple *if it is the behavior of a simple cfwa (resp. cfwBa) over A and K.*

Definition 11.

- *A series $r \in K\langle\langle A^* \rangle\rangle$ is called* almost simple *if $r = \sum_{1 \leq i \leq n} \left(r_1^{(i)} \cdot \ldots \cdot r_{m_i}^{(i)}\right)$ where, for every $1 \leq i \leq n$, $r_1^{(i)}, \ldots, r_{m_i}^{(i)}$ are simple counter-free series over A and K.*
- *A series $r \in K\langle\langle A^\omega \rangle\rangle$ is called* almost simple *if $r = \sum_{1 \leq i \leq n} \left(r_1^{(i)} \cdot \ldots \cdot r_{m_i}^{(i)}\right)$ where, for every $1 \leq i \leq n$, $r_1^{(i)}, \ldots, r_{m_i - 1}^{(i)}$ are simple counter-free series and $r_{m_i}^{(i)}$ is a simple ω-counter-free series over A and K.*

From the above definition and Proposition 4, we get that a finitary (resp. infinitary) almost simple series is a counter-free (resp. an ω-counter-free) series[1]. We shall denote by $ascf(K,A)$ (resp. ω-$ascf(K,A)$) the class of all almost simple counter-free (resp. ω-counter-free) series over A and K.

Theorem 3. *(i)* $sf(K,A) \subseteq ascf(K,A)$.
(ii) $\omega\text{-}sf(K,A) \subseteq \omega\text{-}ascf(K,A)$.

In the sequel, we state the inclusion of the class of almost simple ω-counter-free series into the class ω-$ULtl(K,A)$. For this, we need several notions and technical auxiliary results. Due to space limitations we skip all this stuff.

Proposition 5. *Let $L \subseteq A^+$ be a star-free language and $r \in K\langle\langle A^*\rangle\rangle$ be a letter-step series. Then, for every $\varphi \in ULTL(K,A)$ the infinitary series $(1_L \odot r^+) \cdot \|\varphi\|$ is ω-ULtl-definable.*

Theorem 4. $\omega\text{-}ascf(K,A) \subseteq \omega\text{-}ULtl(K,A)$.

Proof. Clearly it suffices to show that whenever $\mathcal{A}_1, \ldots, \mathcal{A}_{n-1}$ are simple cfwa and $\mathcal{A}_n$ is a simple cfwBa over A and K, then $\|\mathcal{A}_1\| \cdot \ldots \cdot \|\mathcal{A}_n\| \in \omega\text{-}ULtl(K,A)$. We let $r_i = \|\mathcal{A}_i\|$, and denote by k_i the initial weight$\neq 0$ and $k_a^{(i)}$ the weight$\neq 0$ of the transitions of $\mathcal{A}_i$ $(1 \leq i \leq n)$ labelled by $a \in A$. Since K is zero-divisor free, we get that $supp(r_n)$ is an ω-counter-free language, and thus it is also ω-LTL-definable. Hence, there is formula $\varphi \in bLTL(K,A)$ with $\|\varphi\| = 1_{supp(r_n)}$. We let $\varphi_n = k_n \wedge \varphi \wedge \left(\Box\left(\bigvee_{a\in A}\left(k_a^{(n)} \wedge p_a\right)\right)\right)$ and we trivially get $r_n = \|\varphi_n\|$. By construction $\varphi_n \in ULTL(K,A)$. Furthermore, for every $1 \leq i \leq n-1$, the language $supp(r_i) \setminus \{\varepsilon\} \subseteq A^*$ is counter-free hence, star-free. Since

$$r_i|_{A^+} = 1_{supp(r_i)\setminus\{\varepsilon\}} \odot \left(k_i\left(\sum_{a\in A}\left(k_a^{(i)}\right)_a\right)^+\right)$$

for every $1 \leq i \leq n-1$, and

$$r_{n-1}|_{A^+} \cdot r_n = k_{n-1}\left(\left(1_{supp(r_{n-1})\setminus\{\varepsilon\}} \odot \left(\sum_{a\in A}\left(k_a^{(n-1)}\right)_a\right)^+\right) \cdot r_n\right),$$

by applying Proposition 5, we get that

$$\left(1_{supp(r_{n-1})\setminus\{\varepsilon\}} \odot \left(\sum_{a\in A}\left(k_a^{(n-1)}\right)_a\right)^+\right) \cdot r_n \in \omega\text{-}ULtl(K,A)$$

which implies that there exists a $ULTL(K,A)$ formula φ_{n-1}^+ such that

$$\left(1_{supp(r_{n-1})\setminus\{\varepsilon\}} \odot \left(\sum_{a\in A}\left(k_a^{(n-1)}\right)_a\right)^+\right) \cdot r_n = \left\|\varphi_{n-1}^+\right\|.$$

Hence, $r_{n-1}|_{A^+} \cdot r_n = \left\|k_{n-1} \wedge \varphi_{n-1}^+\right\|$. We let

$$\varphi_{n-1} = \left(k_{n-1} \wedge \varphi_{n-1}^+\right) \vee \left((r_{n-1}, \varepsilon) \wedge \varphi_n\right) \in ULTL(K,A)$$

[1] In fact we can define an *almost simple* counter-free weighted (resp. Büchi) automaton, but we do not need it here.

and we have $\|\varphi_{n-1}\| = r_{n-1} \cdot r_n$. Thus $r_{n-1} \cdot r_n \in \omega\text{-}ULtl(K, A)$. We proceed in the same way, and we show that $r_i \cdot \ldots \cdot r_n \in \omega\text{-}ULtl(K, A)$, for every $1 \leq i \leq n-2$, which concludes our proof.

By Theorems 1, 2, 3(ii), and 4 we conclude the main result of our paper.

Theorem 5 (Main theorem).

$$\omega\text{-}ULtl\,(K, A) = \omega\text{-}wqFo(K, A) = \omega\text{-}sf(K, A) = \omega\text{-}ascf(K, A).$$

6 Conclusion

We showed the coincidence of series definable in fragments of the weighted *LTL* and *FO* logic, the class of ω-star-free series, and the class of almost simple ω-counter-free series. Our underlying semiring required to be idempotent, zero-divisor free and totally commutative complete. Recently, in [4], the authors studied weighted automata and weighted *MSO* logics over general structures which play an important role in practical applications. Therefore, the development of our theory in that setup is a challenging perspective.

References

1. Baier, C., Katoen, J.-P.: Principles of Model Checking. The MIT Press (2008)
2. Bollig, B., Gastin, P., Monmege, B., Zeitoun, M.: Pebble weighted automata and transitive closure logics. In: Abramsky, S., Gavoille, C., Kirchner, C., Meyer auf der Heide, F., Spirakis, P.G. (eds.) ICALP 2010. LNCS, vol. 6199, pp. 587–598. Springer, Heidelberg (2010)
3. Diekert, V., Gastin, P.: First-order definable languages. In: Logic and Automata: History and Perspectives. Texts in Logic and Games, vol. 2, pp. 261–306. Amsterdam University Press (2007)
4. Droste, M., Meinecke, I.: Weighted automata and weighted MSO logics for average and long-time behaviors. Inform. and Comput. 220-221, 44–59 (2012)
5. Droste, M., Vogler, H.: Weighted automata and multi-valued logics over arbitrary bounded lattices. Theoret. Comput. Sci. 418, 14–36 (2012)
6. Eilenberg, S.: Automata, Languages and Machines, vol. A. Academic Press (1974)
7. Ésik, Z., Kuich, W.: A semiring-semimodule generalization of ω-regular languages I. J. of Automata Languages and Combinatorics 10, 203–242 (2005)
8. Ésik, Z., Kuich, W.: On iteration semiring-semimodule pairs. Semigroup Forum 75, 129–159 (2007)
9. Kuich, W.: Semirings and formal power series: Their relevance to formal languages and automata theory. In: Rozenberg, G., Salomaa, A. (eds.) Handbook of Formal Languages, vol. 1, pp. 609–677. Springer (1997)
10. Kuperberg, D.: Linear temporal logics for regular cost functions. In: Proceedings of STACS 2011. LIPIcs, vol. 9, pp. 627–636. Schloss Dagstuhl - Leibniz-Zentrum fuer Informatik (2011)
11. Kuperberg, D., Vanden Boom, M.: On the expressive power of cost logics over infinite words. In: Czumaj, A., Mehlhorn, K., Pitts, A., Wattenhofer, R. (eds.) ICALP 2012, Part II. LNCS, vol. 7392, pp. 287–298. Springer, Heidelberg (2012)

12. Kupferman, O., Pnueli, A., Vardi, M.Y.: Once for all. J. Comput. Syst. Sci. 78, 981–996 (2012)
13. Mandrali, E., Rahonis, G.: On weighted first-order logics with discounting (submitted)
14. Thomas, W.: Languages, automata and logic. In: Rozenberg, G., Salomaa, A. (eds.) Handbook of Formal Languages, vol. 3, pp. 389–485. Springer (1997)
15. Vardi, M.Y.: From philosophical to industrial logics. In: Ramanujam, R., Sarukkai, S. (eds.) ICLA 2009. LNCS (LNAI), vol. 5378, pp. 89–115. Springer, Heidelberg (2009)

Linear Induction Algebra and a Normal Form for Linear Operators

Laurent Poinsot

Université Paris 13, Sorbonne Paris Cité, LIPN, CNRS (UMR 7030), France
laurent.poinsot@lipn.univ-paris13.fr
http://lipn.univ-paris13.fr/~poinsot/

Abstract. The set of natural integers is fundamental for at least two reasons: it is the free induction algebra over the empty set (and at such allows definitions of maps by primitive recursion) and it is the free monoid over a one-element set, the latter structure being a consequence of the former. In this contribution, we study the corresponding structure in the linear setting, *i.e.* in the category of modules over a commutative ring rather than in the category of sets, namely the free module generated by the integers. It also provides free structures of induction algebra and of monoid (in the category of modules). Moreover we prove that each of its linear endomorphisms admits a unique normal form, explicitly constructed, as a non-commutative formal power series.

Keywords: Universal algebra, free module, recursion theory, formal power series, infinite sums.

Mathematics Subject Classification (2010): 03C05, 08B20, 18D35.

1 Overview

The set of natural integers is fundamental for at least two reasons: it is the free induction algebra over the empty set (and at such allows definitions of maps by primitive recursion) and it is the free monoid over a one-element set, the latter structure being a consequence of the former. It is possible to define a similar object, with similar properties, in the category of modules over some commutative ring R (with a unit), namely the free R-module V generated by $\mathbb{N}$. We prove that this module inherits from the integers a structure of initial R-linear induction algebra, and also of free R-linear monoid (a usual R-algebra). General definitions of varieties of algebraic structures (in the setting of universal algebra) in the category of R-modules, rather than set-based, are given in section 2 together with some results concerning the relations between a set-theoretic algebra and its R-linear counterpart. These results are applied to V in section 3, and allow us to outline a theory of R-linear recursive functions, and to provide relations between the (free) monoid structure of V and well-known usual algebraic constructions (polynomials, tensor algebra and algebra of a monoid). Finally in section 4 we prove that any R-linear endomorphism of V may be written uniquely as an infinite sum, and so admits a unique *normal form* as a non-commutative formal power series.

T. Muntean, D. Poulakis, and R. Rolland (Eds.): CAI 2013, LNCS 8080, pp. 260–273, 2013.

2 Linear Universal Algebra

In this contribution are assumed known some basic notions about category theory and universal algebra that may be found in any textbooks ([2,11] for instance). We also refer to [5] for notions concerning modules and their tensor product. However some of them are recalled hereafter. The basic categories used are the category $\mathcal{Set}$ of sets (with set-theoretic maps) and the category R-$\mathcal{Mod}$ of modules over some fixed commutative ring R with a unit (and R-linear maps). If C denotes a category and a, b are two objects of this category, then the class of all morphisms from a to b in C is denoted by $C(a, b)$. For instance, if V, W are two R-modules, then R-$\mathcal{Mod}(V, W)$ denotes the set of all R-linear maps from V to W. Let (Σ, α) be a (finitary and homogeneous) signature (also called an algebra type or an operator domain), *i.e.*, a set Σ (the elements of which are referred to as *symbols* of functions) together with a map $\alpha \colon \Sigma \to \mathbb{N}$ called the *arity function*. In what follows we simply denote by Σ a signature (Σ, α), and $\alpha^{-1}(\{n\})$ is denoted by $\Sigma(n)$. The elements of $\Sigma(0) \subseteq \Sigma$ with an arity of zero are called *symbols of constants*. A *Σ-algebra*, or algebra of type Σ, is a pair (A, F) where A is a set and F is a map that associates to each symbol of function f of arity $\alpha(f) = n$ (for each n) an actual map $F(f) \colon A^n \to A$ (we sometimes call F the *Σ-algebra structure map* of A). In particular if $\alpha(c) = 0$, then $F(c)$ is identified to an element of A (which explains the term of symbol of constant). An homomorphism between two algebras $(A, F), (B, G)$ over the same signature Σ is a set-theoretic map $\phi \colon A \to B$ such that for every $f \in \Sigma(n)$, and every $a_1, \cdots, a_n \in A$, $\phi(F(f)(a_1, \cdots, a_n)) = G(f)(\phi(a_1), \cdots, \phi(a_n))$ (in particular for each $c \in \Sigma(0)$, $\phi(F(c)) = G(\phi(c)))$. An *isomorphism* is a homomorphism which is also a bijective map. A *sub-algebra* (B, G) of (A, F) is a Σ-algebra such that the natural inclusion $B \subseteq A$ is a homomorphism of Σ-algebras. A *congruence* $\cong$ on a Σ-algebra (A, F) is an equivalence relation on A such that for every $f \in \Sigma(n)$, if $a_i \cong b_i$, $i = 1, \cdots, n$, then $F(f)(a_1, \cdots, a_n) \cong F(f)(b_1, \cdots, b_n)$. This implies that the quotient set $A/_{\cong}$ inherits a natural structure of Σ-algebra from that of A. It is well-known (see [2]) that such congruences form a lattice, and then for every $R \subseteq A^2$, we may talk about the *least congruence on A generated by R* in an evident way. For any set X there exists a *free Σ-algebra* $\Sigma[X]$ on X. It is constructed by induction as follows (it is a subset of the free monoid $(\Sigma \sqcup X)^*$ over $\Sigma \sqcup X$, and the parentheses to form its elements are only used for readability; see [2]). The base cases: $\Sigma(0) \subseteq \Sigma[X]$ and $X \subseteq \Sigma[X]$, the induction rule: for every n, and every $f \in \Sigma(n)$, if $t_1, \cdots, t_n \in \Sigma[X]$, then $f(t_1, \cdots, t_n) \in \Sigma[X]$, and the closure property: it is the least subset of $(\Sigma \sqcup X)^*$ with these two properties. Its structure of Σ-algebra is the evident one. It is called *free* because for any Σ-algebra (A, F) and any set-theoretic map $\phi \colon X \to A$, there exists a unique homomorphism $\widehat{\phi} \colon \Sigma[X] \to (A, F)$ such that $\widehat{\phi}(x) = \phi(x)$ for every $x \in X$. In category-theoretic terms, this means that the (obvious) forgetful functor from the category of Σ-algebras to $\mathcal{Set}$ admits a left adjoint, and this implies that a free algebra is unique up to a unique isomorphism (we can talk about *the* free algebra).

Example 1. The set $\mathbb{N}$, together with the constant 0 and the usual successor function, is the free induction algebra over the empty set (see for instance [2]) where we call *induction algebra* any algebra over the signature $\mathcal{Ind} = \{0, S\}$ where $0 \in \mathcal{Ind}(0)$ and $S \in \mathcal{Ind}(1)$.

A *variety* of Σ-algebras is a class of algebras closed under homomorphic images, sub-algebras, and direct products. A *law* or *identity* over Σ on the *standard alphabet* $X = \{x_i : i \geq 0\}$ is a pair $(u, v) \in \Sigma[X]^2$ sometimes written as an equation $u = v$. We say that a law (u, v) *holds* in a Σ-algebra (A, F), or that (A, F) *satisfies* (u, v), if under every homomorphism $\Sigma[X] \to (A, F)$ the values of u and v coincide. If E is any set of laws in $\Sigma[X]$, then $\mathcal{V}_{\Sigma,E}$ or simply $\mathcal{V}_E$, is the class of all algebras which satisfy all the laws in E. By the famous Garrett Birkhoff's theorem, $\mathcal{V}_E$ is a variety and any variety arises in such a way.

Example 2. The variety of all monoids is given by $\mathcal{V}_{M,E}$ where $M(0) = \{1\}$, $M(2) = \{\mu\}$, $M(n) = \emptyset$ for every $n \neq 0, 2$, and E consists in the three equations $(\mu(x_1, 1), x_1)$, $(\mu(1, x_1), x_1)$ and $(\mu(\mu(x_1, x_2), x_3), \mu(x_1, \mu(x_2, x_3)))$. The variety of all commutative monoids is obtained in an obvious way.

A *free* algebra over a set X in a variety $\mathcal{V}_{\Sigma,E}$ is a Σ-algebra V_X in the class $\mathcal{V}_{\Sigma,E}$, together with a set-theoretic map $i_X : X \to V_X$, such that for every algebra (A, F) in $\mathcal{V}_{\Sigma,E}$ and every map $\phi : X \to A$, there is a unique homomorphism $\widehat{\phi} : V_X \to (A, F)$ with $\widehat{\phi} \circ i_X = \phi$. Thus the free Σ-algebra $\Sigma[X]$ is easily seen as a free algebra in the variety $\mathcal{V}_{\Sigma,\emptyset}$. In category-theoretic terms, when a variety is seen as a category (whose morphisms are the homomorphisms of algebras), this means that the obvious forgetful functor from the variety to $\mathcal{Set}$ admits a left adjoint. This implies that a free algebra is unique up to a unique isomorphism. Let us see a way to construct it. Let $\cong_E$ be the least congruence of Σ-algebra on $\Sigma[X]$ generated by the relations $\{(\widehat{\sigma}(u), \widehat{\sigma}(v)) : (u, v) \in E,\ \sigma : X \to \Sigma[X]\}$ (recall that $\widehat{\sigma} : \Sigma[X] \to \Sigma[X]$ is the unique homomorphism of Σ-algebras that extends σ). Let $V_X = \Sigma[X]/_{\cong_E}$ together with its structure of quotient Σ-algebra inherited from that of $\Sigma[X]$. Let (B, G) be any Σ-algebra in the variety $\mathcal{V}_{\Sigma,E}$, and $\phi : X \to B$ be a set-theoretic map. It admits a unique homomorphism extension $\widehat{\phi} : \Sigma[X] \to (B, G)$ since $\Sigma[X]$ is free. Because (B, G) belongs to $\mathcal{V}_{\Sigma,E}$ and $\widehat{\phi} \circ \sigma : \Sigma[X] \to (B, G)$ is a homomorphism whenever $\sigma : \Sigma[X] \to \Sigma[X]$ is so, then for each $(u, v) \in E$, $\widehat{\phi}(\sigma(u)) = \widehat{\phi}(\sigma(v))$. Therefore $\widehat{\phi}$ passes to the quotient by $\cong_E$ and defines a homomorphism from V_X to (B, G) as expected.

Example 3. For instance $\mathbb{N}$ with its structure of (commutative) monoid is the free algebra in $\mathcal{V}_{M,E}$ over $\{1\}$, while $\mathbb{N} \setminus \{0\}$ with its multiplicative structure of monoid in the free algebra in the variety of all commutative monoids over the set of all prime numbers.

Up to now, we only describe *set-based* algebras. But it is possible to talk about *linear algebras*. For this let us recall some basic facts about modules and their tensor product (see [5]). Let X be any set. The free R-module generated by X is the R-module RX of all formal sums $\sum_{x \in X} \alpha_x x$ ($\alpha_x \in R$) where all but finitely many coefficients $\alpha_x \in R$ are zero (this is the free R-module with basis X), and for any $x_0 \in X$, we refer to the element $e_{x_0} \in RX$, obtained as the formal sum $\sum_{x \in X} \alpha_x x$ with $\alpha_x = 0$ for every $x \neq x_0$ and $\alpha_{x_0} = 1$ (the unit of R), as the *canonical image of* x_0 *into* RX, and therefore this defines a one-to-one map $e\colon X \to RX$ by $e(x) = e_x$. If W is any over R-module, then any R-linear map $\phi\colon RX \to W$ is entirely defined by its values on the basis X. Let $V_1, \cdots, V_n, W$ be R-modules. A map $\phi\colon V_1 \times \cdots \times V_n \to W$ is said to be *multilinear* (or *bilinear* when $n = 2$) if it is linear in each of its variables when the other ones are fixed. Given a multilinear map $\phi\colon V_1 \times \cdots \times V_n \to W$, there is a unique *linear map* $\psi\colon V_1 \otimes_R \cdots \otimes_R V_n \to V$, where $\otimes_R$ denotes the tensor product over R (see [5]), such that $\psi \circ q = \phi$ (where $q\colon V_1 \times \cdots \times V_n \to V_1 \otimes_R \cdots \otimes_R V_n$ is the canonical multilinear map; the image of $(v_1, \cdots, v_n)$ under q is denoted by $v_1 \otimes \cdots \otimes v_n$). In what follows, ϕ is referred to as the *multilinear map* associated to ψ, and denoted by ψ_0. If $V_1, \cdots, V_n$ are free qua R-modules with basis $(e^{(j)})_{i \in I_j}$, $j = 1, \cdots, n$, then $V_1 \otimes_R \cdots \otimes_R V_n$ also is free with basis $\{\, e^{(1)}_{i_1} \otimes \cdots \otimes e^{(n)}_{i_n} : i_j \in I_j,\ j = 1, \cdots, n \,\}$. Moreover given a linear map $\phi\colon V_1 \to R\text{-}\mathcal{Mod}(V_2, V_3)$, then it determines a unique linear map $\psi\colon V_1 \otimes_R V_2 \to V_3$ (it is obtained from the bilinear map $\phi'\colon V_1 \times V_2 \to V_3$ given by $\phi'(v_1, v_2) = \phi(v_1)(v_2)$).

Lemma 1. *For every sets* $X_1, \cdots, X_n$, $R(X_1 \times \cdots \times X_n)$ *and* $RX_1 \otimes_R \cdots \otimes_R RX_n$ *are isomorphic* R*-modules.*

Proof. (Sketch) It is clear that $R(X_1 \times \cdots \times X_n)$ is identified as a sub-module of $R(RX_1 \times \cdots \times RX_n)$ by $\iota\colon (x_1, \cdots, x_n) \mapsto e(e(x_1), \cdots, e(x_n))$. Let $q \circ \iota\colon R(X_1 \times \cdots \times X_n) \to RX_1 \otimes_R \cdots \otimes_R RX_n$ be the restriction of the canonical multilinear map (it is clearly onto and is easily shown to be R-linear), and $s\colon RX_1 \times \cdots \times RX_n \to R(X_1 \times \cdots \times X_n)$ be the multilinear map given by $s(e(x_1), \cdots, e(x_n)) = e(x_1, \cdots, x_n)$ for every $x_i \in X_i$, $i = 1, \cdots, n$. Therefore it gives rise to a unique linear map $\widetilde{s}\colon RX_1 \otimes_R \cdots \otimes_R RX_n \to R(X_1 \times \cdots \times X_n)$. It is easy to see that $\widetilde{s} \circ q = id$, but q is onto so that it is an R-linear isomorphism (the details are left to the reader). □

From lemma 1, it follows that any set-theoretic map $\phi\colon X_1 \times \cdots \times X_n \to W$ may be extended in a unique way to a linear map $\widetilde{\phi}\colon RX_1 \otimes_R \cdots \otimes_R RX_n \to W$. Following the notations from the proof of lemma 1, $\phi\colon X_1 \times \cdots \times X_n \to W$ is first freely extended to a R-linear map $\phi\colon R(X_1 \times \cdots \times X_n) \to W$, and then $\phi \circ \widehat{s}\colon RX_1 \otimes_R \cdots \otimes_R RX_n \to W$ is the expected linear map $\widetilde{\phi}$. Moreover its associated multilinear map $\widetilde{\phi}_0\colon RX_1 \times \cdots \times RX_n \to W$ is sometimes referred to as the *extension of* ϕ *by multilinearity.* We are now in position to introduce R-linear Σ-algebras and varieties. Let Σ be an operator domain, and R be a commutative ring with a unit. A *R-linear Σ-algebra* is a R-module with a structure of Σ-algebra such that all operations are R-multilinear. More precisely it is a R-module V with a Σ-algebra structure map F such that for each $f \in \Sigma(n)$ $(n \geq 0)$, $F(f)\colon \underbrace{V \otimes_R \cdots \otimes_R V}_{n \text{ factors}} \to V$ is R-linear. Following [3], if V is a R-linear Σ-algebra, let $\mathcal{U}(V)$ denote its underlying (set-theoretic) Σ-algebra (its structure of Σ-algebra is given by the multilinear map $F_0(f)\colon V \times \cdots \times V \to V$ associated to $F(f)$), and if (A, F) is a usual Σ-algebra, let $(RA, \widetilde{F})$ denote the R-linear Σ-algebra made from the free R-module RA on A by extending the Σ-operation $F(f), f \in \Sigma(n)$, of A by multilinearity. More precisely, $\widetilde{F}(f)\colon RA \otimes_R \cdots \otimes_R RA \to RA$ is the unique linear map obtained from lemma 1. It is given by $\widetilde{F}(f)(e(a_1) \otimes \cdots \otimes e(a_n)) = e(F(f)(a_1, \cdots, a_n))$ for each $a_1, \cdots, a_n \in A$ (this map is well defined since $\{ e(a_1) \otimes \cdots \otimes e(a_n) \colon a_1, \cdots, a_n \in A \}$ forms a basis). (According to the above discussion, this is equivalent to a multilinear map $\widetilde{F}_0(f)\colon RA^n \to RA$ with $\widetilde{F}_0(f)(e(a_1, \cdots, a_n)) = e(F(f)(a_1, \cdots, a_n))$.) Actually we obtain a *functorial* correspondence between Σ-algebras and R-linear Σ-algebras: the forgetful functor $\mathcal{U}$ admits a left adjoint given by the construction RA. More precisely, given a R-linear Σ-algebra (W, G), and a homomorphism $\phi\colon (A, F) \to \mathcal{U}(W, G)$, $\widetilde{\phi}\colon (RA, \widetilde{F}) \to (W, G)$, given by $\widetilde{\phi}(e(a)) = \phi(a)$ for each $a \in A$, is the unique extension of ϕ which is a homomorphism of R-linear Σ-algebras (this means that $\widetilde{\phi}$ is R-linear, and $\widetilde{\phi}(\widetilde{F}(f)(x_1 \otimes \cdots \otimes x_n)) = G(f)(\widetilde{\phi}(x_1) \otimes \cdots \otimes \widetilde{\phi}(x_n))$ for every $x_1, \cdots, x_n \in RA$). To determine such a correspondence between varieties and linear varieties we must be more careful due to multilinearity. A law $u = v$ on $\mathcal{X}$ is said to be *regular* when the same elements of $\mathcal{X}$ occur in u and v, and exactly once in both of them. For instance, $\mu(x_1, 1) = x_1$, $\mu(\mu(x_1, x_2), x_3) = \mu(x_1, \mu(x_2, x_3))$ are regular laws. If E is any set of regular equations on $\Sigma[\mathcal{X}]$, and (V, F) is a R-linear Σ-algebra, then we say that (V, F) *satisfies* E when under all homomorphisms $\Sigma[\mathcal{X}] \to \mathcal{U}(V)$, the images of u and of v are equal for each $(u, v) \in E$. If E is any set of regular equations on $\Sigma[\mathcal{X}]$, then there is a very close connection between the variety $\mathcal{V}_{\Sigma,E}$ of Σ-algebras satisfying E, and the variety $\mathcal{V}_{\Sigma,R,E}$ of R-linear Σ-algebras satisfying E: it is easy to see that a R-linear Σ-algebra V will lie in $\mathcal{V}_{\Sigma,R,E}$ if, and only if, $\mathcal{U}(V)$ lies in $\mathcal{V}_{\Sigma,R,E}$. Conversely, according to [3], a Σ-algebra (A, F) will lie in $\mathcal{V}_{\Sigma,E}$ if, and only if, RA lies in $\mathcal{V}_{\Sigma,R,E}$. A *free* R-linear algebra in $\mathcal{V}_{\Sigma,R,E}$ over a set (resp. a R-module, resp. a Σ-algebra in the variety $\mathcal{V}_{\Sigma,E}$) X is a R-linear Σ-algebra V_X in the variety $\mathcal{V}_{\Sigma,R,E}$ with a set-theoretic map (resp. a R-linear map, resp. a homomorphism) $j_X\colon X \to V_X$ (called the *canonical map*) such that for all R-linear algebra W in $\mathcal{V}_{\Sigma,R,E}$ and all set-theoretic map (resp. R-linear map, resp. homomorphism) $\phi\colon X \to W$ there is a unique homomorphism $\widehat{\phi}\colon V_X \to W$ of R-linear algebras such that $\widehat{\phi} \circ j_X = \phi$. Such a free algebra is unique up to a unique isomorphism. As an example, the free R-linear algebra in $\mathcal{V}_{\Sigma,R,E}$ over a R-module W is made as follows. Let us assume that the free R-linear algebra V_W on the underlying *set* W is constructed with the *set-theoretic* map $j_W\colon W \to V_W$ (we see in lemma 2 that it always exists). Let F be the Σ-algebra structure map of V_W (this means that $F(f)$ is a linear map from $V_W \otimes_R \cdots \otimes_R V_W \to V_W$ for each $f \in \Sigma$). Let $\overline{W}$ be the least sub-module of V_W *stable under all* $F(f)$*'s* (this means that the image of $\overline{W} \otimes_R \cdots \otimes_R \overline{W}$ by all $F(f)$'s lies into $\overline{W}$) and that contains the submodule generated by $j_W(w_1 + w_2) - j_W(w_1) - j_W(w_2)$, $j_W(\alpha w) - \alpha j_W(w)$ for every $\alpha \in R$, $w_1, w_2, w \in W$. Then it is easily seen that the quotient module $V_W/\overline{W}$ inherits a structure of R-linear Σ-algebra from that of V_W, and is the expected free algebra (where the canonical map is the composition of the natural epimorphism $V_W \to V_W/\overline{W}$ with the set-theoretic canonical map $W \to V_W$).

Remark 1. These three notions of free algebras (over a set, a module or a Σ-algebra in the variety $\mathcal{V}_{\Sigma,E}$) come from the fact that there are three forgetful functors, and each of them admits a left adjoint.

Lemma 2. *Let E be a set of regular equations. Let X be a set and V_X be the free Σ-algebra over X in the variety $\mathcal{V}_{\Sigma,E}$ with $i_X \colon X \to V_X$. Then, RV_X with $j_X \colon X \to RV_X$ given by $j_X(x) = e(i_X(x))$ is the free R-linear Σ-algebra over X in $\mathcal{V}_{\Sigma,R,E}$. Moreover, RV_X, with the R-linear map $k_{RX} \colon RX \to RV_X$ defined by $h_{RX}(e_x) = e(i_X(x))$ for every $x \in X$, is the free R-linear Σ-algebra (in $\mathcal{V}_{\Sigma,R,E}$) over RX. Finally, let $k_{V_X} \colon V_X \to RV_X$ be the unique homomorphism such that $k_{V_X} \circ i_X = j_X = e \circ i_X$. Then, RV_X with h_{V_X} is free over V_X.*

Proof. (The proof of this lemma is easy for a category theorist or universal algebraist but is given for the sake of completeness.) Let (W, G) be a R-linear Σ-algebra, and $\phi \colon X \to W$ be a set-theoretic map. Then, there exists a unique homomorphism of Σ-algebras $\widehat{\phi} \colon V_X \to \mathcal{U}(W)$ such that $\widehat{\phi} \circ i_X = \phi$. Since RV_X is free with basis V_X over R, there is a unique R-linear map $\psi \colon RV_X \to W$ such that $\psi \circ e = \widehat{\phi}$ (so $\psi \circ j_X = \psi \circ e \circ j_X = \widehat{\phi} \circ j_X = \phi$). Moreover from the above discussion we know that RV_X is a R-linear Σ-algebra of the variety $\mathcal{V}_{\Sigma,R,E}$. It remains to prove that ψ is a homomorphism of Σ-algebra from V_X to W. Let $f \in \Sigma(n)$, and $a_1, \cdots, a_n \in V_X$. Let F be the Σ-algebra structure map of V_X. We have $\psi(\widetilde{F}(f)(e(a_1) \otimes \cdots \otimes e(a_n))) = \psi(e(F(f)(a_1, \cdots, a_n))) = \widehat{\phi}(F(f)(a_1, \cdots, a_n)) = G(f)(\widehat{\phi}(a_1), \cdots, \widehat{\phi}(a_n)) = G(f)(\psi(e(a_1)) \otimes \cdots \otimes \psi(e(a_n)))$, for each $a_1, \cdots, a_n \in V_X$. Now, let $\phi \colon RX \to W$ be any R-linear map (where W is a R-linear Σ-algebra in the variety $\mathcal{V}_{\Sigma,R,E}$). Then, there exists a unique set-theoretic map $\phi_0 \colon X \to W$ such that $\phi_0(x) = \phi(e(x))$ for every $x \in X$. Therefore there exists a unique homomorphism of Σ-algebras $\widehat{\phi_0} \colon V_X \to W$ such that $\widehat{\phi_0} \circ i_X = \phi_0$. Finally, there exists a unique R-linear map, wich is also a homomorphism of Σ-algebras $\psi \colon RV_X \to W$ such that $\psi \circ e = \widehat{\phi_0}$. Then, $\phi_0 = \psi \circ j_X = \psi \circ e \circ i_X = \psi \circ h_{RX} \circ e$. But $\phi \circ e = \phi_0$, and both maps ϕ and $\psi \circ h_{RX}$ are R-linear and equal on basis elements of RX, so that they are equal on RX as expected. Finally, let $\phi \colon V_X \to \mathcal{U}(W)$ be a homomorphism of Σ-algebras. Then, there exists a unique set-theoretic map $\phi_0 \colon X \to W$ such that $\phi_0 = \phi \circ i_X$. Then, there exists a unique homomorphism of Σ-algebras which is a R-linear map $\widehat{\phi_0} \colon RV_X \to W$ with $\widehat{\phi_0} \circ j_X = \phi_0$. Then, $\widehat{\phi_0} \circ k_{V_X} \circ i_X = \widehat{\phi_0} \circ j_X = \phi_0 = \phi \circ i_X$, and since $\widehat{\phi_0} \circ k_{V_X}$ and ϕ are both homomorphisms from V_X to W it follows that their are equal (since V_X is free). □

3 *R*-linear Induction Algebra

3.1 The Initial *R*-linear Induction Algebra

The free R-module $R\mathbb{N}$ over $\mathbb{N}$ is denoted by V. The canonical image of an integer n into V is denoted by e_n so $e_i \neq e_j$ for every $i \neq j$ and $\{ e_n \colon n \in \mathbb{N} \}$ happens to be a basis of V over R. The constant 0 of the signature *Ind* corresponds to e_0, and the successor map $s\colon \mathbb{N} \to \mathbb{N}$ is uniquely extended by R-linearity (no need here of multilinearity) to $U \in R\text{-}\mathcal{Mod}(V,V)$ defined on the basis elements by $Ue_n = e_{n+1}$, $n \in \mathbb{N}$. It is clear that (V, e_0, U) is a R-linear induction algebra, and according to lemma 2, (V, e_0, U) is even the free R-linear induction algebra over the empty set, the free R-linear induction over the zero vector space, and the free R-linear induction over the induction algebra $\mathbb{N}$. We call (V, U, e_0) the *initial* R-linear induction algebra because given another R-linear induction algebra (W, w, S) ($w \in W$, $S \in R\text{-}\mathcal{Mod}(W,W)$), there is a unique R-linear map $\phi\colon V \to W$ such that $\phi(e_0) = w$, and $\phi \circ U = S \circ \phi$. This may be proved directly from the fact that V is free over $(e_n)_{n\in\mathbb{N}}$, and $e_n = U^n(e_0)$ for each $n \in \mathbb{N}$. (Indeed, there is a unique linear map $\phi\colon V \to W$ such that $\phi(e_n) = S^n(w)$.)

Remark 2. It is obvious that $\mathbb{N}$ is the initial induction algebra (since it is freely generated by the empty set). This means that for each induction algebra A, we have a natural isomorphism (see [11] for a precise definition of this notion) of sets $\mathcal{V}_{\mathit{Ind},\emptyset}(\mathbb{N}, A) \cong \mathcal{S}et(\emptyset, A) = \{\emptyset\}$ (where the variety $\mathcal{V}_{\mathit{Ind},\emptyset}$ of all induction algebras is considered as a category). Now, since V is the free R-linear induction algebra on $\mathbb{N}$, for every R-linear induction algebra W, one also has natural isomorphisms (of sets) $\mathcal{V}_{\mathit{Ind},R,\emptyset}(V, W) \cong \mathcal{V}_{\mathit{Ind},\emptyset}(\mathbb{N}, \mathcal{U}(W)) \cong \{\emptyset\}$.

For every $n \in \mathbb{N}$, let V_n be the sub-module of V generated by $(e_k)_{k\geq n}$ (which is obviously free over $(e_k)_{k\geq n}$). It is a R-linear induction algebra on its own (V_n, e_n, U) (since $U\colon V_n \to V_{n+1} \subseteq V_n$). Therefore, for every $n \in \mathbb{N}$, there exists a unique R-linear map, which is a homomorphism of induction algebras, $\mu_n\colon V \to V_n$ such that $\mu_n(e_0) = e_n$ and $\mu_n(e_{k+1}) = \mu_n(Ue_k) = U(\mu_n(e_k))$. It is easy to prove by induction that $\mu_n(e_k) = e_{k+n}$. Now, we define $\overline{\mu}\colon V \to R\text{-}\mathcal{Mod}(V,V)$ by $\overline{\mu}(e_n) = \mu_n$ for each $n \geq 0$. Therefore we obtain a bilinear map $V \times V \to V$ given by $\phi(e_m, e_n) = \overline{\mu}(e_m)(e_n) = \mu_m(e_n) = e_{m+n}$. Finally this leads to the existence of a linear map $\mu\colon V \otimes_R V \to V$ defined by $\mu(e_m \otimes e_n) = e_{m+n}$. A simple calculation shows that μ is associative (in the sense that $\mu(\mu(u \otimes v) \otimes w) = \mu(u \otimes \mu(v \otimes w))$ for every $u, v, w \in V$ and not only for basis elements) and $\mu(v \otimes e_0) = v = \mu(e_0 \otimes v)$ for every $v \in V$. This means that V becomes a monoid, and more precisely an R-algebra (an internal monoid in the category of R-modules, see [11]). We see below another way to build this R-algebra structure on V.

Remark 3. Similarly it is also possible to define the free linear extension of the usual multiplication on $\mathbb{N}$ to a linear map $\mu'\colon V \otimes V \to$ by $\mu'(e_m \otimes e_n) = e_{mn}$, which happens to be associative and has a unit e_1. But e_0 is not an absorbing element: for instance $\mu'((\alpha e_m + \beta e_n) \otimes e_0) = \alpha\mu'(e_m \otimes e_0) + \beta\mu'(e_n \otimes e_0) = (\alpha + \beta)e_0 \neq e_0$ whenever $\alpha + \beta \neq 0$ (in R). It is due to the fact that the equation $x_1 \times 0 = 0$ or $0 \times x_1 = 0$ is not a regular law. Similarly, even if we have $\mu'(e_m \otimes \mu(e_n \otimes e_p)) = \mu(\mu'(e_m \otimes e_n) \otimes \mu'(e_m \otimes e_p))$, the distributivity law does not hold for any $u, v, w \in V$ (again essentially because it is not a regular law).

3.2 A Free Monoid Structure and Its Links with Classical Algebra

We also know that $(\mathbb{N}, +, 0)$ is the free monoid $\{1\}^*$ over $\{1\}$. Therefore, again by lemma 2, (V, μ, e_0) is the free monoid over $\{1\}$, or over the module R, or over the monoid $(\mathbb{N}, +, 0)$, where $\mu\colon V \otimes_R V \to V$ is the R-linear map given by $\mu(e_m \otimes e_n) = e_{m+n}$ (it satisfies $\mu(\mu(u \otimes v) \otimes w) = \mu(u \otimes \mu(v \otimes w))$ for every $u, v, w \in V$, and $\mu(v \otimes e_0) = v = \mu(e_0 \otimes v)$ for every $v \in V$). Therefore (V, μ, e_0) has a structure of commutative R-algebra which is actually the same as that defined in subsection 3.1. Moreover it is nothing else than the usual algebra of polynomials $R[x]$ in one indeterminate (an isomorphism is given by $e_n \mapsto x^n$). The fact that (V, μ, e_0) is the free monoid over R is also re-captured by the fact that $R[x]$ may be seen as the tensor R-algebra generated by $Rx \cong R$ (see [5]). Finally the fact that (V, μ, e_0) is free over the monoid $(\mathbb{N}, +, 0)$ is recovered in the usual algebraic setting by the fact that qua a R-algebra V (and therefore $R[x]$) is isomorphic to the R-algebra of the monoid $\mathbb{N}$.

Remark 4. According to the remark 3, there is no hope to use the multiplication from $\mathbb{N}$ in order to define a structure of ring on V internal to the category of modules.

3.3 Linear Primitive Recursion Operator

Back to the fact that V is the initial R-linear induction algebra, we show here how to define linear maps by primitive recursion in a way similar to the usual clone of primitive recursive functions (see for instance [16]). Recall that given two maps $g\colon \mathbb{N}^k \to \mathbb{N}$ and $h\colon \mathbb{N}^{k+2} \to \mathbb{N}$ it is possible to define a unique map $R(g, h) = f\colon \mathbb{N}^{k+1} \to \mathbb{N}$ by *primitive recursion* as $f(0, n_1, \cdots, n_k) = g(n_1, \cdots, n_k)$ and $f(n + 1, n_1, \cdots, n_k) = h(n_1, \cdots, n_k, n, f(n_1, \cdots, n_k))$ for every $n_1, \cdots, n_k, n \in \mathbb{N}$. If W is a R-module, then $W^{\otimes n}$ is the tensor product $\underbrace{W \otimes_R \cdots \otimes_R W}_{n\ times}$ (so that $W^{\otimes 0} \cong R$). Now, any set-theoretic map $f\colon \mathbb{N}^\ell \to \mathcal{U}(V)$ gives rise to a unique R-linear map $\widehat{f}\colon V^{\otimes \ell} \to V$ by $\widehat{f}(e_{n_1} \otimes \cdots \otimes e_{n_\ell}) = f(n_1, \cdots, n_\ell)$. Therefore given $g\colon \mathbb{N}^k \to V$ and $h\colon \mathbb{N}^{k+2} \to V$, there exists a unique R-linear map $\widehat{R}(g, h)\colon V^{\otimes k+1} \to V$ by $\widehat{R}(g, h)(e_{n_1} \otimes \cdots \otimes e_{n_{k+1}}) = R(g, h)(n_1, \cdots, n_{k+1})$ and thus by $\widehat{R}(g, h)(e_0 \otimes e_{n_1} \otimes \cdots \otimes e_{n_k}) = g(n_1, \cdots, n_k) = \widehat{g}(e_{n_1} \otimes \cdots \otimes e_{n_k})$ and $\widehat{R}(g, h)(e_{n+1} \otimes e_{n_1} \otimes \cdots \otimes e_{n_k}) = h(n_1, \cdots, n_k, n, R(g, h)(n, n_1, \cdots, n_k)) = \widehat{h}(e_{n_1} \otimes \cdots \otimes e_{n_k} \otimes e_n \otimes \widehat{R}(g, h)(e_n \otimes e_{n_1} \otimes \cdots \otimes e_{n_k}))$ for each $n, n_1, \cdots, n_k, n_{k+1} \in \mathbb{N}$. The following result is then proved.

Theorem 1 (Linear primitive recursion). *Let $g \in Set(\mathbb{N}^k, V)$ and $h \in Set(\mathbb{N}^{k+2}, V)$. Then there exists a unique linear map $\phi\colon V^{\otimes k+1} \to V$ such that $\phi(e_0 \otimes e_{n_1} \otimes \cdots \otimes e_{n_k}) = \widehat{g}(e_{n_1} \otimes \cdots \otimes e_{n_k})$ and $\phi(e_{n+1} \otimes e_{n_1} \otimes \cdots \otimes e_k) = \widehat{h}(e_{n_1} \otimes \cdots \otimes e_{n_k} \otimes e_n \otimes \phi(e_n \otimes e_{n_1} \otimes \cdots \otimes e_{n_k}))$ for every $n, n_1, \cdots, n_k \in \mathbb{N}$.*

Remark 5. The two R-linear maps μ and μ' from subsection 3.1 may be obtained by linear primitive recursion.

In order to close this subsection, let us briefly see the corresponding notion of clone of primitive recursive functions in the linear case. Let $f\colon R\text{-}\mathcal{Mod}(V^{\otimes m}, V)$, and $g_1 \cdots , g_m \in R\text{-}\mathcal{Mod}(V^{\otimes n}, V)$, then the *superposition* $\mu(f, g_1, \cdots , g_m)$ in $R\text{-}\mathcal{Mod}(V^{\otimes n}, V)$ is defined by $\mu(f, g_1, \cdots , g_m)(e_{i_1} \otimes \cdots \otimes e_{i_n}) = f(g_1(e_{i_1} \otimes \cdots \otimes e_{i_n}) \otimes \cdots \otimes g_n(e_{i_1} \otimes \cdots \otimes e_{i_n}))$ for every $e_{i_1}, \cdots , e_{i_n} \in V$. For every n, $i = 1, \cdots , n$, we define the *projections* $\pi_i^{(n)} \in R\text{-}\mathcal{Mod}(V^{\otimes n}, V)$ by $\pi_i^{(n)}(e_{j_1} \otimes \cdots \otimes e_{j_n}) = e_{j_i}$ for every $j_1, \cdots , j_n \in \mathbb{N}$. Then the clone of all linear primitive recursive functions is the set of all R-linear maps from $V^{\otimes k}$, for varying k, to V which is closed under superposition, and linear primitive recursion (in the sense that if $g\colon V^{\otimes k} \to V$ and $h\colon V^{\otimes k+2} \to V$ are primitive recursion linear maps, then $\widehat{R}(g_0, h_0)$ is linear primitive recursive, where $g_0\colon \mathbb{N}^k \to V$ and $h_0\colon \mathbb{N}^{k+2} \to V$ are the unique maps such that $g = \widehat{g}_0$ and $h = \widehat{h}_0$), that contains, for every set-theoretic primitive recursive function $f \in \mathbb{N}^k \to \mathbb{N}$, the map $\widetilde{f}\colon V^{\otimes k} \to V$ where $\widetilde{f}(e_{i_1} \otimes \cdots \otimes e_{i_k}) = e_{f(i_1,\cdots,i_k)}$ for all $(i_1, \cdots , i_k) \in \mathbb{N}^k$, and that contains the projections.

4 A Normal Form for *R*-linear Endomorphisms of *V*

In [6] the authors generalize a result from [9] that concerns the decomposition of linear endomorphisms of V (in [6] only the case where R is a field is considered) with respect to a pair of raising and lowering ladder operators. In the present paper, after recalling this result in a more general setting, we show that it may be seen as a strong version of Jacobson's density theorem and that it gives rise to a unique normal form for the endomorphisms of V in a way made precise hereafter.

4.1 Jacobson's Density Theorem

Jacobson's density theorem is a result made of two parts: an algebraic and a topological one. Let us begin with definitions needed for the algebraic part. Let R be a unitary ring (commutative or not). If M is a left R-module, then we denote by $\nu\colon R \to \mathcal{Ab}(M)$ the associated (module) structure map (where $\mathcal{Ab}$ denotes the category of all Abelian groups). This is a ring map since it is a linear representation of R. A left R-module M is said to be a *faithful module* if the structure map is one-to-one, *i.e.*, $\ker \nu = (0)$. A left R-module M is said to be a *simple module* if it is non-zero and it has no non-trivial sub-modules (modules different from (0) and M itself). Finally, a ring R is said to be *(left-)primitive* if it admits a faithful simple left-module. Now, let us turn to the topological part. Given two topological spaces X, Y, we let $\mathcal{Top}(X, Y)$ be the set of all continuous maps from X to Y (here $\mathcal{Top}$ denotes the category of all topological spaces). Let K be a compact subset of X and U be an open set in Y, then we define $V(K, U) = \{ f \in \mathcal{Top}(X, Y)\colon f(K) \subseteq U \}$. The collection of all such sets $V(K, U)$ (with varying K and U) forms a subbasis for the *compact-open topology* on $\mathcal{Top}(X, Y)$. This means that for every non-void open set V in the compact-open topology, and every $f \in V$, there exist compact sets $K_1, \cdots , K_n$ of X and open sets $U_1, \cdots , U_n$ in Y such that $f \in \bigcap_{i=1}^n V(K_i, U_i) \subseteq V$, see [1,7].

Remark 6. Let R be a ring (commutative or not), and let M be a left module over R. Let us assume that M has the discrete topology. Therefore its compact subsets are exactly its finite subsets. Then, the compact-open topology induced by $\mathcal{Top}(M, M) = M^M$ on the sub-space of all R-linear endomorphisms $R\text{-}\mathcal{Mod}(M, M)$ of M is the same as the topology of simple convergence (here $R\text{-}\mathcal{Mod}$ is the category of all left R-modules), *i.e.* for every topological space X, a map $\phi\colon X \to R\text{-}\mathcal{Mod}(M, M)$ is continuous if, and only if, for every $v \in M$, the map $\phi_v : x \in X \mapsto \phi(x)(v) \in M$ is continuous. Moreover with this topology, and R discrete, $R\text{-}\mathcal{Mod}(M, M)$ is a Hausdorff complete topological R-algebra ([17]).

We are now in position to state Jacobson's density theorem (see [8] for a proof).

Theorem 2 (Jacobson's density theorem). *Let R be a unitary ring (commutative or not). The ring R is primitive if, and only if, it is a dense subring (in the compact-open topology) of a ring $\mathbb{D}\text{-}\mathcal{Mod}(M, M)$ of linear endomorphisms of some (left) vector space M over a division ring $\mathbb{D}$ (where M is discrete).*

4.2 Decomposition of Endomorphisms

A direct consequence of Jacobson's density theorem is the following. Let $\mathbb{K}$ be a field of characteristic zero, and $A(\mathbb{K})$ be the *Weyl algebra* which is the quotient algebra of the free algebra $\mathbb{K}\langle x, y\rangle$ in two non-commutative variables by the two-sided ideal generated by $xy - yx - 1$ (this means that although the generators of $A(\mathbb{K})$ do not commute their commutator is equal to 1). (See [10] for more details.) Now, $A(\mathbb{K})$ is a primitive ring by Jacobson's density theorem. Indeed, $A(\mathbb{K})$ admits a faithful representation into $\mathbb{K}\text{-}\mathcal{Mod}(\mathbb{K}[z], \mathbb{K}[z])$ by $[x] \mapsto (P(z) \mapsto zP(z))$ and $[y] \mapsto (P(z) \mapsto \frac{d}{dz}P(z))$ (where $P(z)$ denotes an element of $\mathbb{K}[z]$, $[x], [y]$ are the canonical images of x, y onto $A(\mathbb{K})$, and it is clear that the commutation relation is preserved by this representation), and it is an easy exercise to check that through this representation $A(\mathbb{K})$ is a dense subring of $\mathbb{K}\text{-}\mathcal{Mod}(\mathbb{K}[z], \mathbb{K}[z])$ (under the topology of simple convergence with $\mathbb{K}[z]$ discrete). Nevertheless given $\phi \in \mathbb{K}\text{-}\mathcal{Mod}(\mathbb{K}[z], \mathbb{K}[z])$ and an open neighborhood V of ϕ, Jacobson's density theorem does not provide any effective nor even constructive way to build some $\phi_0 \in A(\mathbb{K})$ such that $\phi_0 \in V$. In [9] the authors show how to build in a recursive way a sequence of operators $(\Omega_n)_{n\in\mathbb{N}}$, $\Omega_n \in A(\mathbb{K})$ for each n, such that $\lim_{n\to\infty} \Omega_n = \phi$. In [6] the authors generalize this result to the case of $\mathbb{K}$-linear endomorphisms of V, with $\mathbb{K}$ any field (of any characteristic), proving that the multiplicative structure of the algebra $\mathbb{K}[z]$ is unnecessary (recall that as $\mathbb{K}$-vector spaces, $V \cong \mathbb{K}[z]$). We now recall this result in a more general setting where a commutative ring R with unit replaces the field $\mathbb{K}$. Let $(e_n)_{n\in\mathbb{N}}$ be a basis of $V = R\mathbb{N}$. We define a R-linear map $D\colon V \to V$ by $D(e_0) = 0$ and $D(e_{n+1}) = e_n$ for every $n \in \mathbb{N}$. (This linear map D may be given a definition by linear primitive recursion as $D = \widehat{R}(0, \pi_2^{(1)})$.) According to [12] (see page 109), for any sequence $(\phi_n)_{n\in\mathbb{N}}$ with $\phi_n \in R\text{-}\mathcal{Mod}(V, V)$, the family $(\phi_n \circ D^n)_{n\in\mathbb{N}}$ is summable in the topology of simple convergence of $R\text{-}\mathcal{Mod}(V, V)$ (where, for every endomorphism ϕ

of V, $\phi^0 = id_V$ and $\phi^{n+1} = \phi \circ \phi^n$). This means that there is an element of $R\text{-}\mathcal{M}od(V,V)$ denoted by $\sum_{n\in\mathbb{N}} \phi_n \circ D^n$, and called the *sum* of $(\phi_n \circ D^n)_{n\geq 0}$, such that for every $v \in V$, $v \neq 0$, $\left(\sum_{n\in\mathbb{N}} \phi_n \circ D^n\right)(v) = \sum_{n=0}^{d(v)} \phi(D^n(v))$, where $d(v)$ is the maximum of all k's such that the coefficient of e_k in the decomposition of v in the basis $(e_n)_{n\geq 0}$ is non-zero.

Remark 7. The above summability of $(\phi_n \circ D^n)_{n\in\mathbb{N}}$ essentially comes from topological nilpotence of D in the topology of simple convergence which means that for every $v \in V$, there exists $n_v \in \mathbb{N}$ (for instance $d(v)$ when $v \neq 0$) such that for every $n \geq n_v$, $D^n(v) = 0$ ($D^n \to 0$ in the topology of simple convergence).

For every polynomial $P(x) = \sum_{n=0}^{m} p_n x^n \in R[x]$, every sequence $\mathbf{v} = (v_n)_{n\in\mathbb{N}}$ of elements of V, and every R-linear endomorphism ϕ of V, we define $P(\mathbf{v}) = \sum_{n=0}^{m} p_n v_n \in V$, and $P(\phi) = \sum_{n=0}^{m} p_n \phi^n \in R\text{-}\mathcal{M}od(V,V)$. It is clear that $P(x) \in R[x] \mapsto P(\mathbf{e}) \in V$ for $\mathbf{e} = (e_n)_{n\in\mathbb{N}}$ defines a linear isomorphism between $R[x]$ and V. Moreover we have $P(\mathbf{e}) = P(U)(e_0)$. Now, let ϕ be given. There exists a sequence of polynomials $(P_n(x))_n$ such that $\phi = \sum_{n\in\mathbb{N}} P_n(U) \circ D^n$ (this means that ϕ is the sum of the summable family $(P_n(U) \circ D^n)_{n\geq 0}$ and it is equivalent to $\phi(e_n) = \sum_{k=0}^{n} P_k(U)(D^k(e_n))$ for each $n \in \mathbb{N}$, because $D^k(e_n) = 0$ for every $k > n$). This can be proved by induction on n as follows. We have $\phi(e_0) = P_0(\mathbf{e}) = P_0(U)(e_0)$ for a unique $P_0(x) \in R[x]$. Let us assume that there are $P_1(x), \cdots, P_n(x) \in R[x]$ such that $\phi(e_n) = \sum_{k=0}^{n} P_k(U) D^k(e_n) = \sum_{k=0}^{n} P_k(U) e_{n-k}$. Let $P_{n+1}(U)(e_0) = P_{n+1}(\mathbf{e}) = \phi(e_{n+1}) - \sum_{k=0}^{n} P_k(U) e_{n+1-k}$ (P_{n+1} is uniquely determined). Then, $\phi(e_{n+1}) = \sum_{k=0}^{n+1} P_k(U) \circ D^k(e_{n+1})$.

Remark 8. This result is outside the scope of Jacobson's density theorem since R is not a division ring, and also more precise since it provides a recursive algorithm to construct explicitly a sequence that converges to any given endomorphism.

Every sequence $(P_n)_n$ defines an endomorphism ϕ given by the sum of $(P_n(U) \circ D^n)_n$, and the above construction applied to ϕ recovers the sequence $(P_n)_n$. The correspondence between ϕ and $(P_n)_n$ as constructed above is functional, and it is actually a R-linear map ($R[x]^{\mathbb{N}}$ is the product R-module), onto and one-to-one.

4.3 A Normal Form for R-linear Endomorphisms of V

Let us consider the following subset of the R-algebra of non-commutative series $R\langle\langle x, y\rangle\rangle$ in two variables (see [5]): $R\langle x, y\rangle\rangle = \{\sum_{n\geq 0} P_n(x) y^n : \forall n,\ P_n(x) \in R[x]\}$. This is a R-sub-module of $R\langle\langle x, y\rangle\rangle$, and a $R[x]$-module with action given by $Q(x) \cdot (\sum_{n\geq 0} P_n(x) y^n) = \sum_{n\geq 0}(Q(x) P_n(x)) y^n = (\sum_{n\geq 0} P_n(x) y^n) \cdot Q(x)$. (We observe that $xy = y \cdot x$ but yx does not belong to $R\langle x, y\rangle\rangle$.) According to the result of subsection 4.2, there exists a R-linear isomorphism $\pi \colon R\langle x, y\rangle\rangle \to R\text{-}\mathcal{M}od(V,V)$ which maps $\sum_{n\geq 0} P_n(x) y^n$ to $\sum_{n\in\mathbb{N}} P_n(U) \circ D^n$.

Remark 9. It is essential that $xy \neq yx$, otherwise $\pi(xy) = U \circ D \neq id_V = D \circ U = \pi(yx)$, and π would be ill-defined.

For any $\phi \in R\text{-}\mathcal{Mod}(V,V)$, the unique $S = \sum_{n \geq 0} P_n(x) y^n \in R\langle x, y\rangle\rangle$ such that $\pi(S) = \phi$ should be called the *normal form* $s(\phi)$ of ϕ for a reason made clear hereafter. We observe that any set-theoretic map $\phi \colon \mathbb{N} \to \mathbb{N}$ also has such a normal form through the natural isomorphism $\mathcal{Set}(\mathbb{N}, \mathbb{N}) \cong R\text{-}\mathcal{Mod}(V,V)$.

Example 4.
1. Let us assume that R contains $\mathbb{Q}$ as a sub-ring. Let us consider the *formal integration* operator $\int$ on V defined by $\int e_n = \frac{e_{n+1}}{n+1}$ for every integer n. Then, $s(\int) = \sum_{n \geq 0} (-1)^n \frac{x^{n+1}}{(n+1)!} y^n$ (by recurrence).
2. Since the commutator $[D, U] = D \circ U - U \circ D = id_V - U \circ D$, we obtain $s([D,U]) = 1 - xy$.

Let $\pi_0(x) = U$, $\pi_0(y) = D$, and $\widehat{\pi} \colon \{x,y\}^* \to R\text{-}\mathcal{Mod}(V,V)$ be the unique monoid homomorphism extension of π_0 (where $R\text{-}\mathcal{Mod}(V,V)$ is seen as a monoid under composition). Let $R\{\{x,y\}\}$ be the set of all series $S = \sum_{w \in \{x,y\}^*} \alpha_w w$ in $R\langle\langle x,y\rangle\rangle$ such that the family $(\alpha_w \widehat{\pi}(w))_{w \in \{x,y\}^*}$ of endomorphisms of V is summable.

Example 5. Let us consider the series $S = \sum_{n \geq 0} y^n x^n \in R\langle\langle x,y\rangle\rangle$. Then, $S \notin R\{\{x,y\}\}$ since $\widehat{\pi}(y^n x^n) = \pi_0(y)^n \circ \pi_0(x)^n = D^n \circ U^n = id_V$ for each n. Whereas $S' = \sum_{n \geq 0} x^n y^n \in R\{\{x,y\}\}$ since $\sum_{n \geq 0} U^n D^n$ is equal to the operator $e_n \mapsto (n+1) e_n$.

From general properties of summability [17], $R\{\{x,y\}\}$ is a sub R-algebra of $R\langle\langle x,y\rangle\rangle$, and the homomorphism of monoids $\widehat{\pi}$ may be extended to an algebra map $\widetilde{\pi} \colon R\{\{x,y\}\} \to R\text{-}\mathcal{Mod}(V,V)$ by $\widetilde{\pi}(\sum_w \alpha_w w) = \sum_w \alpha_w \widehat{\pi}(w)$ which is obviously onto, so that $R\text{-}\mathcal{Mod}(V,V) \cong R\{\{x,y\}\}/\ker \widetilde{\pi}$ (as R-algebras). We have $\widetilde{\pi}(s(\phi)) = \phi$, so that s defines a linear section of $\widetilde{\pi}$. Let $\mathcal{N} \colon R\{\{x,y\}\} \to R\langle x,y\rangle\rangle$ be the R-linear map defined by $\mathcal{N}(S) = s(\widetilde{\pi}(S))$. Then, for every $S, S' \in R\{\{x,y\}\}$, $S \cong S' \bmod \ker \widetilde{\pi}$ (*i.e.*, $\widetilde{\pi}(S) = \widetilde{\pi}(S')$) if, and only if, $\mathcal{N}(S) = \mathcal{N}(S')$. Also it holds that $\mathcal{N}(\mathcal{N}(S)) = S$. The module of all normal forms $R\langle x,y\rangle\rangle$ inherits a structure of R-algebra by $S * S' = \mathcal{N}(SS') = s(\widetilde{\pi}(SS')) = s(\widetilde{\pi}(S) \circ \widetilde{\pi}(S'))$ isomorphic to $R\text{-}\mathcal{Mod}(V,V) \cong R\{\{x,y\}\}/\ker \widetilde{\pi}$.

Example 6. We have $y * x = 1$ while $x * y = xy$, so that $[y,x] = y * x - x * y = 1 - xy$. Let us define the operator ∂ on V by $\partial e_{n+1} = (n+1) e_n$ for each integer n and $\partial e_0 = 0$. Then, we have $s(\partial) = \sum_{n \geq 1} x^n y^{n-1}$. Moreover, $[\partial, U] = \partial \circ U - U \circ \partial = id_V$. It follows that $[s(\partial), x] = 1$. Let $A(R)$ be the quotient algebra $R\langle x,y\rangle$ by the two-sided ideal generated by $xy - yx - 1$, namely the *Weyl algebra over R*. Therefore there exists a unique morphism of algebras $\phi \colon A(R) \to R\langle x,y\rangle\rangle$ such that $\phi(x) = x$ and $\phi(y) = s(\partial)$. Composing with the isomorphism $\pi \colon R\langle x,y\rangle\rangle \to R\text{-}\mathcal{Mod}(V,V)$, we obtain a representation of the algebra $A(R)$ on the module V (x acts on V as U while y acts on V as ∂). When R is a field $\mathbb{K}$ of characteristic zero, then this representation is faithful (see [4]), hence in this case $\mathbb{K}\langle x,y\rangle\rangle$ contains a copy of the Weyl algebra $A(\mathbb{K})$, namely the sub-algebra generated by x and $s(\partial)$.

5 Concluding Remarks and Perspectives

5.1 Free Linear Induction Algebras

Let X be any set. According to section 2, we may define the free R-linear induction algebra V_X on X. It is isomorphic to the direct product of $|X|+1$ copies of V, namely the R-module $V \oplus \bigoplus_{x \in X} V$, this is so because the free induction algebra on X is $\{S^n(0)\colon n \geq 0\} \sqcup \bigsqcup_{x \in X}\{S^n(x)\colon n \geq 0\}$ (where $\bigsqcup$ is the set-theoretic disjoint sum). As an example, take X finite of cardinal say n, then V_X is isomorphic to V^{n+1}. In this finite case, we have $R\text{-}\mathcal{Mod}(V^{n+1}, V^{n+1}) \cong R\text{-}\mathcal{Mod}(V,V)^{(n+1)^2} \cong R\langle x, y\rangle\rangle^{(n+1)^2}$. From subsection 4.3 it follows that any endomorphism of V^{n+1} may be written as a vector of length $(n+1)^2$ or better a $(n+1) \times (n+1)$ matrix with entries some members of $R\langle x, y\rangle\rangle$. More generally, for each integers m, n, we have $R\text{-}\mathcal{Mod}(V^m, V^n) \cong R\text{-}\mathcal{Mod}(V,V)^{mn}$ so that we have obtained a complete description of all linear maps between spaces of the form V^n in terms of the basic operators U and D.

5.2 Links with Sheffer Sequences

It is not difficult to check that we may define a new associative multiplication on $R\langle x, y\rangle\rangle$, and therefore also on $R\text{-}\mathcal{Mod}(V,V)$, by

$$\left(\sum_{n \geq 0} P_n(x) y^n\right) \# \left(\sum_{n \geq 0} Q_n(x) y^n\right) = \sum_{n \geq 0}\left(\sum_{k \geq 0} \langle P_n(x) \mid x^k\rangle Q_k(x)\right) y^n$$

where $\langle P(x) \mid x^k\rangle$ denotes the coefficient of x^k in the polynomial $P(x)$ (so that in the above formula the sum indexed by k is actually a sum with a finite number of non-zero terms for each n), with a two-sided identity $\sum_{n \geq 0} x^n y^n$ (that corresponds to the operator $e_n \mapsto (n+1)e_n$ of V). This product is a generalization of the so-called umbral composition [14]. Let us assume that $\mathbb{K}$ is a field of characteristic zero. Following [15] (see also [13]) a sequence $(p_n(x))_{n \geq 0}$ of polynomials in $\mathbb{K}[x]$ such that the degree of $p_n(x)$ is n for each integer n is called a *Sheffer sequence* if there are two series $\mu(y), \sigma(y) \in \mathbb{K}[[y]]$, where x and y are assumed to be commuting variables, with $\mu(0) \neq 0$, $\sigma(0) = 0$, and $\sigma'(0) \neq 0$ (where σ' denotes the usual derivation of series) such that $\sum_{n \geq 0} p_n(x) \frac{y^n}{n!} = \mu(y) e^{x\sigma(y)} \in \mathbb{K}[[x, y]]$. A series $S = \sum_{n \geq 0} \frac{1}{n!} p_n(x) y^n \in \mathbb{K}\langle x, y\rangle\rangle$ is said to be a *Sheffer series* whenever $(p_n(x))_n$ is a Sheffer sequence. Such series correspond to *Sheffer operators* on V given by $\sum_{n \geq 0} \frac{1}{n!} p_n(U) \circ D^n$. For instance Laguerre's polynomials given by $L_n(x) = \sum_{k=0}^{n} \binom{n}{k} \frac{(-1)^k}{k!} x^k$ form a Sheffer sequence, and thus $\sum_{n \geq 0} \frac{1}{n!} L_n(U) \circ D^n$ is a Sheffer operator. We observe that the above multiplication $\#$ corresponds to the umbral composition of $(P_n(x))_n$ and $(Q_n(x))_n$. Because Sheffer sequences form a group under umbral composition (see [14]), it follows that Sheffer operators and Sheffer series form an isomorphic group under the corresponding umbral composition. The perspectives of our present contribution concern the study of such operators and their combinatorial properties.

References

1. Arens, R.: Topologies for homeomorphism groups. American Journal of Mathematics 68(4), 593–610 (1946)
2. Cohn, P.M.: Skew fields - Theory of general division rings. Encyclopedia of Mathematics and its Applications, vol. 57. Cambridge University Press (1995)
3. Bergman, G.M.: The diamond lemma for ring theory. Advances in Mathematics 29, 178–218 (1978)
4. Björk, J.-E.: Rings of differential operators. North-Holland Publishing Company (1979)
5. Bourbaki, N.: Elements of mathematics - Algebra, ch. 1-3. Springer (1998)
6. Duchamp, G.H.E., Poinsot, L., Solomon, A.I., Penson, K.A., Blasiak, P., Horzela, A.: Ladder operators and endomorphisms in combinatorial physics. Discrete Mathematics and Theoretical Computer Science 12(2), 23–46 (2010)
7. Eilenberg, S.: Sur les groupes compacts d'homéomorphies. Fundamenta Mathematicae 28, 75–80 (1937)
8. Farb, B., Dennis, R.K.: Noncommutative algebra. Graduate Texts in Mathematics, vol. 144. Springer (1993)
9. Kurbanov, S.G., Maksimov, V.M.: Mutual expansions of differential operators and divided difference operators. Dokl. Akad. Nauk. UzSSR 4, 8–9 (1986)
10. Mac Connell, J.C., Robson, J.C., Small, L.W.: Noncommutative Noetherian rings. Graduate studies in mathematics, vol. 30. American Mathematical Society (2001)
11. Mac Lane, S.: Categories for the Working Mathematician. Graduate Texts in Mathematics, vol. 5. Springer (1971)
12. Poinsot, L.: Contributions à l'algèbre, à l'analyse et à la combinatoire des endomorphismes sur les espaces de séries. Habilitation à diriger des recherches en Mathématiques. Université Paris 13, Sorbonne Paris Cité (2011), http://lipn.univ-paris13.fr/~poinsot/HDR/HDR.pdf
13. Poinsot, L.: Generalized powers of substitution with pre-function operators. To be published in Applied Mathematics, special issue on Fractional Calculus Theory and Applications (2013)
14. Roman, S.: The umbral calculus. Pure and Applied Mathematics, vol. 111. Academic Press Inc. (1984)
15. Rota, G.-C., Kahaner, D., Odlyzko, A.: On the foundations of combinatorial theory VIII: Finite Operator Calculus. Journal of Mathematical Analysis and its Applications 42(3), 684–750 (1973)
16. Soare, R.I.: Recursively enumerable sets and degrees. Perspective in Mathematic Logic. Springer (1987)
17. Warner, S.: Topological rings. North-Holland Mathematics Studies, vol. 178. Elsevier (1993)

Author Index